Sustainability Transitions and Industrial Transformation

Markus Steen · Taran Thune · Teis Hansen ·
Tuukka Mäkitie · Koen Frenken
Editors

# Sustainability Transitions and Industrial Transformation

## Multi-Sector Dynamics and Policy Perspectives

palgrave
macmillan

*Editors*
Markus Steen
Department of Technology Management
SINTEF
Trondheim, Norway

Taran Thune
Center for Technology, Innovation and Culture
University of Oslo
Oslo, Norway

Teis Hansen
Department of Technology Management
SINTEF
Trondheim, Norway

Department of Food and Resource Economics
University of Copenhagen
Frederiksberg C, Denmark

Tuukka Mäkitie
Center for Technology, Innovation and Culture
University of Oslo
Oslo, Norway

Department of Technology Management
SINTEF
Oslo, Norway

Koen Frenken
Copernicus Institute of Sustainable Development
Utrecht University
Utrecht, The Netherlands

ISBN 978-981-95-4892-7 ISBN 978-981-95-4893-4 (eBook)
https://doi.org/10.1007/978-981-95-4893-4

This work was supported by TIK Centre of Technology, Innovation and Culture, University of Oslo.



This Palgrave Macmillan imprint is published by the registered company Springer Nature Singapore Pte Ltd.
The registered company address is: 152 Beach Road, #21-01/04 Gateway East, Singapore 189721, Singapore

# Acknowledgements

As Editors of this book, we would first like to thank all our colleagues who have contributed their work. Our gratitude also extends to yet other colleagues who were generous enough to review and provide feedback on individual chapters.

We also wish to thank the Research Council of Norway for funding via grant 295021 to the centre for the study of innovation policy for industrial transformation, sustainability, and digitalization (INTRANSIT). The key ideas behind this book were developed in an INTRANSIT seminar in June 2023, and we are very grateful for all the ideas and suggestions from our INTRANSIT friends who helped shape the book's scope and key themes.

Lastly, we acknowledge timely support offered by the team at Palgrave/ Springer Nature.

# Contents

# About the Editors

**Markus Steen** is Senior Research Scientist at the Department of Technology Management (research group Innovation and Sustainability) in SINTEF Digital. He holds a Ph.D. in economic geography from the Norwegian University of Science and Technology (NTNU), Department of Geography and Social Anthropology, where he is also Associate Professor II. His research focuses on sustainability transitions, innovation, policy, industrial transformation, and regional development. Empirically, his work has centred on a range of sectors in the Nordics and Northern Europe, including energy, aquaculture, manufacturing, and the energy-intensive process industries.

**Koen Frenken** is Full Professor in Innovation Studies at the Copernicus Institute of Sustainable Development at Utrecht University. He won personal grants in the Marie Curie and NWO Vidi and Vici programmes. Theoretical interests include evolutionary economics, institutional sociology, and complexity theory. He works on breakthrough innovation, economic geography, platform economy, and innovation policy.

**Teis Hansen** is Full Professor at the Department of Food and Resource Economics, University of Copenhagen, and Senior Research Scientist at the Department of Technology Management, SINTEF, Trondheim. He is also Member of the Danish Government's expert group on the role of research in the green transition, and a Semper Ardens Accomplish Grant Holder of the Carlsberg Foundation. His research interests are located at the intersections of geography, innovation, sustainability, and policy. Empirically, his current research is focused on understanding the scope for policy to promote rapid transitions across the bioeconomy, the energy system and energy-intensive process industries.

**Tuukka Mäkitie** is Research Scientist at SINTEF Digital, and holds a Ph.D. in innovation studies from University of Oslo, Norway. His research interests include sustainability transitions and green innovation in energy, transport, and manufacturing sectors. He is particularly interested in the processes of novel value chain formation as a key part of the acceleration phase of sustainability transitions.

**Taran Thune** is Professor of Innovation at the Centre for Technology, Innovation and Culture at the University of Oslo and Director of research centre INTRANSIT. INTRANSIT focuses on understanding how industries transform in face of mounting environmental and climate concerns, and address such questions in industries such as maritime transport, energy sectors, aquaculture and agri-food, process industries, and more. Taran is particularly interested in policy questions and has published widely on research and innovation policy topics in a range of journals.

# List of Figures

# List of Tables

# 1

# Introduction: Why This Book and Its Contributions

Markus Steen, Koen Frenken, Teis Hansen, Tuukka Mäkitie, and Taran Thune

## The Need for Transformation

Consider this: coal, oil, and natural gas, all primary sources of greenhouse gas emissions, accounted for 80% of global primary energy use in 2023. This simple statistic underscores that the global energy system

M. Steen (✉) · T. Hansen · T. Mäkitie
Department of Technology Management, SINTEF Digital, Trondheim, Norway
e-mail: markus.steen@sintef.no

K. Frenken
Copernicus Institute of Sustainable Development, Utrecht University, Utrecht, The Netherlands

T. Hansen
Department of Food and Resource Economics, University of Copenhagen, Copenhagen, Denmark

T. Mäkitie · T. Thune
TIK Centre for Technology, Innovation and Culture, University of Oslo, Oslo, Norway



M. Steen et al. (eds.), *Sustainability Transitions and Industrial Transformation*,
https://doi.org/10.1007/978-981-95-4893-4_1

remains reliant on coal, gas, and oil. Transport and mobility are similarly largely based on fossil fuels. In light of the climate crisis, the world is by and large off track in making the necessary progress to reach net-zero by mid-century (IEA, 2023). What is more, biodiversity loss and chemical pollution continue to increase due to massive negative externalities in food, housing, and other sectors. And, while circular economy has become a mainstream concept, only 7% of all material consumed is currently reused or recycled (Circle Economy, 2023).

The world is on an unsustainable course. Achieving the changes deemed necessary to mitigate climate change, halt the destruction of natural land and oceanic habitats, and stagger the loss of biodiversity remain daunting challenges. This is not because we lack knowledge about facts or remedies. On the contrary, we know that fossil carbon emissions need to be curbed, and that we need to shift towards low-carbon energy sources. It is also obvious that the overall strain on land use and natural resources must be reduced, and that we need to replace our current linear economic system with an economy based largely on circular principles. We know that this necessitates changes in both technologies and consumption practices, and that policies and regulations must evolve to incentivize more sustainable production and consumption. While all of this has arguably been understood for decades (cf. the Limits to Growth publication by the Club of Rome in 1972, the UN Brundtland report in 1987, and the first report by the Intergovernmental Panel on Climate Change (IPCC) in 1990), and despite the manifold "global" agreements on the need to more profoundly solve environmental challenges, the current unsustainable course has regretfully proved itself resilient.

The overall picture is therefore gloomy. We are heading towards a much warmer planet with less biodiversity and severe public health effects, with expected dire consequences for humans and the natural world alike. Yet, when zooming in on particular places, regions, and countries, there is considerable heterogeneity, and many accounts of successful transition processes indeed allow for a certain degree of optimism. In recent years there have been many initiatives and efforts in research, policy, and increasingly also in industries to develop and support new sustainable technologies and practices, as well as to phase out unsustainable ones (e.g., coal, oil, landfilling, chlorofluorocarbons, chemical pollutants). These efforts have not been futile. On the contrary, some 'cleantechs', such as wind energy, solar PV, electric vehicles, LED lighting, and heat pumps, are diffusing rapidly, at least in some countries. This growth in cleantechs is in some instances also enabled by new digital technologies.

Over the last couple of decades, the field of sustainability transition studies (or, in short, 'transition studies') has been at the forefront in exploring why change is cumbersome and how transformation towards more environmentally desirable futures can be fostered. This research field emerged in the 1990s on the basis of previous work in among other evolutionary economics, history of technology, and science and technology studies (Berkhout et al., 2004; Markard et al., 2012; Rip & Kemp, 1998). Over the years, transition studies have seen fruitful encounters with additional research fields, such as sociology, political science, management and organization studies, and economic geography, leading to a truly interdisciplinary research area that involves the breadth of social sciences and humanities as well as increasing engagements with engineering and natural scientists.

The field has in particular focused on the production-consumption systems that provide key societal services such as energy, mobility, food, housing, and health, and how these societal functions can be fulfilled more sustainably (Köhler et al., 2019; Markard et al., 2012). A limitation of transition research to date is however that it has been dominated by studies focusing on singular technologies, sectors, or production-consumption systems. While systems like energy, mobility, and food can be meaningfully delineated as an object of study, all systems are ultimately related and connected through flows of energy, materials, trade, data, and actors. Such relations exist both between production-consumption systems (e.g., energy-mobility, energy-food, food-health) and within and across technology value chains (e.g., batteries-mining, food-chemicals, vehicles-steel). Through these interconnections, transition dynamics in one system can affect transition dynamics in other systems through positive and negative feedback mechanisms. Understanding interactions between different systems is thus important to fully grasp system transitions as well as the policies to foster such transitions.

Therefore, new or recomposed analytical and conceptual perspectives are needed, as are empirical studies.

In sum, understanding transitions towards sustainability requires attention to the complexities not only within sectors such as energy or agriculture, but across different production-consumption systems. The potentially enabling role of general-purpose technologies, and particularly digitalization in its many forms, for fostering transitions inside and across industries and value chains also needs further exploration, as claims of the promise of the "twin transition" are frequently heard and the effects are still uncertain (Andersen et al., 2021; Mäkitie et al., 2023; Sareen & Haarstad, 2021). Moreover, to fully understand transformation towards sustainability in sectors and industries, the systems perspective needs to be accompanied with more attention to agency (De Roeck & Van Poeck, 2023; Farla et al., 2012), and particularly the role of firms (Kungl, 2024; Magnusson & Werner, 2023).

Against this brief background, the key aim with this book is to explore new views on multi-system dynamics in transitions, the role of firms and industries in these processes, and novel policy approaches to support sustainability transitions and industrial transformation[1]. The book also deals with empirical contexts that have received scant attention in the sustainability transition field to date. Examples include hard-to-abate process industries, aquaculture, and multi-sector electrification. It also directs attention to cross-sectoral value chains, upstream supplier industries, and multi-purpose or key enabling technologies such as digitalization and electrification. Some chapters take an explicit firm level perspective (e.g., Magnusson & Werner, 2023) to analyse how firms act and the potential impacts on industrial transformations towards sustainability (Andersen et al., 2020). This allows us to bring in literatures from

[1] Many of the contributors to the book are part of or associated with 'The centre for the study of innovation policy for industrial transformation, sustainability and digitalization' (INTRANSIT). INTRANSIT partners are the University of Oslo (TIK Centre for Technology, Innovation and Culture, and Department of Informatics), SINTEF (Department of Technology Management), and Utrecht University (Copernicus Institute of Sustainable Development). The centre is funded by the Research Council of Norway. Additional contributors have been part of conference sessions and workshops focused on multi-system dynamics and the role of organizations in transitions that INTRANSIT researchers have been involved in as organizers.

management research, organizational studies, and industrial dynamics to enrich the field by linking systems and agency-oriented perspectives on industrial transformations and sustainability transitions. Some chapters also in this section focus on digitalization and sustainability. The final set of chapters focus on novel policy approaches that have emerged over the last decade or so, advocating the need for directionality and more holistic coordination across policy domains and scales. Some of these chapters also address the resurgence of industrial policy and how that fits within the new '3rd generation' of innovation and technology policies (Aiginger & Rodrik, 2020; Edler et al., 2023; Schot & Steinmueller, 2018).

This book thus attempts to bridge perspectives on sustainability transitions that emerge from studies of sectoral systems and (increasingly) multi-systems transformations, studies of firms and inter-firm collaboration, and studies of policy development and their impacts. All chapters provide stand-alone contributions, and they are all empirically underpinned, some with comprehensive analysis, while others use empirical material in more illustrative ways.

To interested audiences in academia, industry, and policy settings, the book contributes in four ways:

(1) It offers a broad and interdisciplinary approach to understanding industrial transformation and sustainability transitions;
(2) It provides empirical studies and examples from a diversity of sectors, including ones considered hard to abate;
(3) It emphasizes how sectors are interlinked, and how broad-based technological advances such as digitalization and electrification create multi-system transformations;
(4) It addresses policy implications and policy tools developed to handle these complex transition dynamics.

It is our hope that the book will be found to be relevant among fellow academics, students, and practitioners, and that it will inspire further work on how we can transform our economies and societies towards a sustainable future for humans and nature.

The remainder of this introductory chapter proceeds as follows. We next revisit the sustainability transitions literature. Our purpose is not to provide a comprehensive account of this vibrant and dynamically evolving field, but rather to give readers—especially those that may be unfamiliar with transition studies—sufficient understanding of this research field to grasp the logics behind the choice of key themes and contributions of the book. In this brief review we focus on what we consider to be the transition literature's two main conceptual approaches—the Multi-Level Perspective (MLP) and Technological Innovation Systems (TIS)—and recent developments and highlighted gaps. The key themes of the book—multi-system transition dynamics, firms and industries in transition, and new perspectives on policy in transitions—are then presented, followed by a brief introduction to the chapters in each of these sections. The chapter ends with a brief summary.

## Sustainability Transition Studies

Sustainability transition studies emerged in the 1990s and now constitute a distinct, yet interdisciplinary, scientific field. Markard et al., (2012, p. 956) defined the empirical phenomenon of sustainability transitions as "*long-term, multi-dimensional, and fundamental transformation processes through which established socio-technical systems shift to more sustainable modes of production and consumption*". At the core of the sustainability transitions field is the understanding that current production-consumption systems are configured by a complex interweaving of technological, infrastructural, political, institutional, social, and cultural elements. The key concept is the *socio-technical system*, referring precisely to how technology—while in many ways central to both the problem and solution side of sustainability challenges—co-evolves with the "social" world (actors, institutions, networks) as well as other material (technologies, infrastructures, material resources) elements.

Established socio-technical systems surmount to configurations of interrelated technologies, actors, and institutions 'that work' (Rip & Kemp, 1998). They are stabilized by path dependencies accruing both from mechanisms such as network effects and increasing returns as well as from the interweaving of heterogeneous systemic elements. This means first, that it is normally insufficient to simply substitute one technology with another. Rather, transitions in socio-technical systems tend to rely on changes in multiple technological and non-technological dimensions. For example, the electrification of automobiles requires not only that automobile manufacturers use their resources to develop electric vehicles (EVs): the system around automobility needs to be reconfigured, including complementary technologies and services such as charging stations, availability of service and maintenance providers, and users need to adapt their driving patterns. For transport as a whole to become more sustainable, there is a need for changes also in modal shifts, including from the personal automobile to public and/or shared transportation. The second key implication is that interventions—notably political ones—are necessary in order to shift development trajectories.

The need for policy interventions is further underpinned by another key feature of transition studies, which is the view that stepwise incremental changes are insufficient to bring about necessary change at system level. Not only are radical innovations considered necessary, they also need to occur at the 'systemic' level (Elzen et al., 2004; Grin et al., 2010). According to Geels (2004), therefore, transitions research aims to conceptualize and explain how radical changes can occur in ways that allow systems, such as energy, transport, and food, to fulfil their societal functions. This ambition is most clearly articulated in the Multi-Level Perspective (MLP).

The MLP is an analytical framework that is applied to studying historical processes of technological change in specific socio-technical systems or sectors[2] by focusing on the interaction between different levels of structuration and institutionalization (Geels, 2002, 2005; Verbong & Geels, 2007). At the core of the early MLP was the idea that transitions involved shifts in socio-technical configurations, understood mainly as a substitution process. While it is now firmly acknowledged that transitions transpire in manifold ways (Geels & Schot, 2007), there is

[2] The terms 'sectors' and 'socio-technical systems' are often used interchangeably in the transition field.

consensus that transitions are long-term and complex processes inhibited by path dependencies and the rigidities of established systems (Geels, 2010; Lovio et al., 2011; Markard et al., 2012).

The MLP distinguishes three interrelated analytical levels, with the socio-technical *regime* (meso-level) constituting the key concept. The regime refers to an interrelated and stable configuration of heterogeneous actor networks, established products and technologies, infrastructures, user practices, expectations, norms, and regulations (Geels, 2002; Rip & Kemp, 1998; Smith et al., 2005). Regimes therefore have both social and material dimensions. The *landscape* (macro) level refers to broader societal factors that cannot be directly influenced by regime actors. Examples include global commodity prices, cultural change, pressures arising from increasing environmental concerns, as well as more sudden shocks caused by for example wars, natural disasters, or pandemics. Although new technologies may have many benefits (lower operational costs, less pollution, etc.), they often struggle to develop beyond a nascent phase. A key idea in the MLP is that technological innovation requires '*niches*', protected spaces for development 'free' from the constraints of mainstream market selection, performance standards, and the infrastructural rigidities of established systems (Geels, 2002; Smith & Raven, 2012). This may be market niches, where actors have different preferences and selection criteria than in mainstream markets, or technological niches where resources are provided by public subsidies or private strategic investments (Geels & Kemp, 2007).

The technological innovation system (TIS) approach focuses on the development, upscaling, and implementation of new technologies, and associated institutional and organizational changes (Bergek et al., 2008; Hekkert et al., 2007; Markard et al., 2012). A TIS is commonly defined as (a) "*network(s) of agents interacting in a specific technology area under a particular institutional infrastructure for the purpose of generating, diffusing and utilizing technology*" (Carlsson & Stankiewicz, 1991, p. 111). A key point of departure is that innovation emanates from inter-organizational interactions straddling public and private sectors, with knowledge creation and dissemination as a core process. Interaction is needed to convert ideas into products and services that are useful, implying that the TIS framework draws attention to systemic innovation (Weber & Rohracher, 2012).

Considering emerging technologies, the TIS framework also highlights how "niche-like" conditions are needed to facilitate the necessary inter-organizational interaction and resource formation and mobilization processes. Such early or (semi-)protected markets for instance via public procurement provide early commercial opportunities and a context for the development of different designs, as well as development of value chains and capacities to provide key products and services (Geels et al., 2008). In this way, niches "*help building a constituency behind a new technology, and set in motion interactive learning processes and institutional adaptation (...) that are all-important for the wider diffusion and development of the new technology*" (Kemp et al., 1998, 184). This points to how also institutions and networks are important in TIS formation (Jacobsson & Johnson, 2000). Whether an emerging TIS has enabling or inhibiting conditions can be assessed by analysis of innovation system strengths and weaknesses, or key "TIS functions" as they are often referred to (Bergek et al., 2008; Hekkert et al., 2007; Johnson, 1998). Different versions of these functions have been suggested, but typically include knowledge development and diffusion, influence on the direction of search, market formation, resource mobilization, legitimation, entrepreneurial experimentation, and development of external economies.

## Current Research Challenges

The system perspectives provided by the MLP and TIS have important implications for how transitions are understood, which underlying processes and mechanisms that are considered essential, and the types of insights and recommendations that sustainability transition studies typically provide to policymakers and other stakeholders. It is however clear that the MLP and TIS have clear limitations concerning multi-system dynamics and current transition processes (Andersen & Markard, 2024). Partly, this reflects that both frameworks were developed (i.e., from late 1990s onwards) when the 'sustainable reorientation' of sectors such as energy and mobility were in early stages of development, with for example low diffusion levels of intermittent renewables or EVs. However, precisely for these sectors, we are now in a different (albeit geographically uneven) phase (Markard, 2018), wherein several cleantechs such as solar PV, heat pumps, and EVs are diffusing rapidly. The MLP and TIS have typically also been applied to one sector or technology at a time.

As already argued, there has also been considerably more empirical attention to certain renewable technologies, forms of mobility or food production, than for example process industries, heavy transport, or construction (Truffer et al., 2022). The explanatory power offered by existing theoretical frameworks for understanding innovation and transition dynamics in these other sectors has thus not been extensively tested. Typically, existing theories and frameworks need to be modified or adapted—or even substituted by new theories and frameworks—when confronted with 'other' empirical phenomenon, or different contexts.

An important point of departure for this book is that practically all sectors are increasingly pressured to address environmental issues, including those considered hard-to-abate, such as the energy-intensive process industries and shipping. We would argue that this has been apparent especially since the 2015 COP21 and the adoption of the legally binding international treaty on climate change known as the 'Paris Agreement'. For the targets set in this treaty to become reality, much more stringent emission reduction targets were introduced for sectors that are largely laggards in the realm of climate mitigation. The shipping sector is a case in point. Although shipping emissions were not directly included in the Paris Agreement, the treaty contributed to an increasing pressure on the sector and the International Maritime Organization to begin quantifying emission reduction targets and identify solutions to reaching those targets (Bach & Hansen, 2023).

The upscaling of for example renewable energy and the development and realization of innovations and changing practices to allow for transitions in hard-to-abate sectors implicates pervasive and economy-wide transformation. For example, the decarbonization of transport (road, rail, aviation, shipping) requires that fossil fuels are replaced with fuels and energy carriers based on renewable energy and sustainably sourced regenerative biomass. Such substitution processes in the transport sector therefore have repercussions for sectors that can provide sustainable electrons and molecules, including the energy sector, waste, agriculture, and forestry.

While multi-system dynamics are increasingly acknowledged, they remain under-researched both conceptualized and empirically. We need to better understand how multiple socio-technical systems, sectors, and technologies interact, how new types of resource and knowledge flows between systems are created, the agentic processes that underpin these interactions, and their conditions and outcomes. These changes go well beyond simple transitions in focal sectors, but require 'architectural change' (McMeekin et al., 2019) across multiple sectors and thereby also our economies (Andersen et al., 2026).

Relatedly, there has been limited attention to upstream and downstream industrial transformation processes that are part-and-parcel of sustainability transitions (Andersen et al., 2026). Transitions have repercussions not only for 'end-users', but also for upstream supplier industries and the multiple sectors that are connected through value chains or production networks. The industrial supply chains that provide all the 'nuts and bolts' (i.e., machinery, engines, power control systems, pipelines, cables, storage tanks for gases and liquids, etc.) all need to be adapted or developed anew, while the adoption of new digital technologies (e.g., artificial intelligence) and platforms may lead to novel system configurations. This underscores the need to better understand the role of different types of actors (incumbents, newcomers, suppliers, intermediaries, public organizations) operating across different systems and domains, pointing also to different forms of (cross-) system intermediation and entanglement.

There has furthermore been considerable policy and research attention to the interplay between sustainability and digitalization, often viewing the latter as central to the realization of the former. This has not least been epitomized in the notion of a 'twin transition' (Muench et al., 2022). Yet, the relationship between sustainability and digitalization is complex, as indeed also recognized by Muench et al. (2022). From a transition's point of view, Mäkitie et al. (2023) argued that digitalization can indeed be part-and-parcel of radical socio-technical systems change, such as in the case of transitioning to autonomous electrified transport. Digitalisation may however also simply serve to optimize existing processes and practices, and thereby further entrench existing socio-technical systems (see also, e.g., Andersen et al., 2021; John et al., 2022; Sareen & Haarstad, 2021).

Moving beyond 'singular systems' implies the need for rethinking how we understand innovation dynamics and systems (i.e., beyond spatially, sectorally, or technologically confined), as well as the need for new and more comprehensive policy approaches (such as transformative or mission-oriented policy), and policymaking and practicing capabilities to steer and support transition processes (e.g., Borrás et al., 2024; Edler et al., 2025; Rogge & Reichardt, 2016; Schot & Steinmueller, 2018). This includes attention to broader policy mixes and coordination, both across different policy domains as well as at different governance levels. It also directs attention towards the resurgence of industrial policy, now increasingly framed around aspirations of 'green' industrial development and leadership (Allan et al., 2021; Arcand & Meadowcroft, 2024; Rodrik, 2014).

In this volume, we organize the contributions to these overlapping themes under three main headings: multi-system dynamics in transitions, firms and industries in transitions, and policy in transitions. In the sections that follow, we briefly introduce these themes as well as the different chapter contributions.

## New Perspectives on Multi-System Dynamics in Transitions

The fact that all sectors need to transition towards sustainability (rapidly) implies the need for rethinking how we understand innovation dynamics and system change. As argued by Rosenbloom (2020, p. 336), stretching the scope of analysis is pertinent since "*sustainability challenges and the transformative changes they necessitate stretch well beyond individual socio-technical systems and niche innovation journeys*". A similar plea comes from Andersen and Geels (2023, p. 10), who argued that there is need for "*a shift in research from focusing on single system transitions plus wider interactions towards investigating a 'patchwork' of interacting transitions*".

The agri-food system, for example, is intrinsically linked to systems that provide energy, water, and transport (Hebinck et al., 2021; Klerkx & Begemann, 2020). Transport provides services in many forms to basically all other sectors and is of course interdependent with the extraction and processing of fuels (for energy), as well as other material inputs to manufacture vehicles, aircraft, and ships. Novel energy solutions, such as biogas, also rely on boundary-crossing innovations in, among other, energy, waste management, agriculture and transport (Huttunen et al., 2014).

As different socio-technical systems advance in their transition processes, it becomes increasingly clear that such inter-system linkages can be pervasive, meaning that change in one system has ripple effects in one or several other sectors (Andersen et al., 2026). This includes upstream sectors in technology value chains. For example, the growth in EVs impacts and necessitates changes across a range of sectors involved in EV and battery technology value chains, including mining, processing, manufacturing, and assembly (Gong & Andersen, 2024; Stephan et al., 2017). Other studies have shown how resource scarcities hinder or delay transition processes, such as how limits to biomass availability and processing capacity constrains a shift towards biofuels in shipping (Bach et al., 2021).

Transition scholars have already made important advancements concerning multi-system dynamics, and as recently noted by Andersen and Markard (2024), the theme has strong traction in the transition community.[3] Key themes in this work include how specific socio-technical systems or technological innovation systems interact with other systems. This is reflected in attention to various kinds of interlinkages, interactions, and couplings between different systems. An early example is the work by Konrad et al. (2008) that distinguished between functional and structural couplings across different socio-technical systems.

---

[3] This is witnessed also in recent conferences (e.g., the International Sustainability Transitions conference in Oslo in June 2024), as well as in several journal special issue initiatives.

Scholars have also identified various interaction modes (e.g., competition, symbiosis), both between technologies (Sandén & Hillman, 2011) and systems. This burgeoning literature contributes new theoretical frameworks and empirical analyses beyond the 'usual suspects' (i.e., wind energy, solar PV, EVs) of sustainability transitions. This requires attention to broader processes such as multi-sector electrification (Nykamp et al., 2023) and the shift towards a circular economy (Kanda et al., 2021; Pfeffer et al., 2025), which tend to involve a variety of sectors and (novel) value chains that cut across them. It also requires attention to the relationship between digitalization and sustainability (sometimes discussed as 'twin transitions'), as well as being attentive to how change dynamics are influenced by different contextual conditions.

## Book Contributions on Multi-System Dynamics in Transitions

In this book, Ateş et al. (this volume) explore how electrification and digitalization imply more interconnectedness as well as more complex interactions between different socio-technical systems. Their case study uses the example of energy and mobility systems in the UK. They mobilize the concept of *interlinkages* to examine such interactions, and develop an analytical framework derived from Large Technical Systems (LTS) framework (Hughes, 1983). Their analysis distinguishes functional, technical, and inter-organizational interlinkages and how these have emerged and evolved in response to UK net-zero targets. Functional interlinkages, which refer to dependencies and interactions between different (sub-)systems, include for example smart charging and vehicle-to-grid technologies that enable EVs as flexible energy assets. Technical interlinkages refer to structural connections between technological artefacts or components across different socio-technical systems and include the expansion of digital infrastructure and standardization for interoperability. Organizational interlinkages comprise new partnerships and cross-sector collaborations. Through their empirical analysis the authors also identify governance interlinkages as a central dimension shaping system interactions. These interlinkages for example include coordination mechanisms and regulatory alignment. A key takeaway from this chapter is that digital technologies act as a form of enabling infrastructure rather than constituting separate systems. The authors convincingly argue that LTS offers a valuable complementary lens to existing frameworks and perspectives in transition studies for understanding multi-system dynamics and the governance of increasingly complex transition processes.

Hanson and Mäkitie (this volume) similarly draw attention to cross-system dynamics, focusing on how the hydrogen technological innovation system in Norway is heterogeneously configured by multiple sectors. In so doing, the chapter contributes to a burgeoning strand of work framed within a TIS perspective that emphasizes the need to better understand multi-sectoral context structures (Bergek et al., 2015; Stephan et al., 2017). Globally, hydrogen is currently mainly produced from natural gas, so-called 'grey hydrogen'. Empirically, the chapter analyses two types of alternative production methods compared with grey hydrogen. These—'green' and 'blue' hydrogen—constitute alternative hydrogen value chains each with their specific sectoral configuration notably on the upstream side. Green hydrogen is produced via electrolysis, using renewable energy. Blue hydrogen, on the other hand, is produced from natural gas, but coupled with carbon capture and storage (CCS) technology to remove and sequester $CO_2$. The chapter thus demonstrates how different sectors make up segments of the alternative hydrogen value chains, which in turn have differing sectoral interdependencies, each with idiosyncratic system dynamics, including institutional and actor characteristics. The blue and green hydrogen value chains are found to be confronted with different challenges both on the input side as well as in market formation. For blue hydrogen, the reliance on available $CO_2$ transport and storage infrastructure is a key bottleneck. For green hydrogen, a challenge is access to electricity at affordable prices. For both, developments on the downstream or offtake side—in heavy processing industry and transport both on land and at sea—is slow and cumbersome. The authors argue that an effective governance of the upscaling of hydrogen as a climate change solution and as a green industrial opportunity must therefore consider the specific conditions caused by its sectoral configurations and how they shape innovation processes over time in a given region or country.

Multi-system interactions are also at the core of the analysis by Mignon et al. (this volume) on the couplings between energy, transport, and food systems in the urban context of Gothenburg, Sweden. They reveal a multitude of such cross-system interconnections that can contribute to more sustainable individual systems, highlighting the identification of key multi-system actors. The authors analyse various municipal documents (policy recommendations, guidelines, action plans, and strategies) to map inter-system interactions between key actors, infrastructures/technologies, and institutions. Their analysis reveals a range of planned, emerging, and established technology couplings, implicit and explicit institutional couplings, and multiple key actor couplings. On this basis, the chapter identifies strengths and weaknesses of existing structures that have been put in place to foster multi-system interactions. Among the strengths, structural actor couplings showed the largest number of interlinkages across systems. Few interactions between the food system and the two other systems also highlight some of the challenges of effectively integrating and prioritizing multi-system linkages. It also underscores the difficulties of operationalizing visions for multi-system transitions, even in urban contexts which are often perceived to have conducive conditions for such interactions.

Mäkitie et al. (this volume) take yet a different approach to multi-sector interactions. They provide a first empirical application of a novel heuristic framework to conceptualize the complex interactions between sectors that affect a sustainability transition in a focal socio-technical system. This framework, the multi-sectoral perspective (MSP) (Andersen et al., 2026), conjoins insights from evolutionary economics, the multi-level perspective (MLP), and the technological innovation system (TIS) approach, and outlines three main analytical dimensions. Drawing on the MLP literature, the first key dimension in the MSP are the "horizontal" multi-sectoral interdependencies that exist across production, distribution, and end-use sectors, for example energy and food. These involved sectors all have their own regime and niche dynamics that may affect transition processes. Second, from the TIS approach, the MSP distinguishes “vertical” interdependencies between a focal sector (developing a technology such as electrical vehicles) and various component and material supply sectors, such as batteries and materials in electric mobility. The third analytical dimension of the MSP concerns how adjacent sectors may influence transition processes in a focal sector even in the absence of direct inter-sectoral couplings. Examples of such effects include technology spillovers from adjacent sectors into a focal sector, or competition over scarce resources. The authors apply the MSP to the case of electrification of maritime transport in Norway. The findings suggest that the MSP is useful for distilling the myriad of ‘enablers’ and ‘constrainers’ both within and beyond a focal sector that may impact transition processes.

Löhr (this volume) develops a novel conceptual differentiation or typology of multi-system configurations in the context of decarbonisation. She argues that through decarbonisation efforts that typically span multiple sectors, such as electrification, hydrogen or cross-sector value chains, sectors are coupled, and system boundaries redefined. Löhr proposes five distinct multi-system configurations: unconnected systems, emerging connections, interconnected systems, integrated systems, and new system emergence. For each configuration, the chapter conceptualizes systemic dimensions (technologies, institutions, actors) and defined the strategies of system entanglers. It also discusses shifts in system boundaries, and the dynamic sites in which systems interact. Empirically, the framework is employed in a comparative case study on the role of green hydrogen in the multi-system coupling of electricity, industry, mobility, and heating in Germany. Notably, the chapter advances our conceptual understanding of multi-system dynamics especially with regard to system interfaces and entanglers, as well as how these changes take place in different multi-system contexts.

In the final chapter in this section, Koutsimpogiorgos et al. (this volume) analyse the institutional reactions to the entry of Uber in countries across the globe. Uber’s app can be understood as a disruptive innovation stemming from the tech sector which is applied in the mobility sector. This, in turn, prompts ‘multi-system’ conflicts between digital platform companies that are rooted in the digital regime, and incumbent business rooted in the taxi regime. While there have been many previous case studies on institutional responses by governments and unions to ride-hailing platforms like Uber, this chapter provides the first global analysis of its kind, covering 81 different countries across the world. Doing so, the authors can explain the negative reactions that Uber encountered from prevailing national institutions including rule of law, openness to innovation, labour protection, and labour informality. The main takeaway from the analysis is that the strongest negative reactions occurred in the countries with well-functioning justice systems, showing that, while disruptive by intent, the institutionalization of tech innovations is contingent on national institutions and the role of incumbents.

## New Perspectives on Firms and Industries in Transitions

Alongside increased interest in transformation dynamics in existing industries and the complexities of multi-systems transformations, there has been a surge of interest in understanding agency (Farla et al., 2012) and particularly the role of established firms and incumbents in transition studies (e.g., Kungl, 2024; Magnusson & Werner, 2023; Turnheim & Sovacool, 2019). This also reflects increasing attention to emission-intensive and hard-to-abate sectors, such as heavy transport, petrochemicals, oil and gas, and process industries like steel and cement production. In these sectors, transitions rarely happen by existing systems being outcompeted by radical new solutions, and incumbent firms are likely to play major roles in any transition scenario. Thus, positioned against the "classic" narrative in transition studies of established firms and organizations as being barriers to radical transformation, the now quite broad literature on incumbency has described the varied roles established firms play in transitions, depending on phase in transitions, industrial characteristics, and firm level factors including size, resource base, market position, and diversification strategies (Kungl, 2024; Steen & Weaver, 2017).

An important insight emanating from research on transformation in multiple systems is that large firms typically dominate and orchestrate complex supply and value chains. Hence, their reach, also in sustainability transitions, extends far beyond their organizational boundaries (Bohnsack et al., 2020). This is particularly prevalent in mature industries, such as auto manufacturing or energy, where large lead firms dominate global supply and value chains. Moreover, global digital companies, such as Google and Meta, play orchestrating roles in multiple sectors across the globe, and harness immense economic and political power (Frenken & Fuenfschilling, 2021; Rikap, 2024). As power is increasingly concentrated in large platform companies, the complex relationships *between* firms and industries in sustainability transitions is an important topic for research.

This recognition that incumbents are powerful players not only through market dominance or value chain positions has also spurred attention to the roles they play in system building (Musiolik et al., 2020) and how they actively interpret and shape institutional contexts (Hoogstraaten et al., 2020; Kukk et al., 2016). This has been accompanied with attention to the direct and indirect ways incumbents influence institutional change and stability, including framing of behaviours and product categories, networking and alliance building, lobbying and working proactively with for example regulators. There is now a rich empirical literature that highlights the complex and sometimes paradoxical ways that established firms champion sustainability transitions, while simultaneously working towards maintaining status quo (e.g., Hahn et al., 2018; Smink et al., 2015).

This emerging interest in understanding firms and industries in transitions has brought the sustainability transitions literature closer to management and organization theory. There are currently attempts at cross-fertilization between these research fields, particularly as institutional theory and a general interest in the role of organizations has gained traction in sustainability transitions research (Fuenfschilling & Truffer, 2014; Kungl & Hess, 2021). The sustainability transitions literature has picked up and developed a wide range of concepts from organizational theory, but also the latter field has increasingly become interested in sustainability and how organizations deal with "grand challenges" (e.g., George et al., 2016). It is perhaps still mainly a one-way street, and management studies, other than the area of corporate sustainability research, has had limited engagement with sustainability transitions research. However, more attention to systemic perspectives has also been asked for in the management and organization studies communities that deal with sustainability (Grewatsch et al, 2023).

## Book Contributions on Firms and Industries in Transition

With these developments in mind, the contributions in the section on "industries and firms in transitions" address how firms manoeuvre in complex industrial and institutional landscapes to deal with sustainability concerns. As part of these developments, several of the chapters also address how firms view digitalisation as integral to transformation strategies towards sustainability. Two of the chapters in this section address circular economy as a firm and industry level transformation towards sustainability in manufacturing and transport, and three chapters look specifically at ongoing transformations in the oil and gas (O&G) sector.

The first chapter in this section is written by Ertelt (this volume) who examines the Swedish heavy-duty vehicle sector and its adoption of circular economy principles. Building on insights from institutional theory and the concept "meta-rules", Ertelt proposes an analytical framework to analyse the diffusions and (re)interpretation processes of circular economy meta-rules across different parts of the nascent circular value chain of the Swedish heavy-duty vehicle sector. Based on this analysis, the chapter discusses the broader implications of interpretative flexibility and clashing prioritizations of potential meta-rules for the speed and scope of multi-sector industrial transformations towards more circular practices.

In their chapter, Thune and Mäkitie (this volume) turn attention to industry structures and value chain relations to study incumbent firm behaviours towards sustainability. This chapter attempts to understand mimetic behaviour between lead and follower firms, and to explore whether positions and relationships in value chains may influence how and when firms engage in sustainability reorientation. This question is addressed in a longitudinal case study of two large incumbent firms in the Norwegian O&G value chain over a 15-year period. The chapter discusses how the relationship between the two firms have influenced sustainability reorientation in different ways and finally, discusses the implications of addressing value chains and inter-firm relationships when investigating the role of firms and industries in sustainability transitions.

The chapter by Ullern and Damman (this volume) explores the complexities faced by incumbent firms in the transition to a circular economy. Drawing on paradox theory, the chapter investigates how incumbents navigate opportunities and challenges in transition processes, through a longitudinal qualitative case study of a furniture manufacturing company. The analysis adds a more pluralized view of incumbents by illustrating contradictory yet interrelated aspects of internal and external activities in transition processes. The authors argue that such complexities can meaningfully be understood through paradox theory and implications for research on transition dynamics are discussed.

The two last chapters in this section look at the interplay between digitalization and sustainability in firms' reorientation strategies, both with examples from the energy sector. Hydle and Schupfer (this volume) look at digitalization and sustainability from an ecosystem lens. While past research on digitalization and sustainability has tended to focus on the organizational level, they argue that it is essential to consider entire ecosystems to comprehend how sustainability transformations driven by digitalization can impact entire industries in either incremental or radical manners (John et al., 2022). Empirically, they analyse digitalization and platformization in the O&G industry, seen as a typical case of a hard-to-abate sector. Their findings echo what Mäkitie et al. (2023) denote incremental twin innovation, meaning that digital sustainable innovations have limited transformational effects for sustainability. Hydle and Schupfer reflect on what it would take to move from incremental to radical improvements in O&G.

Finally, Soppe and Schupfer (this volume) explore how incumbents in carbon-intensive sectors increasingly interact with 'digital entrepreneurs' for innovation and sustainability impact. They study corporate venture capital (CVC) investments as a specific form of interaction between incumbents and niche actors. Their findings from a longitudinal case study of a Norwegian oil firm reveal that, although both CVC and digital technologies are widely lauded for their sustainability potential, the incumbent's investments primarily aim to optimize and sustain existing O&G operations. Framed within a corporate narrative of a decarbonized future, the CVC investments in digital entrepreneurs tend to make conventional business operations smarter—that is, more efficient—but not radically greener. Based on the case study, the authors discuss the untapped potential of niche-regime interactions and digital entrepreneurship in advancing sustainability transitions in hard-to-abate sectors.

## New Perspectives on Policy in Transitions

The role of policy in driving or inhibiting transitions has been a central research theme in the sustainability transitions literature. While connected to innovation policy, policies for sustainability transitions require more than a focus on addressing innovation system failures (see Klein Woolthuis et al. (2005) to make societies more innovative *in general*). Policies for transitions entail not only support for development of new, more sustainable alternatives, but also attention to phase-out of unsustainable solutions and a prioritization of radical over incremental change (Alkemade et al., 2011). Given the complexity of transition processes, a central topic in the policy-oriented transition literature is also the need for considering broader mixes of policy instruments and strategies, rather than analysing single instruments (Kern et al., 2019; Rogge, 2019).

In response to this, research over the last decade-and-a-half has led to the development of a range of new and more comprehensive policy approaches, such as 'transformative' or 'mission-oriented' innovation policy, which are often collectively referred to as 3rd generation innovation policy (Kuhlmann & Rip, 2018). As argued by Edler et al., (2025, p. 1), a key backdrop to the introduction of various transformative-oriented policy initiatives is a "*well-justified sense of urgency*" in dealing with the substantial challenges facing our societies, not least related to climate change and the trespassing of various planetary boundaries.

Transformative innovation policy centres on driving radical change towards greater sustainability in socio-technical systems, thereby prioritizing ecological and social well-being rather than economic growth. Core tenets include promoting inclusivity and broad stakeholder engagement, ensuring that diverse perspectives shape transition processes, as well as policies phasing out unsustainable solutions (Schot & Steinmueller, 2018). Transformative innovation policy encourages experimentation and adaptability, supporting flexible frameworks that can respond to emerging needs and uncertainties given the open-ended character of transition processes (Diercks et al., 2019; Grillitsch et al., 2019; Haddad et al., 2022).

In turn, the core to mission-oriented innovation policy is defining measurable and time-bound goals—so-called "missions"—that direct innovation efforts across sectors and disciplines as well as across the public and private spheres (Mazzucato, 2021). Rather than adhering to the traditional policy objective of levelling the playing field between technologies, missions are intended to tilt the playing field towards solutions that address core societal challenges. As such, missions provide guidance for research and innovation investments at a less specific level than technology-specific policies, yet more specific than navigating according to broader objectives such as the sustainable development goals. While it is acknowledged that today's missions are different from the Manhattan or Apollo missions in the sense that these did not have to compete with already existing technologies (Foray et al., 2012), mission-oriented innovation policy does not emphasize phase-out to the same extent as transformative innovation policy.

Connected to the theme of the book, work on transformative and mission-oriented innovation policy has given scarce attention to multi-system interactions, even though early contributions on this topic highlighted the importance of coordination across policy domains (Weber & Rohracher, 2012). From the literature on multi-system interactions, contributions give some attention to the role of and implications for policy, for example Rogge and Goedeking (2024) on electrification of mobility, Käsbohrer et al. (2024) on the interface of residential storage systems and electric vehicles, and Mäkitie et al. (2022) on complementarity formation mechanisms across sectors and technologies. However, the overall picture of the state of research is that there is very little policy-focused transition research with an explicit focus on transitions in multiple sectors.

Further, work on policies for transitions increasingly needs to take interfaces with multiple policy objectives into consideration. Partly this is motivated by the current strong emphasis (e.g., in the European policy context) on competitiveness, but the background is broader than this: security concerns, geopolitics, the rise of technology sovereignty and digitalization (Edler et al., 2023; Kivimaa, 2022; Mäkitie et al., 2023) all reshape the conditions for formulating and implementing transformative and mission-oriented innovation policy.

## Book Contributions on Policy in Transitions

This book includes four chapters that focus on novel transformative-oriented policy approaches. While the chapters differ with regard to scope and empirical context, they all highlight the interface between transformative aims and objectives related to stimulating industrial development and competitiveness.

In the chapter by Andersson and Hellsmark (this volume), the authors introduce a conceptual framework for analysing transformative policy missions. Inspired by an innovation systems approach, the framework distinguishes structural and functional outcomes and links these outcomes back to three aspects: the system in focus (setting the boundaries around the mission), the transformative orientation (the societal problem addressed by the mission), and policy interventions (the specific mission goals and instruments). The chapter uses this framework to analyse the decarbonization of Swedish energy-intensive industry, specifically focusing on the most central policy instrument in this context, the 'Industry Leap'. The results in terms of structural outcomes highlight significant mobilization of incumbent industrial actors to mission-oriented innovation networks. For functional outcomes, the chapter demonstrates that effects are concentrated for specific technologies, in particular carbon capture and hydrogen, while there is limited experimentation with alternative pathways. Overall, the chapter demonstrates how the analysis of transformative policy approaches such as missions can benefit from taking on-board insights from the innovation systems literature.

The second chapter in this section on policy by Elzinga et al. (this volume) also addresses mission-oriented policy. The authors contribute by explicitly focusing on interactions between missions, which has so far been overlooked in the literature. Given that many actors may interact with more than one mission, it becomes important to better understand synergistic and competitive relations between missions. The authors analyse two Dutch missions: electrification and circular economy missions. The results highlight that the different missions do not produce mutually reinforcing interactions. Rather, the chapter points to risks of actor confusion, fragmented resource allocation, and contestation over directionality. Jointly, this may compromise the ability of the different missions to achieve their objectives, hence, pointing towards the necessity of a more integrated approach to managing missions that explicitly takes interrelations into account.

In the chapter by Hansen, Hansen et al. (this volume), the authors address the new generation of industrial policy, which often explicitly focuses on greening of the economy, in particular through support for cleantech industries. The chapter reflects on the transformative capacity of this 'green' industrial policy approach, by bringing it into conversation with the literature on transformative innovation policy. Synergies and conflicts between green industrial policy and transformative innovation policy are considered in the latest wave of policies in the Global North, including the US Inflation Reduction Act and the EU Green Deal. The chapter concludes that the implemented policies to some extent reflect key priorities in the transformative innovation policy approach, such as providing clear directionality towards decarbonisation, explicitly addressing demand articulation, and prioritizing coordination between policy areas. At the same time, this alignment between green industrial policy and transformative innovation policy is conditioned by whether policymakers assess that domestic industries are competitive in emerging green technologies. If that is not the case, green industrial policy can not be expected to take a transformative character.

In the final chapter on policy, John et al. (this volume) similarly turn to policies with multiple purposes, focusing on twin transition policies, i.e., the opportunities for policies to simultaneously stimulate digitalization and sustainability. While policy practice often assumes positive relations between digitalization and sustainability, research demonstrates that this is often not the case. In their contribution, the authors develop a framework that helps guide policy action towards the twin transition by identifying four complementary policy intervention logics. Briefly explained, two approaches are developed: first, a technological approach where policies either stimulate the use of digital technologies to either directly result in positive sustainability gains or focus digital technologies on addressing barriers to sustainable innovation; second, a socio-institutional approach, where policies target barriers that hinder both digital and sustainable innovation, or direct the development of the innovation system around digital technologies in specific user contexts towards sustainable innovation barriers. The different identified policy logics are explored in a case study of the Dutch energy-intensive processing industries. The chapter contributes by providing a more granular view of twin transition policy, thereby disentangling the various options for formulating twin transition policy.

## Contributions and Outlook

As the preceding overview of chapters shows, this book engages conceptually and empirically with key current themes in transition studies and related fields. To briefly summarize, the contributions in this volume revolve around three key themes in relation to sustainability transitions: multi-system dynamics, firms and industries, and policies. Several of the chapters also deal with digitalization, as a second generic megatrend alongside environmental sustainability that drives significant change across different production-consumption systems.

The book as a whole provides a variety of novel as well as adapted analytical and conceptual perspectives to understand multi-system dynamics, industrial transformation, the role of firms and actors, the connections between sustainability and digitalisation, and new policy approaches. The different chapters' conceptual contributions are underpinned by empirical work in a diversity of sectors and geographical contexts.

While extensive, the contributions in this volume are certainly far from exhaustive. On the contrary, the different chapters in this book point towards a range of phenomenon where considerably more research is needed. As such, we see this volume as part of a broader collective endeavour in the scientific community to develop new knowledge about sustainability transitions and their dynamics—how they emerge, unfold, and ways in which they can be nurtured and governed. Because the world is constantly changing—and with that the phenomenon (or objects of study) we are interested in—our theories, concepts, and methodologies also need to be adapted. In light of this, the last chapter in this book (Hansen, Frenken, et al., this volume) addresses our suggestions for further research and broader emerging concerns.

## References

Aiginger, K., & Rodrik, D. (2020). Rebirth of industrial policy and an agenda for the twenty-first century. *Journal of Industry, Competition and Trade, 20*(2), 189–207. https://doi.org/10.1007/s10842-019-00322-3

Alkemade, F., Hekkert, M. P., & Negro, S. O. (2011). Transition policy and innovation policy: Friends or foes? *Environmental Innovation and Societal Transitions, 1*(1), 125–129. https://doi.org/10.1016/j.eist.2011.04.009

Allan, B., Lewis, J. I., & Oatley, T. (2021). Green industrial policy and the global transformation of climate politics. *Global Environmental Politics, 21*(4), 1–19. https://doi.org/10.1162/glep_a_00640

Andersen, A. D., Frenken, K., Galaz, V., Kern, F., Klerkx, L., Mouthaan, M., Piscicelli, L., Schor, J. B., & Vaskelainen, T. (2021). On digitalization and sustainability transitions. *Environmental Innovation and Societal Transitions, 41*, 96–98. https://doi.org/10.1016/j.eist.2021.09.013

Andersen, A. D., & Geels, F. W. (2023). Multi-system dynamics and the speed of net-zero transitions: Identifying causal processes related to technologies, actors, and institutions. *Energy Research & Social Science, 102*, Article 103178. https://doi.org/10.1016/j.erss.2023.103178

Andersen, A. D., & Markard, J. (2026). Multi-system dynamics in sustainability transitions: Introduction and outlook. In J. Wesche & A. Hendriks (Eds.), Introduction to Sustainability Transitions Research (pp. 178–200). Cambridge University Press.

Andersen, A. D., Mäkitie, T., Steen, M., & Wanzenböck, I. (2026). Conceptualizing the relationship between industrial transformation and accelerating sustainability transitions: A multi-sectoral perspective. *Energy Research & Social Science 131,* 104509. https://doi.org/10.1016/j.erss.2025.104509

Andersen, A. D., Steen, M., Mäkitie, T., Hanson, J., Thune, T. M., & Soppe, B. (2020). The role of inter-sectoral dynamics in sustainability transitions: A comment on the transitions research agenda. *Environmental Innovation and Societal Transitions, 34*, 348–351. https://doi.org/10.1016/j.eist.2019.11.009

Andersson, J., & Hellsmark, H. (this volume). The scope and outcomes of transformative policy missions: A conceptual framework and empirical exploration of Swedish industry decarbonization. In M. Steen, K. Frenken, T. Hansen, T. Mäkitie, & T. M. Thune (Eds.), *Sustainability transitions and industrial transformation: Multi-sector dynamics and policy perspectives.* Palgrave.

Arcand, B., & Meadowcroft, J. (2024). Net zero by 2050: The case for green industrial policy. *Environmental Politics,* 1–21. https://doi.org/10.1080/09644016.2024.2375899

Ateş, A., Lovell, K., & Rogge, K. S. (this volume). Understanding multi-system interlinkages between energy and mobility systems and their governance in the UK. In M. Steen, K. Frenken, T. Hansen, T. Mäkitie, & T. M. Thune (Eds.), *Sustainability transitions and industrial transformation: Multi-sector dynamics and policy perspectives.* Palgrave.

Bach, H., & Hansen, T. (2023). Flickering guiding light from the international maritime organisation's policy mix. *Environmental Innovation and Societal Transitions, 47*, 100720. https://doi.org/10.1016/j.eist.2023.100720

Bach, H., Mäkitie, T., Hansen, T., & Steen, M. (2021). Blending new and old in sustainability transitions: Technological alignment between fossil fuels and biofuels in Norwegian coastal shipping. *Energy Research & Social Science, 74*, 101957. https://doi.org/10.1016/j.erss.2021.101957

Bergek, A., Hekkert, M., Jacobsson, S., Markard, J., Sandén, B., & Truffer, B. (2015). Technological innovation systems in contexts: Conceptualizing contextual structures and interaction dynamics. *Environmental Innovation and Societal Transitions, 16*, 51–64. https://doi.org/10.1016/j.eist.2015.07.003

Bergek, A., Jacobsson, S., Carlsson, B., Lindmark, S., & Rickne, A. (2008). Analyzing the functional dynamics of technological innovation systems: A scheme of analysis. *Research Policy, 37*(3), 407–429. https://doi.org/10.1016/j.respol.2007.12.003

Berkhout, F., Smith, A., & Stirling, A. (2004). Socio-technological regimes and transition contexts. In B. Elzen, F. W. Geels, & K. Green (Eds.), *System innovation and the transition to sustainability: Theory, evidence and policy* (pp. 48–75). Edward Elgar.

Bohnsack, R., Kolk, A., Pinkse, J., & Bidmon, C. M. (2020). Driving the electric bandwagon: The dynamics of incumbents' sustainable innovation. *Business Strategy and the Environment, 29*(2), 727–743. https://doi.org/10.1002/bse.2430

Borrás, S., Haakonsson, S., Hendriksen, C., Gerli, F., Poulsen, R. T., Pallesen, T., Somavilla Croxatto, L., Kugelberg, S., & Larsen, H. (2024). The transformative capacity of public sector organisations in sustainability transitions. *Environmental Innovation and Societal Transitions, 53*, 100904. https://doi.org/10.1016/j.eist.2024.100904

Carlsson, B., & Stankiewicz, R. (1991). On the nature, function and composition of technological systems. *Journal of Evolutionary Economics, 1*(2), 93–118. https://doi.org/10.1007/BF01224915

Circle Economy. (2023). *The circularity gap report 2023*. https://www.circularity-gap.world/2023#download

De Roeck, F., & Van Poeck, K. (2023). Agency in action: Towards a transactional approach for analyzing agency in sustainability transitions. *Environmental Innovation and Societal Transitions, 48*, 100757. https://doi.org/10.1016/j.eist.2023.100757

Diercks, G., Larsen, H., & Steward, F. (2019). Transformative innovation policy: Addressing variety in an emerging policy paradigm. *Research Policy, 48*(4), 880–894. https://doi.org/10.1016/j.respol.2018.10.028

Edler, J., Blind, K., Kroll, H., & Schubert, T. (2023). Technology sovereignty as an emerging frame for innovation policy. Defining rationales, ends and means. *Research Policy, 52*(6), 104765. https://doi.org/10.1016/j.respol.2023.104765

Edler, J., Matt, M., Polt, W., Weber, M., Edler, J., Matt, M., Polt, W., & Weber, M. (2025). Transformative mission-oriented innovation policies: Revisiting the role of science, technology and innovation in society. In J. Edler, M. Matt, W. Polt, & M. Weber (Eds.), *Introduction: The motivation for this volume and an overview of its contributions* (pp. 1–14). Edward Elgar Publishing. https://doi.org/10.4337/9781803929521.00006

Elzen, B., Geels, F., & Green, K. (Eds.). (2004). *System innovation and the transition to sustainability: Theory*. Edward Elgar.

Elzinga, R., Negro, S. O., & Janssen, M. (this volume). The more the merrier? A study on multiple mission-specific innovation systems in the Dutch energy sector. In M. Steen, K. Frenken, T. Hansen, T. Mäkitie, & T. M. Thune (Eds.), *Sustainability transitions and industrial transformation: Multi-sector dynamics and policy perspectives*. Palgrave.

Ertelt, S. -M. (this volume). (Re)interpreting circularity? Understanding the contested directionalities of the Swedish heavy-duty vehicle sector towards the circular economy. In M. Steen, K. Frenken, T. Hansen, T. Mäkitie, & T. M. Thune (Eds.), *Sustainability transitions and industrial transformation: Multi-sector dynamics and policy perspectives*. Palgrave.

Farla, J., Markard, J., Raven, R., & Coenen, L. (2012). Sustainability transitions in the making: A closer look at actors, strategies and resources. *Technological Forecasting and Social Change, 79*(6), 991–998. https://doi.org/10.1016/j.techfore.2012.02.001

Foray, D., Mowery, D. C., & Nelson, R. R. (2012). Public R&D and social challenges: What lessons from mission R&D programs? *Research Policy, 41*(10), 1697–1702. https://doi.org/10.1016/j.respol.2012.07.011

Frenken, K., & Fuenfschilling, L. (2021). The rise of online platforms and the triumph of the corporation. *Sociologica, 14*(3), 101–113. https://doi.org/10.6092/issn.1971-8853/1171

Fuenfschilling, L., & Truffer, B. (2014). The structuration of socio-technical regimes—Conceptual foundations from institutional theory. *Research Policy, 43*(4), 772–791. https://doi.org/10.1016/j.respol.2013.10.010

Geels, F. W. (2002). Technological transitions as evolutionary reconfiguration processes: A multi-level perspective and a case-study. *Research Policy, 31*(8–9), 1257–1274. https://doi.org/10.1016/s0048-7333(02)00062-8

Geels, F. W. (2004). From sectoral systems of innovation to socio-technical systems: Insights about dynamics and change from sociology and institutional theory. *Research Policy, 33*(6–7), 897–920. https://doi.org/10.1016/j.respol.2004.01.015

Geels, F. W. (2005). Processes and patterns in transitions and system innovations: Refining the co-evolutionary multi-level perspective. *Technological Forecasting and Social Change, 72*(6), 681–696. https://doi.org/10.1016/j.techfore.2004.08.014

Geels, F. W. (2010). Ontologies, socio-technical transitions (to sustainability), and the multi-level perspective. *Research Policy, 39*(4), 495–510. https://doi.org/10.1016/j.respol.2010.01.022

Geels, F. W., Hekkert, M. P., & Jacobsson, S. (2008). The dynamics of sustainable innovation journeys. *Technology Analysis & Strategic Management, 20*(5), 521–536. https://doi.org/10.1080/09537320802292982
Geels, F. W., & Kemp, R. (2007). Dynamics in socio-technical systems: Typology of change processes and contrasting case studies. *Technology in Society, 29*(4), 441–455. https://doi.org/10.1016/j.techsoc.2007.08.009
Geels, F. W., & Schot, J. (2007). Typology of sociotechnical transition pathways. *Research Policy, 36*(3), 399–417. https://doi.org/10.1016/j.respol.2007.01.003
George, R. A., Siti-Nabiha, A. K., Jalaludin, D., & Abdalla, Y. A. (2016). Barriers to and enablers of sustainability integration in the performance management systems of an oil and gas company. *Journal of Cleaner Production, 136*, 197–212. https://doi.org/10.1016/j.jclepro.2016.01.097
Gong, H., & Andersen, A. D. (2024). The role of material resources for rapid technology diffusion in net-zero transitions: Insights from EV lithium-ion battery Technological Innovation System in China. *Technological Forecasting and Social Change, 200*, 123141. https://doi.org/10.1016/j.techfore.2023.123141
Grewatsch, S., Kennedy, S., & Bansal, P. (2023). Tackling wicked problems in strategic management with systems thinking. *Strategic Organization, 21*(3), 721–732. https://doi.org/10.1177/14761270211038635
Grillitsch, M., Hansen, T., Coenen, L., Miörner, J., & Moodysson, J. (2019). Innovation policy for system-wide transformation: The case of strategic innovation programmes (SIPs) in Sweden. *Research Policy, 48*(4), 1048–1061. https://doi.org/10.1016/j.respol.2018.10.004
Grin, J., Rotmans, J., & Schot, J. (2010). *Transitions to sustainable development: New directions in the study of long term transformative change*. Routledge.
Haddad, C. R., Nakić, V., Bergek, A., & Hellsmark, H. (2022). Transformative innovation policy: A systematic review. *Environmental Innovation and Societal Transitions, 43*, 14–40. https://doi.org/10.1016/j.eist.2022.03.002
Hahn, T., Figge, F., Pinkse, J., & Preuss, L. (2018). A paradox perspective on corporate sustainability: Descriptive, instrumental, and normative aspects. *Journal of Business Ethics, 148*(2), 235–248. https://doi.org/10.1007/s10551-017-3587-2
Hansen, T., Frenken, K., Mäkitie, T., Steen, M., & Thune, T. M. (this volume). Multi-system dynamics, industrial transformation and transition policy: Future research priorities. In M. Steen, K. Frenken, T. Hansen, T. Mäkitie, & T. M. Thune (Eds.), *Sustainability transitions and industrial transformation: Multi-sector dynamics and policy perspectives*. Palgrave.

Hansen, T., Hansen, U. E., & Steen, M. (this volume). The transformative capacity of green industrial policy. In M. Steen, K. Frenken, T. Hansen, T. Mäkitie, & T. M. Thune (Eds.), *Sustainability transitions and industrial transformation: Multi-sector dynamics and policy perspectives*. Palgrave.

Hanson, J., & Mäkitie, T. (this volume). The sectoral configurations of the hydrogen innovation system in Norway. In M. Steen, K. Frenken, T. Hansen, T. Mäkitie, & T. M. Thune (Eds.), *Sustainability transitions and industrial transformation: Multi-sector dynamics and policy perspectives*. Palgrave.

Hebinck, A., Klerkx, L., Elzen, B., Kok, K. P. W., König, B., Schiller, K., Tschersich, J., van Mierlo, B., & von Wirth, T. (2021). Beyond food for thought-Directing sustainability transitions research to address fundamental change in agri-food systems. *Environmental Innovation and Societal Transitions, 41*, 81–85. https://doi.org/10.1016/j.eist.2021.10.003

Hekkert, M. P., Suurs, R. A. A., Negro, S. O., Kuhlmann, S., & Smits, R. E. H. M. (2007). Functions of innovation systems: A new approach for analysing technological change. *Technological Forecasting and Social Change, 74*(4), 413–432. https://doi.org/10.1016/j.techfore.2006.03.002

Hoogstraaten, M. J., Frenken, K., & Boon, W. P. C. (2020). The study of institutional entrepreneurship and its implications for transition studies. *Environmental Innovation and Societal Transitions, 36*, 114–136. https://doi.org/10.1016/j.eist.2020.05.004

Hughes, T. P. (1983). *Networks of power: Electrification in western society, 1880–1930*. The Johns Hopkins University Press.

Huttunen, S., Kivimaa, P., & Virkamäki, V. (2014). The need for policy coherence to trigger a transition to biogas production. *Environmental Innovation and Societal Transitions, 12*, 14–30. https://doi.org/10.1016/j.eist.2014.04.002

Hydle, K. M., & Schupfer, H. (this volume). Digitalization and platformization for sustainability: The case of the oil and gas industry. In M. Steen, K. Frenken, T. Hansen, T. Mäkitie, & T. M. Thune (Eds.), *Sustainability transitions and industrial transformation: Multi-sector dynamics and policy perspectives*. Palgrave.

IEA. (2023). *Tracking clean energy progress 2023*. https://www.iea.org/reports/tracking-clean-energy-progress-2023

Jacobsson, S., & Johnson, A. (2000). The diffusion of renewable energy technology: An analytical framework and key issues for research. *Energy Policy, 28*(9), 625–640. https://doi.org/10.1016/s0301-4215(00)00041-0

John, N., Wesseling, J., & Frenken, K. (this volume). Navigating twin transitions: A policy framework to align digital innovation with sustainability, applied to the Dutch processing industry. In M. Steen, K. Frenken, T. Hansen, T. Mäkitie, & T. M. Thune (Eds.), *Sustainability transitions and industrial transformation: Multi-sector dynamics and policy perspectives*. Palgrave.

John, N., Wesseling, J. H., Worrell, E., & Hekkert, M. (2022). How key-enabling technologies' regimes influence sociotechnical transitions: The impact of artificial intelligence on decarbonization in the steel industry. *Journal of Cleaner Production, 370*, 133624. https://doi.org/10.1016/j.jclepro.2022.133624

Johnson, A. (1998). *Functions in innovation system approaches*. Department of industrial dynamics. Chalmers University of Technology.

Kanda, W., Geissdoerfer, M., & Hjelm, O. (2021). From circular business models to circular business ecosystems. *Business Strategy and the Environment, 30*(6), 2814–2829. https://doi.org/10.1002/bse.2895

Kemp, R., Schot, J., & Hoogma, R. (1998). Regime shifts to sustainability through processes of niche formation: The approach of strategic niche management. *Technology Analysis & Strategic Management, 10*(2), 175–198. https://doi.org/10.1080/09537329808524310

Kern, F., Rogge, K. S., & Howlett, M. (2019). Policy mixes for sustainability transitions: New approaches and insights through bridging innovation and policy studies. *Research Policy, 48*(10), 103832. https://doi.org/10.1016/j.respol.2019.103832

Kivimaa, P. (2022). Policy and political (in)coherence, security and Nordic-Baltic energy transitions. *Oxford Open Energy, 1*, 1–14. https://doi.org/10.1093/ooenergy/oiac009

Klein Woolthuis, R., Lankhuizen, M., & Gilsing, V. (2005). A system failure framework for innovation policy design. *Technovation, 25*(6), 609–619. https://doi.org/10.1016/j.technovation.2003.11.002

Klerkx, L., & Begemann, S. (2020). Supporting food systems transformation: The what, why, who, where and how of mission-oriented agricultural innovation systems. *Agricultural Systems, 184*, 102901. https://doi.org/10.1016/j.agsy.2020.102901

Konrad, K., Truffer, B., & Voss, J. P. (2008). Multi-regime dynamics in the analysis of sectoral transformation potentials: Evidence from German utility sectors. *Journal of Cleaner Production, 16*(11), 1190–1202. https://doi.org/10.1016/j.jclepro.2007.08.014

Koutsimpogiorgos, N., Frenken, K., Herrmann, A. M., & Punt, M. B. (this volume). Pulling the brake: Worldwide institutional reactions to Uber's entry. In M. Steen, K. Frenken, T. Hansen, T. Mäkitie, & T. M. Thune (Eds.), *Sustainability transitions and industrial transformation: Multi-sector dynamics and policy perspectives*. Palgrave.

Kuhlmann, S., & Rip, A. (2018). Next-generation innovation policy and grand challenges. *Science and Public Policy, 45*(4), 448–454. https://doi.org/10.1093/scipol/scy011

Kukk, P., Moors, E. H. M., & Hekkert, M. P. (2016). Institutional power play in innovation systems: The case of Herceptin®. *Research Policy, 45*(8), 1558–1569. https://doi.org/10.1016/j.respol.2016.01.016

Kungl, G. (2024). Challenges of the current discourse on incumbent firms in sustainability transitions. *Energy Research & Social Science, 108*, 103367. https://doi.org/10.1016/j.erss.2023.103367

Kungl, G., & Hess, D. J. (2021). Sustainability transitions and strategic action fields: A literature review and discussion. *Environmental Innovation and Societal Transitions, 38*, 22–33. https://doi.org/10.1016/j.eist.2020.10.004

Käsbohrer, A., Hansen, T., & Zademach, H.-M. (2024). Multi-system interactions and institutional work: Actor interactions at the interface of residential storage systems and electric vehicles in Germany. *Environmental Innovation and Societal Transitions, 51*, 100844. https://doi.org/10.1016/j.eist.2024.100844

Köhler, J., Geels, F. W., Kern, F., Markard, J., Onsongo, E., Wieczorek, A., Alkemade, F., Avelino, F., Bergek, A., Boons, F., Fünfschilling, L., Hess, D., Holtz, G., Hyysalo, S., Jenkins, K., Kivimaa, P., Martiskainen, M., McMeekin, A., Mühlemeier, M. S.,…Wells, P. (2019). An agenda for sustainability transitions research: State of the art and future directions. *Environmental Innovation and Societal Transitions, 31*, 1–32. https://doi.org/10.1016/j.eist.2019.01.004

Lovio, R., Mickwitz, P., & Heiskanen, E. (2011). Path dependence, path creation and creative destruction in the evolution of energy systems. In R. Wüstenhagen & R. Wuebker (Eds.), *Handbook of research on energy entrepreneurship* (pp. 274–301). Edward Elgar.

Löhr, M. (this volume). Conceptualising multi-system configurations: The case of hydrogen in Germany. In M. Steen, K. Frenken, T. Hansen, T. Mäkitie, & T. M. Thune (Eds.), *Sustainability transitions and industrial transformation: Multi-sector dynamics and policy perspectives*. Palgrave.

Magnusson, T., & Werner, V. (2023). Conceptualisations of incumbent firms in sustainability transitions: Insights from organisation theory and a systematic literature review. *Business Strategy and the Environment*, *32*(2), 903–919. https://doi.org/10.1002/bse.3081

Markard, J. (2018). The next phase of the energy transition and its implications for research and policy. *Nature Energy*, *3*(8), 628–633. https://doi.org/10.1038/s41560-018-0171-7

Markard, J., Raven, R., & Truffer, B. (2012). Sustainability transitions: An emerging field of research and its prospects. *Research Policy*, *41*(6), 955–967. https://doi.org/10.1016/j.respol.2012.02.013

Mazzucato, M. (2021). *Mission economy: A moonshot guide to changing capitalism*. Penguin.

McMeekin, A., Geels, F. W., & Hodson, M. (2019). Mapping the winds of whole system reconfiguration: Analysing low-carbon transformations across production, distribution and consumption in the UK electricity system (1990–2016). *Research Policy*, *48*(5), 1216–1231. https://doi.org/10.1016/j.respol.2018.12.007

Mignon, I., Onufrey, K., & Thomson, L. (this volume). Mapping multi-system interactions in urban contexts—A study of the energy, transport and food systems in the municipality of Gothenburg. In M. Steen, K. Frenken, T. Hansen, T. Mäkitie, & T. M. Thune (Eds.), *Sustainability transitions and industrial transformation: Multi-sector dynamics and policy perspectives*. Palgrave.

Muench, S., Stoermer, E., Jensen, K., Asikainen, T., Salvi, M., & Scapolo, F. (2022). *Towards a green and digital future*. Publications Office of the European Union, Issue. https://data.europa.eu/doi/10.2760/977331

Musiolik, J., Markard, J., Hekkert, M., & Furrer, B. (2020). Creating innovation systems: How resource constellations affect the strategies of system builders. *Technological Forecasting and Social Change*, *153*, 119209. https://doi.org/10.1016/j.techfore.2018.02.002

Mäkitie, T., Andersen, A. D., & Steen, M. (this volume). The multi-sectoral interactions of electrification in Norwegian coastal shipping. In M. Steen, K. Frenken, T. Hansen, T. Mäkitie, & T. M. Thune (Eds.), *Sustainability transitions and industrial transformation: Multi-sector dynamics and policy perspectives*. Palgrave.

Mäkitie, T., Hanson, J., Damman, S., & Wardeberg, M. (2023). Digital innovation's contribution to sustainability transitions. *Technology in Society*, *73*, 102255. https://doi.org/10.1016/j.techsoc.2023.102255

Mäkitie, T., Hanson, J., Steen, M., Hansen, T., & Andersen, A. D. (2022). Complementarity formation mechanisms in technology value chains. *Research Policy, 51*(7), 104559. https://doi.org/10.1016/j.respol.2022.104559

Nykamp, H., Andersen, A. D., & Geels, F. W. (2023). Low-carbon electrification as a multi-system transition: A socio-technical analysis of Norwegian maritime transport, construction, and chemical sectors. *Environmental Research Letters, 18*(9), Article 094059. https://doi.org/10.1088/1748-9326/acf67a

Pfeffer, D., Reike, D., & Bening, C. R. (2025). Analyzing policy mixes for the circular economy transition: The case of recycled plastics in electronics. *Environmental Innovation and Societal Transitions, 56*, 100982. https://doi.org/10.1016/j.eist.2025.100982

Rikap, C. (2024). Varieties of corporate innovation systems and their interplay with global and national systems: Amazon, Facebook, Google and Microsoft's strategies to produce and appropriate artificial intelligence. *Review of International Political Economy, 31*(6), 1735–1763. https://doi.org/10.1080/09692290.2024.2365757

Rip, A., & Kemp, R. (1998). Technological change. In S. Rayner & E. L. Malone (Eds.), *Human choice and climate change—Resources and technology*. Battelle Press.

Rodrik, D. (2014). Green industrial policy. *Oxford Review of Economic Policy, 30*(3), 469–491. https://doi.org/10.1093/oxrep/gru025

Rogge, K. S. (2019). Policy mixes for sustainable innovation: Conceptual considerations and empirical insights. In F. Boons & A. McMeekin (Eds.), *Handbook of sustainable innovation* (pp. 165–185). Edward Elgar Publishing.

Rogge, K. S., & Goedeking, N. (2024). Challenges in accelerating net-zero transitions: Insights from transport electrification in Germany and California. *Environmental Research Letters, 19*(4), Article 044007. https://doi.org/10.1088/1748-9326/ad2d84

Rogge, K. S., & Reichardt, K. (2016). Policy mixes for sustainability transitions: An extended concept and framework for analysis. *Research Policy, 45*(8), 1620–1635. https://doi.org/10.1016/j.respol.2016.04.004

Rosenbloom, D. (2020). Engaging with multi-system interactions in sustainability transitions: A comment on the transitions research agenda. *Environmental Innovation and Societal Transitions, 34*, 336–340. https://doi.org/10.1016/j.eist.2019.10.003

Sandén, B. A., & Hillman, K. M. (2011). A framework for analysis of multi-mode interaction among technologies with examples from the history of alternative transport fuels in Sweden. *Research Policy, 40*(3), 403–414. https://doi.org/10.1016/j.respol.2010.12.005

Sareen, S., & Haarstad, H. (2021). Digitalization as a driver of transformative environmental innovation. *Environmental Innovation and Societal Transitions, 41*, 93–95. https://doi.org/10.1016/j.eist.2021.09.016

Schot, J., & Steinmueller, W. E. (2018). Three frames for innovation policy: R&D, systems of innovation and transformative change. *Research Policy, 47*(9), 1554–1567. https://doi.org/10.1016/j.respol.2018.08.011

Smink, M. M., Hekkert, M. P., & Negro, S. O. (2015). Keeping sustainable innovation on a leash? Exploring incumbents' institutional strategies. *Business Strategy and the Environment, 24*(2), 86–101. https://doi.org/10.1002/bse.1808

Smith, A., & Raven, R. (2012). What is protective space? Reconsidering niches in transitions to sustainability. *Research Policy, 41*(6), 1025–1036. https://doi.org/10.1016/j.respol.2011.12.012

Smith, A., Stirling, A., & Berkhout, F. (2005). The governance of sustainable socio-technical transitions. *Research Policy, 34*(10), 1491–1510. https://doi.org/10.1016/j.respol.2005.07.005

Soppe, B., & Schupfer, H. (this volume). Making oil and gas smarter, not greener: Fossil fuel incumbents' corporate venture capital investments in digital entrepreneurs during sustainability transitions. In M. Steen, K. Frenken, T. Hansen, T. Mäkitie, & T. M. Thune (Eds.), *Sustainability transitions and industrial transformation: Multi-sector dynamics and policy perspectives*. Palgrave.

Steen, M., & Weaver, T. (2017). Incumbents' diversification and cross-sectorial energy industry dynamics. *Research Policy, 46*(6), 1071–1086. https://doi.org/10.1016/j.respol.2017.04.001

Stephan, A., Schmidt, T. S., Bening, C. R., & Hoffmann, V. H. (2017). The sectoral configuration of technological innovation systems: Patterns of knowledge development and diffusion in the lithium-ion battery technology in Japan. *Research Policy, 46*(4), 709–723. https://doi.org/10.1016/j.respol.2017.01.009

Thune, T. M., & Mäkitie, T. (this volume). Captains of industry? Value chain relations and sustainability reorientation of firms. In M. Steen, K. Frenken, T. Hansen, T. Mäkitie, & T. M. Thune (Eds.), *Sustainability transitions and industrial transformation: Multi-sector dynamics and policy perspectives*. Palgrave.

Truffer, B., Rohracher, H., Kivimaa, P., Raven, R., Alkemade, F., Carvalho, L., & Feola, G. (2022). A perspective on the future of sustainability transitions research. *Environmental Innovation and Societal Transitions, 42*, 331–339. https://doi.org/10.1016/j.eist.2022.01.006

Turnheim, B., & Sovacool, B. K. (2019). Forever stuck in old ways? Pluralising incumbencies in sustainability transitions. *Environmental Innovation and Societal Transitions*. https://doi.org/10.1016/j.eist.2019.10.012

Ullern, E. F., & Damman, S. (this volume). Navigating paradoxes in sustainability transitions: how incumbent firms (can) foster circular economy. In M. Steen, K. Frenken, T. Hansen, T. Mäkitie, & T. M. Thune (Eds.), *Sustainability transitions and industrial transformation: Multi-sector dynamics and policy perspectives*. Palgrave.

Verbong, G., & Geels, F. (2007). The ongoing energy transition: Lessons from a socio-technical, multi-level analysis of the Dutch electricity system (1960–2004). *Energy Policy, 35*(2), 1025–1037. https://doi.org/10.1016/j.enpol.2006.02.010

Weber, K. M., & Rohracher, H. (2012). Legitimizing research, technology and innovation policies for transformative change: Combining insights from innovation systems and multi-level perspective in a comprehensive 'failures' framework. *Research Policy, 41*(6), 1037–1047. https://doi.org/10.1016/j.respol.2011.10.015

# 2

# Understanding Multi-System Interlinkages Between Energy and Mobility Systems and Their Governance in the UK

Aslı Ateş, Katherine Lovell, and Karoline S. Rogge

## Introduction

In the race to decarbonize the systems integral to our lives, countries have prioritized strategies to meet net-zero targets under the Paris Agreement. While energy generation has seen significant progress by shifting from fossil fuels to renewables (IEA, 2023), decarbonizing transport, heating, industry, and the built environment requires additional effort.

A. Ateş (✉) · K. Lovell · K. S. Rogge
Science Policy Research Unit–SPRU, University of Sussex, Brighton, UK
e-mail: a.ate@sussex.ac.uk; ates.as@outlook.com

K. Lovell
e-mail: katherine.a.lovell@durham.ac.uk

K. S. Rogge
e-mail: K.Rogge@sussex.ac.uk; karoline.rogge@isi.fraunhofer.de


M. Steen et al. (eds.), *Sustainability Transitions and Industrial Transformation*,
https://doi.org/10.1007/978-981-95-4893-4_2

Against this backdrop, electrification[1] is increasingly acknowledged as a key decarbonisation strategy for these sectors (Andersen & Geels, 2023).

Like many countries, the UK has made significant progress in decarbonising the energy sector with the aim of full decarbonisation of electricity sector by 2035 (DESNZ & BEIS, 2021). In the first quarter of 2024, renewables accounted for approximately 50.9% of the UK's electricity mix (DESNZ, 2024). More recently, the transport[2] sector has started to undergo major changes aimed at reducing greenhouse gas emissions through electrification. By the end of February 2024, there were more than one million electric cars, with BEVs constituting 15.8% of the market share for new car registrations (SMMT, 2024). Parallel to this, the EV charging infrastructure has also expanded rapidly. At the end of June 2024, there were 64,775 public charging points across 33,829 locations (Zap-Map, 2024).

This decarbonisation progress has been shaped significantly by the smart and flexible systems pathway for the UK (BEIS & Ofgem, 2017, 2021) which aims to provide reliable, low-carbon and affordable energy while responding to increasing demand arising from the electrification of other sectors, such as through the transition to battery-electric vehicles in the transport sector, or from the electrification of heating. For this, the government's focus has been on designing and supporting structures and mechanisms that could enable uses of the storage capacity

K. S. Rogge
Innovation Research ISI, Fraunhofer Institute for Systems, Karlsruhe, Germany

K. Lovell
Department of Engineering, Durham University, Durham, United Kingdom

[1] IEA defines electrification as *replacing technologies or processes that use fossil fuels, like internal combustion engines and gas boilers, with electrically powered equivalents, such as electric vehicles or heat pumps*. https://www.iea.org/energy-system/electricity/electrification.

[2] While official sources often refer to "transport" when discussing greenhouse gas (GHG) emissions and decarbonization efforts, this paper adopts a broader conceptualization by referring to "mobility systems". The term "mobility" encompasses not only the movement of vehicles but also wider socio-technical transformations that include changes in infrastructure, user behaviours, and digital technologies that enable new modes of transportation.

of battery-electric vehicles to better integrate the energy and mobility systems (DESNZ & DfT, 2023).

The UK's electrification strategy is an example of the emerging phenomenon of higher interactions between different systems referred to as multi-system interactions. Scholars have attributed increasingly high importance to this phenomenon due to its potential to contribute to the acceleration of sustainability transitions (Andersen & Geels, 2023; Andersen & Markard, 2024; Markard et al., 2020; Schot & Kanger, 2018). This study aims to contribute to this emerging research stream by investigating the UK's energy and mobility systems as a case study that follows a low-carbon electrification pathway (Andersen & Geels, 2023). For this, we develop an analytical framework derived from the large technical systems (LTS) literature. The analytical lens focuses on classifying and describing the interactions between systems through three types of *interlinkages*: functional, technical, and organizational. With this, we aim to answer the following research question:

> How have the UK's energy and mobility systems' interactions been evolving for net-zero transitions?

The following sections present our literature review, analytical framework, methodology, case study, results, and discussion of implications for emerging multi-system interactions research.

## Literature Review

To establish the theoretical foundation for our study, this section briefly examines the existing body of literature on multi-system interactions in sustainability transitions, followed by an exploration of relevant concepts from the literature on large technical systems (LTS).

## Multi-System Interactions in Transition Studies

Over the past decade, research on multi-system interactions in sustainability transitions has advanced considerably, reflecting the growing complexity of challenges such as electrification, digitalisation, and decarbonisation, which increasingly span multiple socio-technical systems (Andersen & Geels, 2023; Markard et al., 2020; Rosenbloom, 2020). This has led to a greater focus on how systems like energy, mobility, and food interact during transition processes (Kanger, 2021).

The conceptual foundations of multi-system transitions research so far have been shaped by two core approaches: the Multi-Level Perspective (MLP) and Technological Innovation Systems (TIS). Raven (2007) was among the first to apply the MLP to investigate multi-regime interactions by tracing how the waste and electricity regimes in the Netherlands co-evolved into an integrated biomass energy system. Similarly, Konrad et al. (2008) used MLP to examine multi-regime interactions among utility regimes such as electricity, gas, water, sanitation, and telecommunications in Germany. Their work identified both functional couplings (such as input–output relationships, for example, water supply requiring energy inputs) and structural couplings (such as shared infrastructures, actors, or institutions); it also showed how changes within and across these domains can contribute to system innovation and shifting regime boundaries. Papachristos et al. (2013) and Papachristos and Adamides (2016) further mapped types and directions of multi-regime interactions.

The TIS approach, while originally focused on innovation processes within single systems, has increasingly been applied to multi-system contexts (Bergek et al., 2015), contributing important insights into cross-sectoral learning and innovation (Bento et al., 2021; Malhotra et al., 2019). These studies demonstrate that factors such as technological complexity, knowledge specificity, and sectoral relatedness shape learning patterns and determine which firms enter emerging TIS and how they contribute resources and capabilities. As technologies mature, these inter-sectoral interactions intensify, broadening the actor base and strengthening value chains.

Apart from investigating multi-system phenomena through the lens of regimes, systems, and sectors, research has also applied this logic to technologies. For example, Andersen & Markard (2020) analysed the high-voltage direct current value chain and distinguished intra-sector, inter-sector and cross-sector interactions. They show that as transitions mature, these technological interplays become too complex for MLP or TIS alone to capture and so introduce a "multi-technology map" to complement existing frameworks, particularly in advanced transition phases. Recent interdisciplinary research has deepened understanding of actor roles and governance by identifying "system entanglers" (van der Vleuten, 2019) as key actors in multi-system interactions (Löhr & Chlebna, 2023) and highlighting the importance of cross-sectoral coordination (Olbrich et al., 2024). Despite growing conceptual advances and interdisciplinary contributions, the literature on multi-system interactions remains fragmented, often focusing on specific interaction types and offering high-level descriptions. Scholars call for frameworks that clarify directionality and mutual influence among systems, and for integrating broader social science theories to unpack the agency, power relations, and causal mechanisms underpinning multi-system interactions (Andersen & Geels, 2023; Andersen & Markard, 2024; Bakhuis et al., 2024).

## Multi-System Interactions in Large Technical Systems

Prior to reviewing the LTS literature on multi-system interactions, it is useful to underscore how LTS scholars conceptualize socio-technical systems differently from mainstream transition studies. While most transition research adopts a sectoral or technology-driven lens derived from MLP and TIS, the LTS approach is explicitly temporal and historically grounded. LTS conceives system boundaries as fluid configurations that are continually negotiated and renegotiated in response to new challenges or moments when control over the system by "system builders" is contested (Hughes, 1979, 1983; Summerton, 1999). Moreover, LTS traces the evolution of systems through successive phases—originally defined as invention, development, growth, momentum, and consolidation (Hughes, 1987) and subsequently extended to include reconfiguration, contestation, stagnation, and decline (Bolton & Foxon, 2015;

Sovacool et al., 2018; Summerton, 1994). This emphasis on temporal stages and boundary dynamics enables LTS to capture how socio-technical systems—and their interactions with other systems—unfold and transform over time.

Thereby, the LTS literature has long examined multi-system phenomena through their empirical works, which scholars have conceptualized as a form of reconfigurations, drawing on Summerton (1994). In doing so, LTS scholars have investigated various dimensions of the dynamics of interactions between large technical systems, such as technical, functional, organizational, spatial, and physical dimensions, and have employed a range of terminologies including interlinkages, interconnections, integrations, and interdependencies (Braun & Joerges, 1993; Robischon, 1994; Summerton, 1999; Zimmerman, 2001). We therefore argue that the LTS literature can provide important insights for multi-system transition research in general, and multi-system interactions in particular.

The LTS approach is rooted in Hughes (1983) seminal work which explains how initially fragmented electricity utility infrastructures, in both the US and UK, evolved into more consolidated systems through organizational integration and technical interconnections, involving technological adoption, standardization, and organizational integration. Building on Hughes' foundations, Summerton (1999) introduced the concept of inter-*organizational interlinkages* as a mode of system interaction and classifying them into three strategies: full integration, territorial coupling of autonomous systems, and long-term alliances. Her work examined how Swedish electricity systems were transformed by market dynamics and deregulation, which challenged existing control structures. Robischon (1994) added further understanding of these dynamics to the literature through his exploration of organizational and technical interactions in the case of the unification of the German telephone service. The concept of *integration* was *c*entral to his investigation. The technical integration necessitated significant reconfiguration, including a digital overlay network to ensure compatibility and connectivity between the relatively advanced West German and "outdated" East German systems. Organizationally, it entailed restructuring and adopting West German regulatory and organizational structures, to enable for service integration.

Other LTS studies have investigated the complexity of the evolution of large technical systems and their interactions. For example, Braun and Joerges (1993) examined the development and functioning of Transnational Organ Transplant Systems (TOTS) in Europe through a co-evolutionary perspective. Their work explored how large technical systems evolve in response to each other, creating complex, interconnected *second-order* systems. These second-order systems showcase the complexity and interconnection in modern medical and technological infrastructures. Another example is the study by Zimmermann (2001) who investigated the growing dependence of urban areas on interconnected infrastructure networks, including transportation, energy, communications, water supply, and wastewater systems. Zimmermann discussed how these networks are increasingly interdependent, which amplifies their complexity and potential for disruptions such as cascading failures, increased vulnerability, and unpredictability of disruptions. Zimmermann explored functional and spatial interconnectedness, noting how operational dependencies between systems can lead to cascading failures.

We argue that the LTS perspective may offer valuable insights for unpacking multi-system dynamics, particularly by emphasizing the interactions between systems across different historical contexts. First, a rich body of empirical research within the LTS tradition has traced how socio-technical systems evolve from their emergence to maturity, often exhibiting distinct characteristics at different stages. For instance, Hughes demonstrated that early power systems in the nineteenth and early twentieth centuries were highly decentralized, with small utilities operating their own infrastructure. Over time, technical and organizational interlinkages drove consolidation and centralization (see Sect. 3.2). By examining these historical trajectories, we can derive lessons for understanding emerging multi-system interactions, recognizing that the processes of decentralization, integration, and consolidation observed in the past may inform how new system interdependencies develop today. Second, the LTS perspective views system boundaries as fluid, shifting in response to new challenges and moments when system control is contested (Hughes, 1983; Summerton, 1999). Summerton's work in the 1990s, for instance, highlights how electricity systems were transformed by market dynamics and deregulation, which challenged existing control structures. In response, electricity companies developed new forms of organizational interlinkages. This demonstrates that system boundaries depend on where control is exerted, supporting the idea that technological or sectoral boundaries are not fixed but continually redefined. This dynamic perspective enabled LTS scholars to more effectively capture the unfolding interactions between systems and their surroundings including their evolving interactions with other socio-technical systems over time.

## Analytical Framework

To examine the interactions between the energy and mobility systems in the UK's net-zero transition, we develop and apply an analytical framework that draws on the LTS literature. Our framework differentiates between three key types of interlinkages: functional, technical, and organizational.

### Functional Interlinkages

Functional interlinkages refer to the dependencies and interactions between different (sub-)systems that enable them to operate cohesively (Summerton, 1994; Zimmerman, 2001). Functional interlinkages can take the form of resource flows, such as electricity powering transport, but also can encompass organizational, cultural, and economic dimensions (Braun & Joerges, 1993). The nature and effectiveness of functional interlinkages are not static; they can be reshaped by a variety of factors, including technological innovation, regulatory shifts, market developments, and organizational adaptation (Hughes, 1983; Braun & Joerges, 1993; Summerton, 1999).

For example, Braun and Joerges' (1993) study of the European TOTS system demonstrates how functional interlinkages can shift over time. Initially, TOTS experienced operational turbulence due to its dependence on rapid transport and real-time communication across borders.

As inter-organizational networks developed and standardized practices—such as harmonized donor-recipient coordination and pre-identification of potential donors—were introduced, these functional interlinkages evolved. These changes allowed the system to transition from a reactive, post-incident model to a more proactive and reliable network, strengthening the connections between diverse organizations and infrastructures across nations and significantly improving overall system effectiveness and reliability.

By understanding these functional interlinkages and dependencies, we can shed light on the co-evolutionary processes between different socio-technical systems, as they can help indicate potential trajectories that such systems might follow for cohesive operation and adaptation.

## Technical Interlinkages

Technical interlinkages refer to the structural connections between technological artefacts or components across different socio-technical systems that enable or constrain system changes. These interlinkages are necessary for functional interlinkages, and manifest through shared technological infrastructures, standardization of components and interfaces, adoption of cross-system innovations, and technical interdependencies in resource use or service provision.

Hughes (1983) provides a compelling historical example of technical interlinkages driving system change in describing the fragmented nature of London's electrical system:

> Greater London had sixty-five electrical utilities, seventy generating stations averaging only 5,285 kw. in capacity, forty-nine different types of supply systems, ten different frequencies, thirty-two voltage levels for transmission and twenty-four for distribution, and about seventy different methods of charging and pricing. Londoners who could afford electricity toasted bread in the morning with one kind, lit their offices with another, visited associates in nearby office buildings using still another variety, and walked home along streets that were illuminated by yet another kind. (Hughes, 1983, p. 227)

This fragmented system eventually evolved into a more consolidated and efficient network through technical interlinkages, including the adoption of new technologies, standardization of components, and advancements in technical specifications. A similar process occurred in the US, where Samuel Insull facilitated the interlinking of small, isolated electricity systems by adopting alternating current (AC), high-capacity power lines and substations, combined with the standardization of operations (Hughes, 1983).

By examining technical interlinkages, we can gain insight into how multi-system interactions are structured at the interfaces between different socio-technical systems. These interlinkages, such as shared infrastructures and standardization, not only facilitate integration but can also reveal points of tension or differentiation, helping us anticipate the challenges and opportunities that arise as systems interact and co-evolve.

## Organizational Interlinkages

Organizational interlinkages refer to the managerial approaches that enable the coordination and integration of various subsystems and entities within the larger system, and in turn, across systems. These interlinkages manifest through corporate structures, management practices, strategic partnerships, and financial mechanisms. Based on the work of Hughes (1983) and Summerton (1999), we categorize these organizational interlinkages into three primary sub-types: couplings of autonomous systems, vertical and horizontal integration, and partnerships and alliances. The coupling of autonomous systems involves the connection of independent systems across geographical or political boundaries, enabling them to function cooperatively while maintaining their autonomy. Vertical and horizontal integration involves the consolidation of various stages of production and distribution processes within a single organizational framework, or the expansion of control across different geographical areas and related sectors, allowing for greater control over the value chain and potentially increased efficiency and market power. Partnerships and alliances involve strategic collaborations between organizations to leverage shared resources, knowledge, and capabilities while maintaining a degree of autonomy, thus allowing entities to combine complementary strengths and improve overall system performance.

Within this framework, particular attention is given to the prominent role of actors—such as incumbents, new entrants, and other stakeholders—in shaping organizational interlinkages. The framework explicitly considers how the concrete actions, decisions, and collaborations of these actors actively construct, reshape, and govern multi-system interlinkages.

## Methodology

Our study employed a qualitative case study approach, which is particularly appropriate for investigating complex phenomena within their real-world contexts (Eisenhardt, 1989; Yin, 2018). This methodological choice aligns with our aim to explore the multifaceted dynamics of multi-system transitions.

We conducted semi-structured interviews between October 2022 and November 2023 with 19 purposively sampled experts representing four distinct actor types (de Haan & Rotmans, 2018; Wittmayer et al., 2017): business, policy, civil society, and academia (for an overview see Annex 1). We selected semi-structured interviews for their flexibility, allowing us to explore the dimensions of our analytical framework while remaining receptive to emergent ideas and perspectives (Longhurst, 2003). For data analysis, we employed a hybrid approach, combining inductive (bottom-up) and deductive (top-down) strategies (Fereday & Muir-Cochrane, 2006). This approach facilitated the identification of emergent themes from the data while also paying attention to our analytical framework. In addition, we considered other relevant concepts, such as system boundaries and their interactions, specific policies and technologies, and system governance structures. For our analysis we used MaxQDA 24 to aid the systematic coding and interpretation of textual data (Kuckartz & Rädiker, 2019). Our analytical process was iterative, involving multiple rounds of coding and interpretation to ensure the robustness and credibility of our findings, with ongoing discussions between co-authors to refine the coding structure (Saldaña, 2021).

To complement the interview data and enhance the validity of our findings, we also conducted an analysis of relevant policy documents, industry-related materials and news, using an innovative AI-assisted approach: retrieval augmented generation method.[3] This method allowed us to efficiently extract and synthesize information from a large corpus of documents, providing additional context and corroborating evidence gathered from the interviews (Bowen, 2009). We employed triangulation, comparing data from interviews and the AI-assisted document analysis to enhance the validity and reliability of our results (Flick, 2018; Miles & Huberman, 1994). This mixed-method approach enabled us to cross-verify findings and develop a comprehensive understanding of the interactions between energy and mobility systems. In the following section, we provide detailed background on the case context to further situate our analysis.

## Case Background

The UK's smart and flexible systems pathway is a central component of the national strategy to achieve net-zero emissions by 2050. In response to the Climate Change Act (2008), which established legally binding carbon budgets and a target of net-zero by 2050 (Climate Change Committee, 2008), and the energy trilemma "policy paradigm" of decarbonising energy and mobility systems while maintaining system reliability and ensuring affordability for consumers (Bolton & Foxon, 2013). Initially, policy efforts focused on decarbonising the electricity sector, which achieved a 78.4% reduction in greenhouse gas emissions by 2024 through the rapid deployment of renewables, particularly wind power (UK Government, 2024). However, the intermittent nature of

---

[3] This novel method was created following the approach explained in the following video: https://youtu.be/kvdVduIJsc8?si=Ilt4VXGrgvWd-q35.

renewables—where wind alone accounted for 33.8% of electricity generation in 2024 in the UK—has necessitated a fundamental reorganization of the energy system. In response, electrification and digitalisation have emerged as key enablers for decarbonisation by connecting historically siloed systems, particularly transport and heating.

The building blocks of the smart and flexible systems pathway were first introduced in the early 2010s with the launch of smart grid and smart metre programmes intended to improve energy efficiency and enabling and accelerating low-carbon energy technology adoption (BEIS & Ofgem, 2017). These programmes then contributed to the creation of a more comprehensive and integrated policy approach, with the "Smart Systems and Flexibility Plan", first introduced in 2017 and updated in 2021. The plan outlined a vision for a decentralized, net-zero electricity system powered by low-carbon energy sources, with flexibility enabling consumers to shift demand and to store renewable energy (BEIS & Ofgem, 2021). The plan identified key areas of action to realize this vision, including reforming markets to reward flexibility, removing regulatory barriers, and deploying smart technologies like energy storage and demand-side response.

In parallel, the UK government has launched the "Flexibility Innovation Programme" in 2021, committing £65 million to large-scale flexibility solutions, including the V2X Innovation Programme, which supports bidirectional EV charging to unlock electric vehicles as grid-balancing assets (DESNZ, 2023). To further reinforce this integrated approach, the government released the Electric Vehicle Smart Charging Action Plan. This plan supports the development and adoption of vehicle-to-grid (V2G) and vehicle-to-everything (V2X) technologies, promotes data-driven and digital charging solutions, and encourages innovative business models. It also mandates that all new private charge points have smart functionality for interoperability, incentivizes off-peak charging, and aims to address barriers to public and long-duration smart charging (DESNZ & DfT, 2023).

Moreover, the UK's Energy Digitalisation Strategy, launched in 2021, is being actively implemented as a coordinated programme of actions for the smart and flexible systems pathway. The strategy sets out a clear agenda for leadership and coordination across government, regulators, and industry, establishing regulatory frameworks that incentivize digitalization and innovation. It mandates the development of digital tools and infrastructure (Open Energy, Data Spine), and standards to enable interoperability and secure data sharing (Data Best Practice, Consumer Consent Dashboard), and it requires energy network companies to regularly update and align their digitalization strategies with best practice guidance. Key delivery bodies such as the Energy Data Taskforce and the Energy Digitalisation Taskforce have been instrumental in driving progress, recommending measures such as the creation of a consumer consent dashboard, the introduction of a digital spine for plug-and-play interoperability, and mandated carbon data monitoring to support emissions transparency and policy decisions. Through these coordinated actions, the strategy aims to support the integration of millions of low-carbon assets such as solar PV, electric vehicles, and heat pumps.

In summary, the UK's smart and flexible systems pathway demonstrates a multi-system transition characterized by the alignment of technological, institutional, and infrastructural change. The case traces the evolution from single system-specific decarbonization—initially focused on energy—towards a more integrated approach that foregrounds the integration of mobility, and heating systems. This transition has been shaped by a set of coordinated policy instruments such as the Climate Change Act, the Smart Systems and Flexibility Plan, the Flexibility Innovation Programme, and the Energy Digitalisation Strategy. The UK's electricity mix consisting of an increasing share of intermittent renewables has necessitated a focus on flexibility, while digitalisation and electrification have emerged as key drivers of increasing interactions across multiple systems.

## Results

In the following sections, we present our findings on the three key types of interlinkages-functional, technical, and organizational—that structure the interactions between the UK's energy and mobility systems.

## Findings on Functional Interlinkages

Drawing from our data sources, we unpack the dynamics of functional interlinkages between energy and mobility systems in the UK, focusing on their historical evolution and current drivers. With increasing pressure on the UK government for climate action, the energy trilemma[4] became central to net-zero transition debates in the early 2010s (I-15). Under smart grid and smart metre programmes, digitalization emerged as the key enabler for a decarbonised energy system that maintained reliability and affordability (I-10/I-15) while the interaction between electric vehicles (EVs) and the grid was primarily seen as an energy demand challenge (I-1/I-2).

From 2016, coordination efforts by BEIS (Department for Business, Energy, and Industrial Strategy) and Ofgem (Office of the Gas and Electricity Markets, the energy regulator) began shifting this passive, one-way dependency into a more functionally interdependent interaction. Within the smart and flexible systems pathway, EVs, with their battery storage capabilities, were envisioned as playing a key role through providing energy flexibility for grid reliability—becoming *batteries on wheels*(I-4). This shift has evolved through two main developments: load shifting via smart charging and bidirectional electricity flows through V2X technologies.

The first function, load shifting, has been widely adopted since the 2020s, mainly provided by energy suppliers. These services shift EV charging to off-peak hours using smart tariffs and digital infrastructure like smart metres to communicate consumption and pricing signals (I-4/I-6). Load shifting reduces fossil fuel use, flattens demand curves, and optimizes renewables, improving grid stability (I-6). This interlinkage reduces the need for additional capacity investments, lowering the cost of net-zero transitions and benefiting consumers (I-5/I-6/I-18).

The second function involves bidirectional electricity flow enabled by V2X technologies. These technologies allow EVs to act as mobile energy storage units, feeding electricity back to homes, workplaces, and the

[4] The energy trilemma refers to the challenge of balancing three key objectives: ensuring energy security, achieving environmental sustainability, and maintaining affordability for consumers and businesses.

grid. While V2X is not yet fully commercialized, its potential to support net-zero targets—particularly by enhancing grid stability given the UK's reliance on intermittent renewables—has led the government to invest over £12 million in innovation and demonstration projects (DESNZ & DfT, 2023). If commercialized, V2X would make EVs both electricity consumers and providers, supplying power back to the grid or buildings when demand is high (I-4/I-10).

However, significant governance and technological challenges remain. Our interviewees noted that the absence of robust data governance frameworks hinder efforts to achieve greater flexibility through deeper integration of mobility (I-1/I-4). For example, electricity actors are encouraged to improve data sharing, but automotive actors often seek to retain proprietary control (I-1/I-4). This misalignment limits electricity actors' ability to enhance grid flexibility and may create barriers to the government's smart system and flexibility plans (I-1). Attempts to address these issues, such as Ofgem's consultations on expanding data-sharing obligations and using smart metre systems for charging infrastructure (I-1/I-5), have been limited by regulatory gaps and the political influence of some mobility actors. The problem is compounded by Ofgem's remit being limited to energy, while OZEV[5] operates with constrained authority between BEIS and DfT. In the words of one interviewee:

> when I was working in BEIS on electric vehicle charge points, the DfT had the power to regulate charge points to be smart, but not cars to be smart. And so if the technology companies put the smartness in the car, we didn't have a rule that allowed us to regulate cars. We could only regulate charge points. And so that even if we knew what we wanted to do, we don't have the power to regulate it for cars. So you don't even have the rules that let you do cars because you were just an energy department. So you got an energy related power that actually now needs to apply to cars. (I-9)

[5] The Office for Zero Emission Vehicles (OZEV) is a cross-government team, part of the Department for Transport and the Department for Energy Security and Net Zero, responsible for supporting the UK's transition to zero-emission vehicles through policy, funding, and development of charging infrastructure.

DfT has sought to address data governance challenges through the Future of Mobility: Urban Strategy and the Connected Vehicle Data Research Project. While these efforts have articulated principles for data sharing and highlighted the potential of vehicle data for decarbonization, they have not yet resulted in binding obligations for automotive actors. Interviews confirmed that critical issues around vehicle data governance—such as access to real-time charging status, battery health, or location data—are still largely shaped internationally, particularly through forums like the UNECE World Forum for Harmonization of Vehicle Regulations (WP.29), which sets global standards for vehicle cybersecurity and interoperability (I-14/I-15). At the national level, the main regulatory action has been mandating smart functionality for charge points under the Automated and Electric Vehicles Act, ensuring that infrastructure does not become a barrier to future flexibility and integration.

In summary, functional interlinkages in the UK's smart and flexible systems pathway have enabled the mobility system to act as a load shifter, supporting grid stability and optimizing renewables. The transition is moving towards bidirectional interdependency, where mobility not only relies on the energy system for electricity but can also supply electricity back to the grid—a valuable function given the UK's intermittent electricity mix. These interlinkages can enhance the cost-efficiency of net-zero transitions. However, persistent governance challenges, especially around data access and regulatory alignment, continue to shape their development and may either prompt new organizational interlinkages or constrain advanced functional integration if unaddressed.

## Findings on Technical Interlinkages

The technical interlinkages between energy and mobility systems in the UK have evolved significantly. Coordinated by Ofgem, BEIS, DfT, and the Energy Data and Digitalisation Taskforces, these efforts aimed to enable flexibility by making system components visible, dynamically controllable, and responsive. The focus has been on building digital infrastructure, enhancing digital capabilities (especially for energy actors), standardizing smart system components, and expanding charging infrastructure.

We find three main developments enabling the interlinkages between energy and mobility systems through digital infrastructure: the open energy platform, the automatic asset registration (ARR), and smart metres. The open energy platform, supported by the flexibility fund, created a digital ecosystem for data sharing and improved visibility of system resources (I-1/I-3). The AAR Programme enhanced real-time visibility, control, and responsiveness for decentralized energy sources, including EV chargepoints (I-2/I-15/I-17/I-18). Smart metres, initially for demand reduction, are now also used to monitor and manage EV demand on the electricity system (I-2/I-10/I-13/I-16).

Building digital capability has been critical for technical interlinkages. Although data sharing has been part of the UK's smart grid vision since the 2010s, practical progress was slow due to industry reluctance and low digital capability (I-2/I-6/I-18). Interviewees highlighted the Energy Data Taskforce (EDT, 2018) and Energy Digitalisation Taskforce (EDiT, 2021) as key drivers of change. The EDT's 'Data Best Practice Guidance' (2020) provided a framework for improved data governance, focusing on openness, interoperability, and quality (I-1/I-2/I-8/I-18). Ofgem's 2021 mandate required all network companies to publish digital strategies under RIIO-2 (I-2), enhancing data availability and integration capacity, though interviewees noted further progress is needed (I-13/I-16/I-17).

Expansion and standardization of charging infrastructure has also strengthened technical interlinkages. The Electric Vehicles (Smart Charge Points) Regulation (June 2022) mandated smart functionalities for all new private charge points (UK Government, 2021). PAS 1878 and 1879 standards ensure interoperability and prevent barriers from non-smart appliances (I-5/I-18). The Automated and Electric Vehicles Act (2018, amended 2023) extended smart charging requirements to public charge points, including contactless payment, real-time pricing, a 99% reliability standard, open data (OCPI), and roaming (UK Government, 2018). These requirements mainly target charge point manufacturers, not automotive manufacturers.

This regulatory focus reflects the perception that Original Equipment Manufacturers (OEMs) are politically influential actors that cannot be easily pressured to comply with technical mandates, particularly when commercial value is at stake. This is a dynamic not necessarily shared by other actor groups. As one interviewee noted:

> The guidance was published in February or March of that year. So my members had maybe six months, but probably four really to in some cases, completely [to] retool their devices and make something different. Now, imagine if you'd done that to Toyota and Ford and Jaguar, Land Rover - like the government would have... well.... maybe it wouldn't have fallen, but there would have been [..] very important people queued up outside Number ten - very angry indeed. (I–15)

In summary, efforts to create and strengthen technical interlinkages between energy and mobility systems in the UK have centred on building and repurposing digital infrastructure, enhancing digital capabilities—especially among energy system actors—and standardizing system components to ensure smart functionality, all while expanding both private and public charging infrastructure. These coordinated actions have been crucial in making system components more visible, interoperable, and adaptable for transition towards a smart and flexible systems pathway. However, the persistent asymmetry in the political power and the governance gaps, particularly regarding the limited influence over OEMs, continues to shape both the possibilities and the limitations of technical interlinkages between the UK's energy and mobility systems.

## Findings on Organizational Interlinkages

We observe a diverse range of connections that correspond to Summerton's (1999) three primary typologies of organizational interlinkages: couplings of autonomous systems, vertical and horizontal integration, and partnerships and alliances. These interlinkages are not only evidence of growing coordination between energy and mobility system actors, but also reflect how commercial strategies, regulatory pressures, and the pursuit of new value streams are driving new forms of collaboration and competition across system boundaries.

First, on the coupling of autonomous systems, a prime example is the Zap-Pay platform. This platform allows users to search, access, and pay for charging across multiple networks using a single app, effectively coupling various autonomous charging systems. Zap-Map has integrated numerous major charge point operators in the UK (e.g., GeniePoint, char.gy, ESB Energy), providing access to over 95% of public charging points (Zap-Map, 2023). Furthermore, the development of roaming agreements between different charge point operators in the UK represents another form of coupling of autonomous systems. For instance, the partnership between NewMotion and ChargePoint allows users of either network to access charging points on both networks (NewMotion, 2021). These types of organizational interlinkages enable the interconnection of distinct charging networks while preserving the operational autonomy.

Second, vertical and horizontal integration strategies have been prominently employed by energy suppliers and EV charging infrastructure providers seeking to capture greater value and influence in the evolving charging network. For instance, EDF Energy's acquisition of Pod Point, the largest EV charging provider, demonstrates vertical integration by controlling both energy supply and EV charging infrastructure (I-5). Another example is Octopus Energy's launch of Octopus Electric Vehicles, expanding from energy supply into EV leasing and charging solutions by leveraging their proprietary software for dynamic pricing and smart charging (I-6). On the other hand, we also find horizontal integration by large organizations, such as Shell's acquisition of NewMotion and Greenlots for its Recharge charging network brand (Shell, 2021). These integration strategies are motivated by the desire to secure new revenue streams and position themselves as key players in the transitions towards net-zero.

Third, we observe strategic partnerships and alliances as key forms of organizational interlinkages, enabling actors to pool resources and expertise across sectoral boundaries. For example, BP's collaboration with Volkswagen Group to roll out ultra-fast charging stations across Europe combines BP's energy distribution capabilities with Volkswagen's automotive expertise (BP, 2021). Similarly, Nissan's partnership with E.ON Next to provide home charging installations and EV-specific energy tariffs demonstrates the alliances between energy and mobility systems actors (EON, 2018).

To summarize, in our analysis of organizational interlinkages we observe a range of connections that correspond to Summerton's (1999) three primary typologies referring to couplings, integration, and partnerships. The observed heterogeneity in organizational interlinkages illustrates the complex web of relationships spanning across the energy and mobility systems, reflecting the dynamics of boundary re-setting and negotiation.

## Discussion and Conclusions

In this chapter, we have examined the multi-system interlinkages between energy and mobility systems within the context of the UK's pursuit of decarbonisation through a smart and flexible systems pathway. Conceptually, we developed and applied an analytical framework derived from the large technical systems (LTS) literature, focusing on functional, technical, and organizational interlinkages. We show how these interlinkages and their governance have evolved over time. We identified governance-related interlinkages—such as the roles of DESNZ, Ofgem, and the Energy Data and Digitalisation Taskforces in building digital infrastructure and creating coordination channels—which have not been a focus in the LTS literature. Therefore, we suggest these governance interlinkages as a new additional type of interlinkage, complementing functional, technical, and organizational ones.

Building on LTS's treatment of systems has helped us to better capture how energy and mobility systems co-evolve through both technological innovations (e.g., smart charging, V2X) and organizational strategies, such as the formation of strategic alliances between firms like BP and Volkswagen in the development of charging infrastructure. Unlike frameworks such as MLP or TIS, which tend to focus on single system dynamics or niche innovations, LTS provides a more flexible conceptualization of system evolution by recognizing that system boundaries are fluid and can shift over time through interlinkages between technical, organizational, and functional structures. This flexibility allowed us to investigate how new interlinkages explain multi-system interactions in UK's net-zero transitions. However, LTS is often more descriptive than prescriptive, which can limit its utility for policymakers. Nonetheless, as our research shows, LTS can complement other transition frameworks by offering a broader perspective on system interactions, particularly where multiple systems adapt and co-evolve.

More specifically, the concept of interlinkages provides a complementary perspective for examining the different ways systems relate to one another—whether through the roles they play for each other (functional), the technologies and infrastructure that connect them (technical), or the structures and collaborations that bring actors together (organizational). These forms of interlinkages are not mutually exclusive; rather, they often overlap and reinforce each other, shaping the evolution and interaction of multi-system transitions. Functional interlinkages typically reflect the shared vision and direction that actors in multiple systems ascribe to their interactions, essentially describing how systems might serve each other's needs (for example, EVs acting as grid assets through smart charging and V2X technologies). In contrast, technical and organizational interlinkages describe the concrete mechanisms and structures through which these functional visions are realized, such as the development of digital infrastructure, the establishment of data standards, or the formation of partnerships and integrated business models.

An important overarching finding is that digitalization has emerged as a critical enabler of the interlinkages between the UK's energy and mobility systems. Significant efforts have been dedicated to creating technical interlinkages through developing digital infrastructure, mandating standardization and digital capacity building by targeted institutional changes. Additionally, digitalisation has shaped organizational interlinkages by fostering new partnerships, alliances and integration strategies among energy and mobility system actors.

A key question raised by our findings is how to conceptualize the role of digitalisation in the electrification of transport. Our empirical data for the UK suggest that digitalisation should neither be viewed as a separate socio-technical system nor merely as part of the landscape (as defined in the MLP). Instead, digital technologies function as a nested system within both energy and mobility systems, but they are shaped differently by each. In the UK's energy system, digitalisation is more centralized and governed at the national level through initiatives like smart grids. In contrast, within the UK's mobility system digitalisation is less centralized, and is influenced by international governance structures and private sector initiatives. Therefore, rather than viewing digitalisation as a third system or simply an enabling technology like previous general-purpose technologies (Bresnahan & Trajtenberg, 1995), we argue that it should be understood as an enabling infrastructure that facilitates interactions between systems, such as in our case between the UK's energy and mobility systems. Digital technologies such as smart charging, vehicle-to-grid technologies, and smart metres enable new functionalities that allow these two systems to interact more efficiently without constituting a separate socio-technical system.

Based on the findings and limitations of our research, we identify several avenues for future investigation. While this study focused on functional, technical, and organizational interlinkages, the LTS literature also highlights physical and spatial interlinkages, which merit further exploration. Additionally, we propose that governance interlinkages—introduced here as a new type—should be a priority for future work. Expanding research to include these dimensions will deepen our understanding of multi-system interlinkages and, in particular, advance knowledge on how to effectively govern and accelerate multi-system transitions.

**Acknowledgements** Aslı Ateş and Karoline Rogge gratefully acknowledge funding from the European Research Council (ERC) under the European Union's Horizon 2020 research and innovation programme (grant agreement No 852730). We acknowledge the use of artificial intelligence (AI) tools to assist with editing, restructuring, and improving the clarity of this manuscript. All substantive content, analysis, and interpretations remain our own.

**Funding** Aslı Ateş reports financial support was provided by European Research Council. Karoline Rogge reports financial support was provided by European Research Council.

## Annex 1—Details of the Interviews

| Name | Actor type | Date | Time |
|---|---|---|---|
| Interview 1 | Business | 13/10/2022 | 01:22:46 |
| Interview 2 | Business | 10/11/2022 | 01:24:53 |
| Interview 3 | Business | 02/12/2022 | 01:22:30 |
| Interview 4 | Business | 12/07/2023 | 00:58:02 |
| Interview 5 | Business | 12/07/2023 | 00:58:27 |
| Interview 6 | Business | 12/10/2023 | 00:41:07 |
| Interview 7 | Business | 26/07/2023 | 01:01:51 |
| Interview 8 | Business | 21/11/2023 | 01:08:37 |
| Interview 9 | Policy | 27/01/2023 | 01:12:55 |
| Interview 10 | Policy | 09/11/2023 | 00:50:29 |
| Interview 11 | Research | 03/11/2022 | 00:35:13 |
| Interview 12 | Research | 13/01/2023 | 00:57:41 |
| Interview 13 | Civil Society/Association | 23/11/2022 | 01:28:39 |
| Interview 14 | Civil Society/Association | 23/11/2022 | 01:22:33 |
| Interview 15 | Civil Society/Association | 13/07/2023 | 01:03:25 |
| Interview 16 | Civil Society/Association | 08/08/2023 | 00:39:37 |
| Interview 17 | Civil Society/Association | 23/10/2023 | 01:01:46 |
| Interview 18 | Civil Society/Association | 03/11/2023 | 01:26:51 |
| Interview 18 | Civil Society/Association | 20/11/2023 | 01:11:09 |
| Interview 19 | Civil Society/Association | 17/11/2023 | 01:09:14 |
| | | Average | 01:05:53 |

## References

Andersen, A. D., & Geels, F. W. (2023). Energy research & social science multi-system dynamics and the speed of net-zero transitions: Identifying causal processes related to technologies, actors, and institutions. *Energy Research & Social Science, 102*(June), Article 103178. https://doi.org/10.1016/j.erss.2023.103178

Andersen, A. D., & Markard, J. (2020). Multi-technology interaction in socio-technical transitions: How recent dynamics in HVDC technology can inform transition theories. *Technological Forecasting and Social Change*, 151, 119802.

Andersen, A. D., & Markard, J. (2024). Multi-system dynamics in sustainability transitions: Introduction and outlook. In *Cambridge open engage*. Cambridge University Press, 1–20. https://doi.org/10.33774/coe-2024-x6x8n. This content is a preprint and has not been peer-reviewed.

Bakhuis, J., Linda, M., Barbour, N., & Louis, J. (2024). Technological forecasting & social change frameworks for multi-system innovation analysis from a sociotechnical perspective: A systematic literature review. 201(February). https://doi.org/10.1016/j.techfore.2024.123266.

Bento, N., Fontes, M., & Barbosa, J. (2021). Inter-sectoral relations to accelerate the formation of technological innovation systems: Determinants of actors' entry into marine renewable energy technologies. *Technological Forecasting and Social Change, 173*(August). https://doi.org/10.1016/j.techfore.2021.121136.

Bergek, A., Hekkert, M., Jacobsson, S., Markard, J., Sandén, B., & Truffer, B. (2015). Technological innovation systems in contexts: Conceptualizing contextual structures and interaction dynamics. *Environmental Innovation and Societal Transitions, 16*, 51–64. https://doi.org/10.1016/j.eist.2015.07.003

Ronan, B., & Foxon, T. (2013). Negotiating the energy policy 'Trilemma'—An analysis of UK energy governance from a socio-technical systems perspective. *Theorising Governance Change for a Sustainable Economy*, 1–24.

Ronan, B., & Foxon, T. J. (2015). Infrastructure transformation as a socio-technical process—Implications for the governance of energy distribution networks in the UK. *Technological Forecasting and Social Change, 90*(PB), 538–50. https://doi.org/10.1016/j.techfore.2014.02.017.

Bowen, G. A. (2009). Document analysis as a qualitative research method. *Qualitative Research Journal, 9*(2), 27–40. https://doi.org/10.3316/QRJ0902027

BP. (2021). *Volkswagen group and Bp to join forces to expand ultra-fast electric vehicle charging in Europe*, (July 18, 2024) https://www.bp.com/en/global/corporate/news-and-insights/press-releases/vw-group-and-bp-join-forces-ultra-fast-ev-charging-europe.html

Braun, I., & Joerges, B. (1993). How to recombine large technical systems: The case of European organ transplantation. In J. Summerton (ed.), *Changing large technical systems*. CRC Press, 25–51. https://doi.org/10.1201/9780429046575.

Bresnahan, T. F., & Trajtenberg, M. (1995). General purpose technologies 'engines of growth'? *Journal of Econometrics, 65*(1), 83–108. https://doi.org/10.1016/0304-4076(94)01598-T

BEIS, O. (2017). *Upgrading our energy system: Smart systems and flexibility plan.* https://www.gov.uk/government/publications/upgrading-our-energy-system-smart-systems-and-flexibility-plan

BEIS, O. (2021). *Transitioning to a net zero energy system: Smart systems and flexibility plan 2021.* https://www.gov.uk/government/publications/transitioning-to-a-net-zero-energy-system-smart-systems-and-flexibility-plan-2021

DESNZ. (2023). Flexibility innovation programme, (July 18, 2024) https://www.gov.uk/government/publications/flexibility-innovation

Department for Energy Security and Net Zero. 2024. *Energy trends June 2024*. https://assets.publishing.service.gov.uk/media/667c119d7d26b2be17a4b3a2/Energy_Trends_June_2024.pdf

DESNZ & DfT. (2023). Electric vehicle smart charging action plan, July 18, 2024, https://www.gov.uk/government/publications/electric-vehicle-smart-charging-action-plan

DESNZ & BEIS. (2021). *Net zero strategy: Build back greener*. https://www.gov.uk/government/publications/net-zero-strategy

Eisenhardt, K. M. (1989). Building theories from case study research. *Academy of Management Review, 14*(4), 532–550.

EON. (2018). *EMobility: E.ON and Nissan enter into strategic partnership*, July 18, 2024, https://www.eon.com/en/about-us/media/press-release/2018/emobility-eon-and-nissan-enter-into-strategic-partnership.html

Fereday, J., & Muir-Cochrane, E. (2006). Demonstrating rigor using thematic analysis: A hybrid approach of inductive and deductive coding and theme development. *International Journal of Qualitative Methods, 5*(1), 80–92.

Flick, U. (2018). *Doing triangulation and mixed methods*, (ed.) U. Flick. Sage.

de Haan, F. J., & Rotmans, J. (2018). A proposed theoretical framework for actors in transformative change. *Technological Forecasting and Social Change, 128*, 275–286. https://doi.org/10.1016/j.techfore.2017.12.017

Hughes, T. (1983). *Networks of power: Electrification in western society, 1880–1930.* The John Hopkins University Press.

Hughes, T. (1979). The electrification of America: The system builders. *Technology and Culture, 20*(1), 124. https://doi.org/10.2307/3103115

Hughes, T. (1987). The evolution of large technological systems. In W. E. Bijker, T. P. Hughes, and T. Pinch (Eds.), *The social construction of technological systems: New directions in the sociology and history of technology* (pp. 51–82). MIT Press.

IEA. (2023). *Tracking clean energy progress 2023*, July 14, 2024, https://www.iea.org/reports/tracking-clean-energy-progress-2023

Kanger, L. (2021). Rethinking the multi-level perspective for energy transitions: From regime life-cycle to explanatory typology of transition pathways. *Energy Research and Social Science, 71*. https://doi.org/10.1016/j.erss.2020.101829.

Konrad, K., Truffer, B., & Voß, J. P. (2008). Multi-regime dynamics in the analysis of sectoral transformation potentials: Evidence from German utility sectors. *Journal of Cleaner Production, 16*(11), 1190–1202. https://doi.org/10.1016/j.jclepro.2007.08.014

Kuckartz, U., & Rädiker, S. (2019). *Analyzing qualitative data with MAXQDA*. Springer.

Löhr, M., & Camilla, C. (2023). Multi-system interactions in hydrogen-based sector coupling projects: System entanglers as key actors. *Energy Research and Social Science, 105*. https://doi.org/10.1016/j.erss.2023.103282.

Longhurst, R. (2003). Semi-structured interviews and focus groups. *Key Methods in Geography, 3*(2), 143–156.

Malhotra, A., Schmidt, T. S., & Huenteler, J. (2019). The role of inter-sectoral learning in knowledge development and diffusion: Case studies on three clean energy technologies. *Technological Forecasting and Social Change, 146*(February), 464–487. https://doi.org/10.1016/j.techfore.2019.04.018

Markard, J., Geels, F. W., & Raven, R. (2020). Challenges in the acceleration of sustainability transitions. *Environmental Research Letters, 15*(8), Article 081001. https://doi.org/10.1088/1748-9326/ab9468

Miles, M. B., & Huberman, A. M. (1994). Qualitative data analysis: An expanded sourcebook. In *Qualitative data analysis: An expanded sourcebook* (2nd ed.). Sage Publications.

Olbrich, S., Dierk, B., & Philipp, S. (2024). Policy mixes for net-zero energy transitions: Insights from energy sector integration in Germany. *Energy Research and Social Science, 118*. https://doi.org/10.1016/j.erss.2024.103822.

Papachristos, G., & Adamides, E. (2016). A Retroductive systems-based methodology for socio-technical transitions research. *Technological Forecasting and Social Change, 108*, 1–14. https://doi.org/10.1016/j.techfore.2016.04.007

Papachristos, G., Sofianos, A., & Adamides, E. (2013). System interactions in socio-technical transitions: Extending the multi-level perspective. *Environmental Innovation and Societal Transitions, 7*, 53–69. https://doi.org/10.1016/j.eist.2013.03.002

Raven, R. (2007). Co-evolution of waste and electricity regimes: Multi-regime dynamics in the Netherlands (1969–2003). *Energy Policy, 35*(4), 2197–2208. https://doi.org/10.1016/j.enpol.2006.07.005
Robischon, T. (1994). Transformation through Integration: The unification of the German telephone service. *Flux, 15*(1), 25–36. https://doi.org/10.3917/flux.p1994.10n15.0025.
Rosenbloom, D. (2020). Engaging with multi-system interactions in sustainability transitions: A comment on the transitions research agenda. *Environmental Innovation and Societal Transitions, 34*, 336–340. https://doi.org/10.1016/J.EIST.2019.10.003
Saldaña, J. (2021). *The coding manual for qualitative researchers* (4th ed.). Sage.
Schot, J., & Kanger, L. (2018). Deep transitions: Emergence, acceleration, stabilization and directionality. *Research Policy, 47*(6), 1045–1059. https://doi.org/10.1016/j.respol.2018.03.009
Shell. (2021). *New motion & greenlots rebranded as shell recharge solutions*, July 18, 2024, https://www.shell.com/what-we-do/renewable-power/renewable-power-news-releases/shell-to-rebrand-newmotion.html
SMMT. (2024). *UK reaches million EV milestone as new car market grows*. https://www.smmt.co.uk/2024/02/uk-reaches-million-ev-milestone-as-new-car-market-grows
Sovacool, B. K., Lovell, K., & Ting, M. B. (2018). Reconfiguration, contestation, and decline: Conceptualizing mature large technical systems. *Science Technology and Human Values, 43*(6), 1066–1097. https://doi.org/10.1177/0162243918768074
Summerton, J. (1994). *Changing large technical systems*. J. Summerton (Ed.). Westview Press.
Summerton, J. (1999). Power plays: The politics of interlinking systems. In *The governance of large technical systems*, 109–29. https://doi.org/10.4324/9780203016893-13.
van der Vleuten, E. (2019). Radical change and deep transitions: Lessons from Europe's infrastructure transition 1815–2015. *Environmental Innovation and Societal Transitions, 32*(July 2017), 22–32. https://doi.org/10.1016/j.eist.2017.12.004.
Wittmayer, J. M., Avelino, F., van Steenbergen, F., & Loorbach, D. (2017). Actor roles in transition: Insights from sociological perspectives. *Environmental Innovation and Societal Transitions, 24*, 45–56. https://doi.org/10.1016/j.eist.2016.10.003
Yin, R. (2018). *Case study research and applications: Design and methods*. SAGE. https://doi.org/10.1016/b978-0-12-809568-3.00016-4.

Zap-Map. (2023). *Evyve becomes tenth network to go live on Zap-Pay*, July 18, 2024, https://www.zap-map.com/news/evyve-tenth-zap-pay-partner

Zap-Map. (2024). *EV Charging Statistics 2024*, July 17, 2024, https://www.zap-map.com/statistics/

Zimmerman, R. (2001). Social implications of infrastructure network interactions. *Journal of Urban Technology, 8*(3), 97–119. https://doi.org/10.1080/106307301753430764

# 3

# The Sectoral Configurations of the Hydrogen Innovation System in Norway

**Jens Hanson and Tuukka Mäkitie**

## Introduction

This chapter explores hydrogen innovation and value chain dynamics in Norway. Hydrogen itself is not a novel product, and a range of different production methods exist. The most developed production processes that so far have covered main shares of hydrogen demand are based on natural gas (grey hydrogen) (Nikolaidis & Poullikkas, 2017). Low- or zero-carbon hydrogen offers decarbonization opportunities, especially in sectors where direct electrification is difficult. While hydrogen is the simplest element, it has complex and different value chains. Green hydrogen is produced via electrolysis closely linked to the electricity

J. Hanson (✉) · T. Mäkitie
Department of Technology Management, SINTEF Digital, Trondheim, Norway
e-mail: jens.hanson@sintef.no

T. Mäkitie
TIK Center for Technology, Innovation and Culture, University of Oslo, Oslo, Norway


M. Steen et al. (eds.), *Sustainability Transitions and Industrial Transformation*,
https://doi.org/10.1007/978-981-95-4893-4_3

sector, while blue hydrogen comes from natural gas reformation and carbon capture in the oil and gas sector. Ongoing experimentation with hydrogen is happening across various user sectors, including heavy-duty transport and industry. This indicates that complex interactions across multiple sectors shape hydrogen value chains (Andersen et al., 2026). Different sectors thus make up segments of hydrogen value chains which in turn have differing sectoral interdependencies, each with idiosyncratic system dynamics, including its institutional and actor characteristics. An effective governance of the upscaling of hydrogen as a climate change solution and as a green industrial opportunity must therefore consider the specific conditions caused by its sectoral configurations (Hanson, 2018; Mäkitie et al., 2018; Stephan et al., 2017) and how they shape innovation processes over time in a given region or country (Andersson et al., 2021).

Norway has a long history with hydrogen production. Notably, Hydro opened a hydrogen production facility using electrolysis in 1929 in Vemork (Norsk Hydrogenforum, 2023). Additionally, Norway has good conditions for production of different types of hydrogen, due to the presence of key resources such as renewable electricity (green hydrogen) and natural gas (blue/grey hydrogen) as well as CCS to enable blue hydrogen given Norway's history and strong position in CCS. Norway is also a front runner in the application of hydrogen technologies in the maritime sector. Norway is thus an interesting country to explore the evolution of hydrogen innovation, and especially how innovation processes have been shaped by multi-sectoral interconnections.

In this chapter, we assess the innovation dynamics of the hydrogen technological innovation system (TIS) in Norway and identify its central sectoral configurations in terms of key production segments (green and blue hydrogen) and user segments. We thus seek to identify the key multi-sectoral conditions for upscaling of hydrogen as a climate solution and industrial opportunity for Norway, to create understanding of how context-specific sectoral configurations may affect hydrogen's innovation dynamics.

## Theoretical Background and Methods

We employ the TIS framework—a key approach to study the emergence and diffusion of novel technologies and industries. The TIS approach provides a systemic view on social structures, innovation processes, and agency (Bergek et al., 2008; Hekkert et al., 2007). Recognizing that TISs are influenced by different context dimensions, including sectors, we include a multi-sectoral approach to TIS dynamics by exploring the sectoral configuration of the hydrogen TIS (Stephan et al., 2017: 710).

Systemically assessing institutional and organizational processes conditioning the development of innovations is important for identifying the central drivers and bottlenecks related to the emergence and development of novel technologies (Bergek, 2019). We understand a TIS as the dynamic network of actors and institutions related to the generation, diffusion, and use of a technology, such as production, transport, storage, and use of green and blue hydrogen over time (Carlsson & Stankiewicz, 1991: 111).

A TIS is influenced by its wider environment (Bergek et al., 2015), including competing or complementing innovations (Markard & Hoffmann, 2016) and geographical and political context, for example preexisting industrial structures and policy frameworks (Bergek et al., 2015). Crucially, for our case, a TIS may be affected and connected to various sectors providing, e.g., energy, input materials, components, and application contexts for the novel technology. For example, the economics of green hydrogen production is strongly influenced by the price and availability of electricity (Mäkitie, Hanson et al., 2022).

The connection between TIS and sectors therefore is a key analytical building block, and we focus on the sectoral configuration of hydrogen value chains, which *"refers to the number and types of sectors linked via the value chain of a TIS"* (Stephan et al., 2017: 710). For novel technologies to develop and mature, technology value chains that span production, distribution, and use need to form. These value chains typically involve a range of different sectors representing different sets of resources, components, and user areas, which interlink as part of a technology value chain (Mäkitie, Hanson et al., 2022). Stephan et al. (2017) for instance show

how a wide variety of sectors, from chemicals, mining, to transportation equipment are connected in the lithium-ion battery value chain. Dependent on the case and technology, a technology value chain is thus composed of a specific sectoral configuration. Given TISs emergent properties, full and functioning value chains may not exist, and the enrolment of the various sectors necessary to form a value chain may thus be seen as a process of TIS formation. It is therefore relevant to look at how the various sectors being part of the sectoral configuration of a TIS influence it's dynamics. We group sectors in three main segments being part of hydrogen value chain: (1) enabling sectors—those that supply key inputs, for instance electricity or natural gas; (2) hydrogen production and distribution—sectors that produce and/or provide infrastructure and transport of hydrogen; and (3) hydrogen use—sectors that represent important hydrogen use areas.

We analyse how sectoral configurations influence hydrogen TIS functions, i.e., the "functionality" of actor networks and institutions in driving the generation, diffusion, and use of the focal technology (Bergek et al., 2008; Hekkert et al., 2007). We focus on six key "functions", or innovation system processes, each evolving over time through the actions of actors (Table 3.1). The development and performance of TIS functions are interconnected, each often further driving or hindering other functions. For example, maturing knowledge development may lead to heightened expectations around a technology, and the attraction of more resources, such as private investments (resource mobilization) (Suurs & Hekkert, 2009). Assessing TIS functions reveals key factors that promote or hinder innovation (Bergek et al., 2008). This evaluation provides insights into what may facilitate the scaling and maturation of new innovations, which in turn is relevant for policymakers, managers, and researchers.

We conducted a qualitative case study using both qualitative and numerical data. Our primary material consisted of 17 semi-structured interviews from August to November 2023 via Teams. Our informants represent Norwegian companies in the hydrogen value chain, interest organizations, public authorities, research financiers, and NGOs. Interviews were supplemented with secondary data sources like statistics,

**Table 3.1** Technological innovation system functions

| TIS function | Description |
|---|---|
| Knowledge development and diffusion | Development and diffusion of knowledge regarding the focal technology. Considers both depth and breadth of knowledge |
| Influence on the guidance of search | Inducing and pressuring factors for actors to enter the TIS. Mechanisms influencing the direction of innovation in terms, e.g., technologies, applications, and markets |
| Entrepreneurial experimentation | Reduction of uncertainty through experimentation with various versions of the technology, and in different applications and markets |
| Market formation | Opening of early markets, as well as articulation and creation of demand |
| Legitimation | Processes of forming social acceptance and compliance with, and formation of, institutions (regulations, norms, and cognitions) |
| Resource mobilization | Mobilization and creation of human and financial resources, infrastructure, and other complementary assets |

*Source* Adapted from Bergek et al. (2008)

research, regulatory databases, parliamentary transcripts, policy documents, and scientific literature. The interviews followed a guide designed to analyse TIS dynamics in Norwegian hydrogen technologies, focusing on TIS function performance, policy, actor networks, and key contextual developments. Using TIS function definitions (Table 3.1) as qualitative codes, we analysed interview transcripts to identify relevant functions.

## Hydrogen Innovation System Dynamics

The Norwegian hydrogen TIS is populated by actors across the value chain, ranging from key enabling sectors, hydrogen production to use sectors. In 2021, the hydrogen industry was assessed to employ about 815 man years with a turnover of about 1.5 billion NOK, of which 70% was attributed to export (Menon, 2022). Norwegian technology supply companies, such as NEL, have for instance had a long-standing global presence. Although many actors are present, hydrogen is still seen as a nascent industry with growth potential, and full and functioning value chains are seen to be lacking.

Both policymakers and industry articulate hydrogen as a decarbonization solution as well as an industrial opportunity for Norway. In terms of climate policy, the Climate Change Act sets the basic framework for decarbonization in Norway (Ministry of Climate and Environment, 2017), outlining a goal of 90–95% carbon emission reduction, compared with 1990, by 2050. Some Norwegian sectors have additional climate mitigation targets. For example in shipping, the Government's Action Plan for Green Shipping established a target of 50% carbon emission reduction in domestic maritime transport by 2030 compared with the 2005 level (Norwegian Ministries, 2019). To support achieving such targets, the government has announced to gradually increase its national fee on carbon emissions to approximately NOK 2000 per carbon tonne by 2030. This is aimed to incentivize gradual change towards low- and zero-emission technologies in transport (Ministry of Climate and Environment, 2021).

Norway has established its own hydrogen strategy, outlining hydrogen as a top-priority decarbonization solution in the country (Norwegian Ministries, 2020). This document outlines hydrogen as particularly important for decarbonization of "hard-to-abate" sectors, including heavy and energy-intensive processing industries, and heavy goods transport both on land and sea, thereby suggesting a set of key use sectors.

## Knowledge Development and Diffusion

Norway has a long history in hydrogen production, going back to the 1920s, with the basic principles of especially green hydrogen production and hydrogen storage being well-known and mature. However, the potential of hydrogen as a decarbonization solution requires further knowledge development, for example, in terms of higher cost-efficiency and use in new application areas, such as in maritime transport and aviation.

In our interviews, the Norwegian hydrogen TIS was described to have high technological competences. Norway has been an international front runner in the use of hydrogen as an energy carrier in maritime transport. It was nevertheless recognized that this was still technologically and regulatorily somewhat immature area due to limited practical experiences. This "marinification" of hydrogen and ammonia[1] was nevertheless seen to be able to draw on prior experiences from more mature applications such as in land-based industrial processes, understanding the technical properties of ammonia, and experiences with liquified natural gas (LNG) as a marine fuel. Other knowledge needs included better understanding of, and solution development for, hydrogen as a part of an intermittent renewable electricity system. Importantly, and regardless of the application area, a key theme for further knowledge development was the need for increased cost- and energy-effectiveness and safety of hydrogen technologies.

An ascending trend in hydrogen knowledge development is visible in granted research projects by both the Research Council of Norway (RCN) and the European Union (see Fig. 3.1). Especially since 2015, grants awarded to hydrogen projects have been growing, which is an indicator of relevant research funding being made available, but also the willingness of Norwegian organizations to explore hydrogen and its applications.

While the above-discussed research projects are often led by research organizations, the RCN's SkatteFunn programme allows privately owned

[1] Production of ammonia ($NH_3$) requires hydrogen, and is thus relevant for hydrogen value chains.

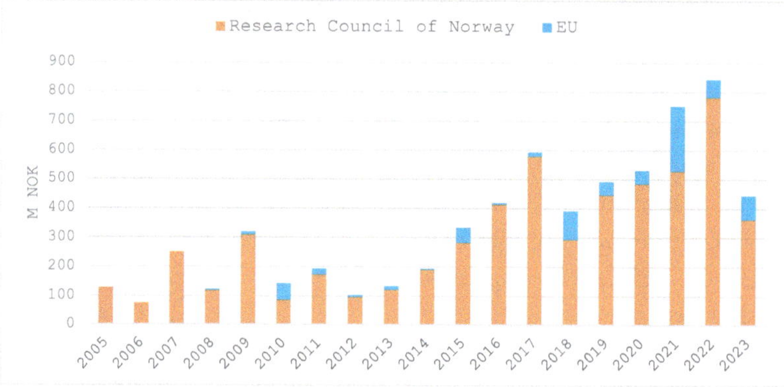


**Fig. 3.1** Total value of hydrogen-related research projects started in each year during 2005–2023, awarded by Research Council of Norway and EU. In EU projects, the value shows the share of the project awarded for Norwegian partners (*Source* Project Bank of Research Council of Norway, search term "hydrogen")

companies to deduct 19% of their self-financed innovation projects. SkatteFunn projects thus offer a perspective on more applied projects. The awarded tax deductions on hydrogen projects saw a notable increase from 2018 onwards (see Fig. 3.2). Notably, while the energy sector is typically the key sector for hydrogen projects, the maritime sector emerged as the second main application domain, since 2018. This further suggests that the more applied knowledge development around hydrogen technologies is focused primarily on maritime and energy applications, where the latter was particularly focused on green and blue hydrogen production.

## Direction of Search

We identified three key factors that motivate hydrogen activities. First, *climate policy* drives the perception of hydrogen as a decarbonization solution for sectors where direct electrification is less feasible. Policy thus represents a motivation for the enrolment of user sectors, which include maritime transport, land-based heavy transport as well as some industrial processes. Second, for firms, the expectations of future market

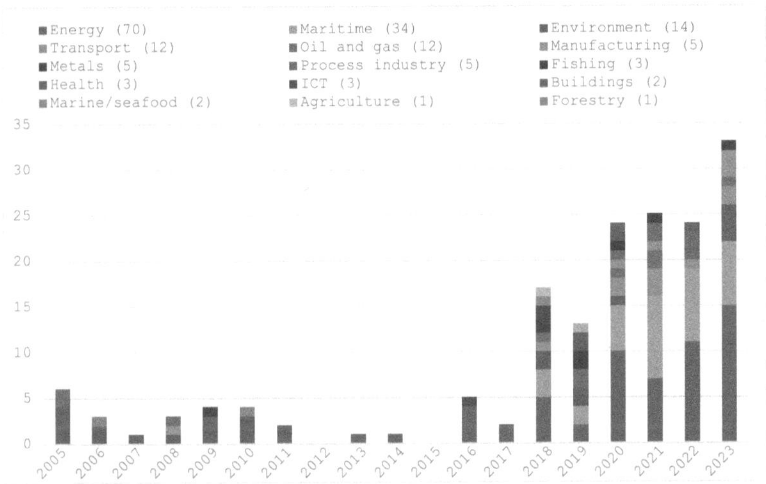


**Fig. 3.2** Number of hydrogen-related SkatteFunn projects started in each year during 2005–2023 in different sectors. The total number of started projects is in parentheses in the legend (*Source* Project Bank of the Research Council of Norway, search term "hydrogen")

opportunities, potentially facilitated by climate policy, combined with the potential for reuse and further development of key Norwegian areas of expertise are a further motivation. Third, the presence of key enabling sectors such as natural gas (blue) and renewable energy (green) are seen as key motivations for entering hydrogen.

In green hydrogen production, access to affordable electricity is critical, illustrating the strong interdependence between enabling and production sectors. However, the Norwegian electricity situation is of concern, particularly with regard to large-scale hydrogen production, where the industry is arguing for a need for more capacity in renewable energy to realize production plans. This relates to both the generation and the distribution (grids), where some actors already have gotten applications for grid access declined.

Similarly, for actors engaged in blue hydrogen, it is not only the presence of gas resources, but also the existing infrastructure, related knowledge bases such as in system engineering, and experiences with health, safety, and environment practices, that make firms see opportunities in hydrogen. Norwegian experience and activities related to carbon capture and storage (Steen et al., 2024) also function as drivers towards hydrogen. Finally, firms with petroleum activities also see hydrogen as a potential diversification opportunity, again with the motivation to reuse existing capabilities.

## Experimentation

Experimentation both in hydrogen production as well as in several user sectors is ongoing. A recent mapping identified an increase in projects that seek to deploy hydrogen solutions: from 50 in 2021 to 129 in 2023 (Hydrogenforum, 2023, see Table 3.2). We categorized these projects and identified projects across the hydrogen value chains, with a predominance in production-related projects (45). We observe a clear majority of projects being focused on green hydrogen (51). This may reflect the fact that green hydrogen production is more feasible in smaller volumes, as opposed to blue hydrogen, which is only perceived as feasible at large scale. Given the market uncertainties discussed under market formation, it is therefore not surprising that the number of green hydrogen projects exceeds that of blue hydrogen. In terms of user sectors, we observe variety with most projects targeted at the maritime sector, heavy industry, and onshore transport.

While the overall number of projects is increasing, interviewees do argue that many projects are in early phases of development and there is a need for projects to move to actual realization. Uncertainties and perceived risks linked to market formation are mentioned as a key factor hampering intensified experimentation. Public support for experimentation is also legitimated out of the concerns related to immature markets and bringing down technology risks and uncertainty. Table 3.3 displays major hydrogen deployment projects supported by Enova, a public financing agency that awards grants for projects that advance transitions towards a low-emission society. These are linked primarily to green hydrogen production and the maritime sector, but also include blue hydrogen as well as industry, truck transport, and construction.

**Table 3.2** Hydrogen projects in Norway categorized into value chain segments; type of hydrogen; type of distribution, and type of user sector

| | Number of projects (excl. 43 R&D) | 83 |
|---|---|---|
| Value chain segment | Distribution | 15 |
| | Production | 45 |
| | Use | 16 |
| | Distribution technology | 0 |
| | Production technology | 3 |
| | Use technology | 4 |
| Type of hydrogen[a] | Green | 51 |
| | Blue | 5 |
| Type of distribution | Offshore | 5 |
| | Onshore | 29 |
| User sector[b] | Construction vehicles | 3 |
| | Industry | 8 |
| | Maritime | 38 |
| | Trucks | 10 |
| | Mixed | 15 |
| | Other | 5 |

[a]Note that these include projects across the value chain where a clear identification of green or blue hydrogen production, distribution, or use was possible.
[b]Note that projects in all value chain segments, where a specified user sector was identified, were categorized into the different user sectors. This means that for example production projects that target a specific user sector are included in the user sector category
*Source* Project database developed by Hydrogenforum (2023)

**Table 3.3** Awarded deployment projects by Enova to hydrogen during 2016–2022, grants of NOK 10 million or more

| Value chain segment | Number of projects | Total value of grants (Millions of NOK) |
|---|---|---|
| Green hydrogen production | 6 | 679 |
| Maritime use | 5 | 506 |
| Industry use | 2 | 272 |
| Blue hydrogen production | 1 | 77 |
| Truck use | 1 | 20 |
| Construction use | 1 | 12 |

*Source* Enova

The maritime sector is often mentioned in interviews as a user sector where Norway has a particularly strong competitive advantage given its long traditions, industry, and knowledge bases. This opens for experimentation with hydrogen and ammonia not only in terms of integration in vessels, but also related to onshore infrastructure, fuels, standardization and development of regulations.

However, even if the maritime sector presents a unique opportunity for experimentation in Norway, it is typically reported that risks associated with technology choices hinder the realization of hydrogen or ammonia projects. Ship owners need to consider investments with long-time horizons for vessels and typically want to avoid technology lock-in to solutions where there is uncertainty.

## Market Formation

Expectations of future development of markets are mentioned as a key motivation for actors to enter hydrogen, particularly in light of the aforementioned emission reduction potential in hard-to-abate user sectors. However, weak market formation is mentioned as a key source of uncertainty for realization of projects. This goes for both hydrogen production and user sectors and reflects the oft-mentioned chicken-and-egg challenge that has characterized the situation for hydrogen in Norway over time (Damman et al., 2021). Interviewees articulate some regret in resorting to this description, but markets remain a key source of uncertainty, particularly for actors targeting activities in Norway. Costs of hydrogen are currently seen as a barrier for potential users, and interviewees argue that prices for green hydrogen are intimately tied to electricity prices, which in turn are not expected to decrease. Enabling sectors, in this case electricity, thus influence market formation. For instance, in the case of hydrogen in the maritime sector, interviewees argue that technology risks, as discussed under experimentation, and the associated higher vessel and fuel costs provide key barriers towards investment decisions, thus stalling user sector development.

Interviewees suggest there are expectations that the carbon price may rise in carbon markets, such as the European Emission Trading System, to the extent that hydrogen becomes competitive with fossil fuels, while hydrogen innovation and knowledge development in parallel contributes to price decreases, but that there are different visions of when that would occur. However, current carbon prices are too low to incentivize hydrogen investments, and actors point towards the need for policy tools that contribute to risk reduction to leverage investments. Actors are quite unanimous on the need for public support for early-phase market formation, which in turn is expected to trigger vital learning processes and cost reductions.

Public procurement has been used in the ferry segment to introduce the world's first LNG and electric vessels, Glutra in 2000, and Ampere in 2015, respectively (Bach et al., 2021; Sæther & Moe, 2021). Following this success, the same instrument has been used to introduce hydrogen technologies to the shipping market. The objectives of supporting hydrogen technologies specifically has been to foster other key technologies besides electric vessels as low-and zero-carbon energy sources in maritime transport, to gain experiences that would support further regulatory and project development, and to incentivize industry to further invest in technology development. Indeed, the National Public Roads Administration specified the use of hydrogen as a propulsion energy as a requirement in the tender for ferry routes Hjelmeland-Nesvik-Skipavik in Rogaland and Bodø-Røst-Værøy-Moskenes in Nordland. The former development contract was awarded to Norled, resulting in the hydrogen-vessel MF Hydra to be put into operation in 2023. The latter route to Lofoten was awarded to Torghatten Nord in 2021.

## Resource Mobilization

Technology development in the formative phase of radically novel innovations like use of hydrogen as an energy carrier does not typically lead to an immediate payback on investment and is therefore often reliant on public support to gain momentum. Respondents in interviews described that the access to public capital (Research Council of Norway, Enova, SIVA, Innovation Norway, and EU) in technology development has been good over the last years. Two research centres, FME Hydrogeni and FME

HyValue, act as prime recent examples of major research funding by Research Council of Norway and industrial actors targeting the whole hydrogen value chain, and especially key Norwegian user sectors like maritime transport. Regarding demonstration, for example the Pilot-E programme has granted demonstration support for hydrogen technologies, such as hydrogen vessels and hydrogen production (Pilot-E, 2023). Moreover, Enova has awarded more than NOK 1,1 billion to the establishment of five hydrogen green production facilities and a number of vessels, again illustrating resources for both production and user sectors (E24, 2022).

However, to mature and upscale technologies, and to build the needed industrial capacity, much larger and typically private investments are eventually needed. As mentioned previously, public financing for early-phase research and demonstration of hydrogen technologies has been made available in Norway.

In terms of access to private capital, respondents reported high variation. Private risk capital was reported to have been particularly strongly available in 2020–2021. This period was marked by low interest rates and stock market listing of several dedicated hydrogen companies. By late 2023, however, it was described that the flow of private capital had somewhat dried up. Moreover, also interest rates have by 2025 risen to higher levels, making investments in hydrogen facilities and technologies less attractive.

Our interviews indicate that general uncertainty in 2023 surrounding the hydrogen market and framework conditions were key barriers for private investors, such as shipowners, to commit investments in hydrogen projects. Despite this, actors described that several hydrogen companies often nevertheless continued to have the backing and patience of large institutional funds and industrial companies as owners due to the expected market formation in the future.

In terms of the availability of competences and human capital, previous studies have suggested that access to qualified workforce is a barrier for Norwegian hydrogen companies (Normann et al., 2023; Winje et al., 2022). Key competences in hydrogen technologies are related to, e.g., process engineering (gases) and electrochemistry which are required also, for example, in the oil and gas industry. The demand for the same competences occasionally leads to competition over employees between the oil and gas sector and the novel "green" industries, such as hydrogen.

## Legitimation

Hydrogen as a decarbonization solution has attracted notable attention from policymakers in recent years. In terms of legislation, hydrogen-specific regulations do not yet exist in many new application areas, such as in maritime transport. This is due to the yet limited experiences with hydrogen applications, which challenges the definition and delineation of hydrogen-specific regulations. Existing regulatory frameworks are used when possible. For example, hydrogen infrastructure of more than 5 tonnes of hydrogen is evaluated through a "Major Accident Regulation", requiring stringent safety considerations (Ministry of Justice and Public Security, 2016). Permitting processes of hydrogen facilities were described to be lengthy and encompassing. Lack of specific regulations was thus seen to hinder implementation of hydrogen technologies in terms of designing hydrogen technologies, as well as making contracting difficult, for instance, in terms of defining responsibilities between suppliers and end-users.

Meanwhile, Norwegian national regulators such as the Norwegian Maritime Authority participate actively in international regulatory development for hydrogen and ammonia. This occurs, for example in the International Maritime Organization, which has the ambition to establish some form of hydrogen regulations by 2024. Experiences from early hydrogen vessels in Norway, such as MF Hydra, were mentioned by informants as valuable in this process. Indeed, in a situation when fixed regulations are not yet in place, the Norwegian Maritime Authority and other regulatory bodies have expressed a wish for early involvement in companies' alternative fuel projects. In this way, the regulatory bodies can constructively contribute to their permission processes and support technology development.

In a report presenting results from a 2020 survey, it was suggested that the general population in Norway has relatively limited knowledge regarding hydrogen. Nevertheless, around half of the population answered to be willing to pay more for hydrogen as a fuel than what they are currently paying (Opinion, 2020). Recently, hydrogen has gained more attention in the mainstream media, as shown in Fig. 3.3. Due to this increased spotlight, also public awareness of hydrogen can be expected to have risen during the last few years. Meanwhile, interviewees mention that the explosion of the hydrogen tanking station in Sandvika (Oslo) in 2019 has negatively influenced public perceptions of hydrogen.

The attention to hydrogen has also increased in debates in the Norwegian parliament. The discussion transcripts of parliamentary debates during early 2020s reveal that political parties have had a relatively large consensus in believing that hydrogen, including blue and green hydrogen as well as ammonia production, has notable future economic opportunities for Norway and Norwegian companies. Moreover, hydrogen technologies were broadly argued to be important decarbonization solutions, especially in maritime and heavy-duty land transport. However, during 2021–2023 electricity prices in Southern and Western Norway have been high, and various industries across the country have struggled to secure grid capacity to cover their growing power needs. This has raised concerns in the Parliament regarding electricity use of hydrogen production, especially in the context of remote areas with limited existing power

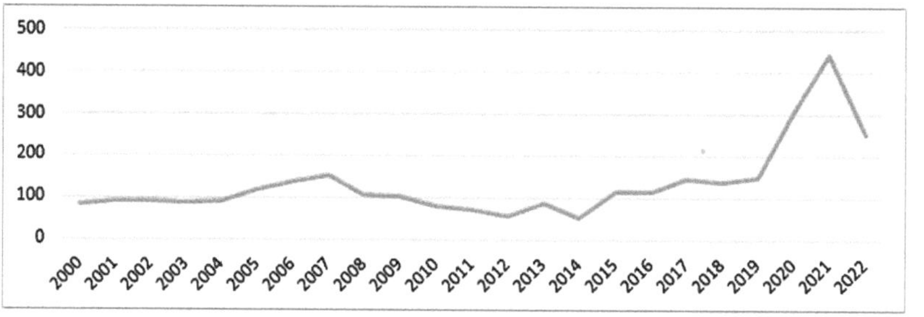


**Fig. 3.3** The total number of articles in Norwegian newspapers Adresseavisen, Aftenposten, Bergens Tidende, Dagbladet, Klassekampen, and VG mentioning "hydrogen" during 2000–2022 (*Source* Retriever/A-tekst)

grids. The debates also featured very different views on to what extent, and in which form, the state should support and subsidize hydrogen. Examples of such debates were whether state ownership in large companies such as Statkraft (utility) should be used to create an industrial locomotive, investing heavily in hydrogen, or whether this should be left to privately managed market forces, and whether contracts-for-difference should be employed to reduce uncertainty regarding hydrogen prices.

## Sectoral Configurations and Innovation Dynamics

The TIS dynamics observed in the previous section point to a number of key sectors across enabling, supply and use sectors, which together make up the sectoral configuration of hydrogen value chains in Norway. Figure 3.4 summarizes these sectors. The functional analysis revealed that the Norwegian hydrogen TIS has specifically focused on several user sectors and two main types of hydrogen production. Between these, there have been notable differences in the performance of the innovation dynamics in different parts of the hydrogen value chain. Meanwhile, the different parts have also suffered from some of the same challenges.

In terms of knowledge development and diffusion, there were activities surrounding both green and blue hydrogen production, representing two different production value chains with differing enabling sectors. Moreover, knowledge development and diffusion has been particularly strong in the shipping sector in Norway. Hydrogen innovation in Norway has thus had rather specific directionality towards certain segments. On the production side, the domestic enabling sectors (renewable electricity and natural gas resources) and CCS development have been shaping the direction of search towards these technologies. Similarly, the large and technologically advanced maritime industry in Norway has engaged in exploring hydrogen solutions, which has created an emphasized directionality towards shipping as a key user sector.

In terms of experimentation on the supply side, green hydrogen production had the highest number of projects, but there were also blue

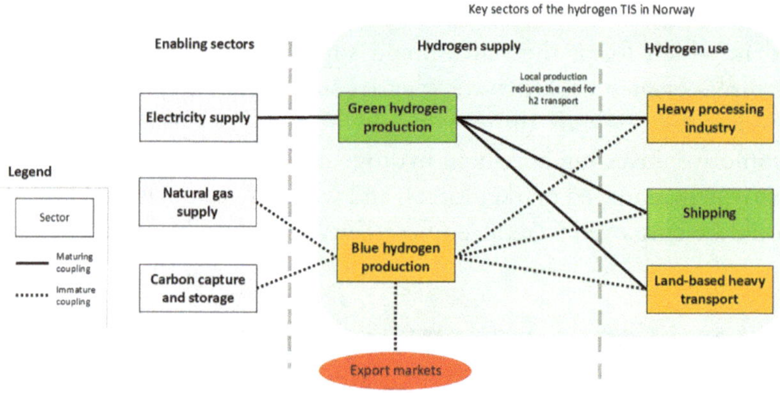


**Fig. 3.4** Key sectoral configuration of the hydrogen TIS in Norway. Green colour indicates high level of activities, yellow average level of activities, and red as low level of activities

hydrogen projects. On the user side, the shipping sector again dominated, but there were activities also in heavy processing industry and land-based heavy transport. Market formation in hydrogen, however, was a major challenge for all involved sectors. Similarly, all parts of the hydrogen value chain seem to be affected by fluctuations in the availability of resources, e.g., in terms of private capital and human resources. In public policymaking, hydrogen has enjoyed somewhat high legitimacy as a decarbonization solution especially in shipping, exemplified by the aforementioned public procurement programmes and support programme. However, private actors in all parts of the hydrogen value chains call for more extensive public support, such as contracts-for-difference, to increase the momentum in the hydrogen TIS in Norway.

As summarized in Fig. 3.4, both green and blue hydrogen supply have been actively explored in Norway. Green hydrogen production is, however, more mature, with proven technological solutions ready, and a few full-scale projects are being realized. Couplings between user sectors are maturing through concrete projects. Currently, blue hydrogen remains unrealized, relying on the advancement of CCS value chains.

Collaborations with user sectors are still in their early stages, as commercial projects have not yet materialized. CCS is emerging in Norway, which may create opportunities for blue hydrogen in future, but the time schedule of this is unclear, as the expansion of the domestic CCS value chain suffers from a number of uncertainties, including market formation (Steen et al., 2024). Blue hydrogen production will also likely have a strong direction towards export markets, but such couplings are yet far from being realized, not least due to immature $CO_2$ transportation networks.

In terms of user sectors, shipping has been the most central sector with a high number of experimentation projects, albeit this has yet resulted in few materialized vessels. Hydrogen vessels have indeed been slower in development and adoption than battery-electric vessels in Norway (Bach et al., 2020; Mäkitie, Steen et al., 2022), but especially through hydrogen-based fuels such as ammonia and methanol, hydrogen has higher potential as a decarbonization solution in the shipping sector due to their applicability in longer sea voyages (Bergek et al., 2023). Also, heavy processing industry (especially metal and chemical) and land-based heavy transport (especially trucks) have had some projects in Norway. Other potential user applications, such as aviation, are yet to attract more attention.

Summing up, we have been able to identify some of the key drivers and barriers for further industry development and upscaling of hydrogen solutions in Norway. In terms of drivers, we observe some general trends that indicate the strengthening of the innovation system overall, as well as some innovation system functions (innovation processes) individually. These include: (i) a strong existing knowledge base surrounding hydrogen and its continuous development over time, (ii) increasing availability of public funding for research, SkatteFunn and deployment grants targeting especially niches, such as maritime sector, where Norwegian companies have competitive advantages, and (iii) a continued growth in hydrogen projects. The activity in experimentation with hydrogen projects is particularly prevalent for green hydrogen. Taken together this indicates, so far, a strengthening of knowledge development as well as experimentation linked to blue and green hydrogen over time, albeit with a majority of green hydrogen projects. A key motivation is the expectation of hydrogen becoming a key climate solution, particularly in areas where direct electrification is less feasible, alongside intensified climate policy and need for emission reductions. Indeed, these are key drivers for many actors to experiment with the hydrogen field.

However, we also see notable barriers, which slow down the innovation system development overall. Some of these barriers span both blue and green hydrogen. While the *expectations* of market formation motivate to experiment with hydrogen, the yet weak realization of markets constitutes a key barrier for Norwegian companies. We observe this manifesting in the hesitancy of making investment decisions for deployment projects, which may already have come some way in terms of knowledge development via R&D and/or pilot activities, and even received public financing. We also observe hurdles related to mobilization of private capital recently, connected to higher interest rates, uncertainties in terms of market development, and future policy support schemes. Moreover, due to the yet limited practical experiences with hydrogen solutions in new user sectors, also regulatory development is yet limited in, for example, the maritime sector, creating a challenge for the actors who seek to bring some of the first hydrogen projects to reality.

Weak market formation is thus seen as a key barrier affecting the overall dynamics of the hydrogen TIS. If overcome, actors expect this to untangle some of the other challenges related to realization of full-scale projects and potentially leverage more private capital. As of late 2023, hydrogen actors were quite unanimous in their problem description for hydrogen upscaling in Norway (lacking markets), and the solution to this problem (contracts-for-difference). However, as discussed under legitimation, the extent and type of public support for hydrogen upscaling is a political choice. For Norwegian industry and policymakers, it remains central to follow the international hydrogen innovation developments to capitalize and develop the natural comparative advantages of Norwegian industry in hydrogen production, transport, and use. These advantages include both fossil and renewable resources, and technological knowledge from various sectors in both blue and green hydrogen value chains.

## Concluding Discussion

This book chapter has assessed the sectoral configurations and innovation dynamics of the hydrogen technological innovation system (TIS) in Norway. We have discussed how these form key conditions for upscaling of hydrogen, which thus affect the further development, diffusion, and use of hydrogen as a low- and zero-emission solution, and the related hydrogen industry development in Norway. Using TIS and sectoral configurations of innovation systems as two key analytical building blocks (Hekkert et al., 2007; Stephan et al., 2017), we observed how enabling, supply, and use sectors are enrolled as part of the sectoral configuration of the Norwegian hydrogen TIS. We found that the sectoral configuration has shaped directionality towards certain segments. For hydrogen supply this is linked to how renewable electricity enables green hydrogen, while natural gas and CCS enable blue hydrogen. In terms of hydrogen use, especially the maritime sector with its advanced knowledge bases combined with the need to decarbonize, has represented a key user sector. We thus see a directionality towards certain production and user segments, forming early-phase building blocks of the sectoral configuration of hydrogen value chains in Norway. Based on the analysis of TIS dynamics, such as knowledge development and experimentation, we thus see that activities have been clustered around certain user sectors (particularly maritime) and two main supply sectors (green and blue hydrogen), each with different enabling sectors (electricity and natural gas/CCS), albeit with a focus on green hydrogen so far.

The focus on green hydrogen represents a particular sectoral configuration of the hydrogen value chain, with a strong interdependence between the electricity sector as an enabling sector and hydrogen supply sectors, where the capacity for upscaling and prices have been noted as a challenge going forward. This illustrates the link between value chain interdependencies and TIS functions. For instance, market formation, which was seen as a key barrier, is influenced by several factors, including enabling sector idiosyncrasies such as electricity prices which impact hydrogen costs (supply sector), which in turn influence user sector adoption decisions and expectations.

Incorporating a sectoral configuration perspective on TIS dynamics is thus useful for identifying segments of emergent value chains, or interdependencies between segments, of TIS where key processes are developing well or where strengthening is needed. This not only underlines the importance of understanding the sectoral context for TIS (Bergek et al., 2015), in terms of its sectoral configuration (Stephan et al., 2017: 710), but also reflects how TISs indeed are linked to multi-system dynamics with different sectors making up the value chains of TIS (Andersen et al., 2026). Our analysis has focused on the case of hydrogen in Norway, with yet limited development. Future studies could look at hydrogen across differing geographies, with differing sectoral configurations, for instance in terms of enabling sectors. Additionally, future research could look at more mature TISs, for instance those that have a wider diversity and scale of user sectors, to analyse how this influences the sectoral configuration of TIS in terms of enabling and supply sector upscaling.

## References


Andersen, A. D., Mäkitie, T., Steen, M., & Wanzenböck, I. (2026). Integrating industrial transformation and sustainability transitions research through a multi-sectoral perspective. Energy Research & Social Science, 131, 104509. https://doi.org/10.1016/j.erss.2025.104509

Andersson, J., Hellsmark, H., & Sandén, B. (2021). The outcomes of directionality: Towards a morphology of sociotechnical systems. *Environmental Innovation and Societal Transitions, 40*, 108–131. https://doi.org/10.1016/j.eist.2021.06.008

Bach, H., Bergek, A., Bjørgum, Ø., Hansen, T., Kenzhegaliyeva, A., & Steen, M. (2020). Implementing maritime battery-electric and hydrogen solutions: A technological innovation systems analysis. *Transportation Research Part d: Transport and Environment, 87*, 102492. https://doi.org/10.1016/j.trd.2020.102492

Bach, H., Mäkitie, T., Hansen, T., & Steen, M. (2021). Blending new and old in sustainability transitions: Technological alignment between fossil fuels and biofuels in Norwegian coastal shipping. *Energy Research & Social Science, 74*, 101957. https://doi.org/10.1016/j.erss.2021.101957

Bergek, A. (2019). Technological innovation systems: A review of recent findings and suggestions for future research. In F. Boons & A. McMeekin (Eds.), *Handbook of sustainable innovation* (pp. 200–218). Edward Elgar Publishing. https://doi.org/10.4337/9781788112574

Bergek, A., Hansen, T., Hanson, J., Mäkitie, T., & Steen, M. (2023). Complexity challenges for transition policy: Lessons from coastal shipping in Norway. *Environmental Innovation and Societal Transitions, 46*, 100687. https://doi.org/10.1016/j.eist.2022.100687

Bergek, A., Hekkert, M., Jacobsson, S., Markard, J., Sandén, B., & Truffer, B. (2015). Technological innovation systems in contexts: Conceptualizing contextual structures and interaction dynamics. *Environmental Innovation and Societal Transitions, 16*, 51–64. https://doi.org/10.1016/j.eist.2015.07.003

Bergek, A., Jacobsson, S., Carlsson, B., Lindmark, S., & Rickne, A. (2008). Analyzing the functional dynamics of technological innovation systems: A scheme of analysis. *Research Policy, 37*(3), 407–429. https://doi.org/10.1016/j.respol.2007.12.003

Carlsson, B., & Stankiewicz, R. (1991). On the nature, function and composition of technological systems. *Journal of Evolutionary Economics, 1*(2), 93–118. https://doi.org/10.1007/BF01224915

Climate Change Act. (2017). https://lovdata.no/dokument/NLE/lov/2017-06-16-60

Damman, S., Sandberg, E., Rosenberg, E., Pisciella, P., & Graabak, I. (2021). A hybrid perspective on energy transition pathways: Is hydrogen the key for Norway? *Energy Research & Social Science, 78*, 102116. https://doi.org/10.1016/j.erss.2021.102116

E24. (2022). *Enova gir milliardstøtte til hydrogen.* Retrieved 20 Fenruary 2023 from https://e24.no/det-groenne-skiftet/i/v5q0xl/enova-gir-milliardstoette-til-hydrogen

Hanson, J. (2018). Established industries as foundations for emerging technological innovation systems: The case of solar photovoltaics in Norway. *Environmental Innovation and Societal Transitions, 26*, 64–77. https://doi.org/10.1016/j.eist.2017.06.001

Hekkert, M. P., Suurs, R. A. A., Negro, S. O., Kuhlmann, S., & Smits, R. E. H. M. (2007). Functions of innovation systems: A new approach for analysing technological change. *Technological Forecasting & Social Change, 74*(4), 413–432. https://doi.org/10.1016/j.techfore.2006.03.002

Hydrogenforum. (2023). *Det norske hydrogenlandskapet kartlagt*. https://www.hydrogen.no/aktuelt/nyheter/det-norske-hydrogenlandskapet-er-kartlagt

Markard, J., & Hoffmann, V. H. (2016). Analysis of complementarities: Framework and examples from the energy transition. *Technological Forecasting and Social Change, 111*, 63–75. https://doi.org/10.1016/j.techfore.2016.06.008
Menon. (2022). *VERDIEN AV DEN NORSKE HYDROGENNÆRINGEN*. https://www.menon.no/wp-content/uploads/2022-134-Verdien-av-den-norske-hydrogennaeringen-1.pdf
Ministry of Climate and Environment. (2021). *Klimaplan for 2021–2030.*
Mäkitie, T., Andersen, A. D., Hanson, J., Normann, H. E., & Thune, T. M. (2018). Established sectors expediting clean technology industries? The Norwegian oil and gas sector's influence on offshore wind power. *Journal of Cleaner Production, 177*, 813–823. https://doi.org/10.1016/j.jclepro.2017.12.209
Mäkitie, T., Hanson, J., Steen, M., Hansen, T., & Andersen, A. D. (2022). Complementarity formation mechanisms in technology value chains. *Research Policy, 51*(7), 104559. https://doi.org/10.1016/j.respol.2022.104559
Mäkitie, T., Steen, M., Saether, E. A., Bjørgum, Ø., & Poulsen, R. T. (2022). Norwegian ship-owners' adoption of alternative fuels. *Energy Policy, 163*, 112869. https://doi.org/10.1016/j.enpol.2022.112869
Ministry of Justice and Security (2016). Forskrift om tiltak for å forebygge og begrense konsekvensene av storulykker i virksomheter der farlige kjemikalier forekommer (storulykkeforskriften). https://lovdata.no/dokument/SF/forskrift/2016-06-03-569
Ministry of Climate and Environment (2017). Act relating to Norway's climate targets (Climate Change Act). https://lovdata.no/dokument/NLE/lov/2017-06-16-60
Nikolaidis, P., & Poullikkas, A. (2017). A comparative overview of hydrogen production processes. *Renewable and Sustainable Energy Reviews, 67*, 597–611. https://doi.org/10.1016/j.rser.2016.09.044
Normann, H.E., Steen, M., Mäkitie, T., Klitkou, A., Børing, P., Solberg, E., Brynthe Lund, H., Wardeberg, M. & Wanderås Fossum, L. (2023). Kompetanse for grønn omstilling. En gjennomgang av forskningslitteratur og arbeidslivets kompetansebehov knyttet til miljø- og klimautfordringer. https://www.nifu.no/rapporter/kompetanse-for-gronn-omstilling-en-gjennomgang-av-forskningslitteratur-og-arbeidslivets-kompetansebehov-knyttet-til-miljo-og-klimautfordringer/

Norsk Hydrogenforum. (2023). *Den norske hydrogenhistorien*. Retrieved 8 December 2023 from https://www.hydrogen.no/om-hydrogen/den-norske-hydrogenhistorien

Norwegian Ministries. (2019). *Handlingsplan for grønn skipsfart*. Retrieved from https://www.regjeringen.no/no/dokumenter/handlingsplan-for-gronn-skipsfart/id2660877/

Norwegian Ministries. (2020). *The Norwegian Government's hydrogen strategy towards a low emission society*.

Opinion. (2020). Kjennskapsundersøkelse Hydrogen - Presentasjon 2. november 2020. https://www.hydrogen.no/files/images/archive/presentasjon-hydrogenundersokelse.pdf

Pilot-E. (2023). *Pilot-E. Raskere fra idé til marked*. Retrieved 26 October 2023 from https://www.enova.no/pilot-e/

Steen, M., Andersen, A. D., Finstad, J., Hansen, T., Hanson, J., Jordal, K., Mäkitie, T., Nordholm, A., Ryghaug, M., & Santoalha, A. (2024). CCS technological innovation system dynamics in Norway. *International Journal of Greenhouse Gas Control, 136*, 104171. https://doi.org/10.1016/j.ijggc.2024.104171

Stephan, A., Schmidt, T. S., Bening, C. R., & Hoffmann, V. H. (2017). The sectoral configuration of technological innovation systems: Patterns of knowledge development and diffusion in the lithium-ion battery technology in Japan. *Research Policy, 46*(4), 709–723. https://doi.org/10.1016/j.respol.2017.01.009

Suurs, R. A. A., & Hekkert, M. P. (2009). Cumulative causation in the formation of a technological innovation system: The case of biofuels in the Netherlands. *Technological Forecasting and Social Change, 76*(8), 1003–1020. https://doi.org/10.1016/j.techfore.2009.03.002

Sæther, S. R., & Moe, E. (2021). A green maritime shift: Lessons from the electrification of ferries in Norway. *Energy Research & Social Science, 81*, 102282. https://doi.org/10.1016/j.erss.2021.102282

Winje, E., Hernes, S., Foseid, H., Schöpfer, A., & Jakobsen, E. (2022). Verdien av den norske hydrogennæringen. Status og fremtidsutsikter. https://menon.no/prosjekter/verdien-av-den-norske-hydrogennaeringen

# 4

# Mapping Multi-System Interactions in Urban Contexts—A Study of the Energy, Transport, and Food Systems in the Municipality of Gothenburg

Ingrid Mignon, Ksenia Onufrey, and Linus Thomson

## Introduction

Recognizing the relevance of a multi-system perspective, a growing body of work within sustainability transitions studies has begun to employ multi-system analyses (Andersen & Geels, 2023; Andersen et al., 2023; Ohlendorf et al., 2023; Rosenbloom, 2019), promoting a closer examination of the fluidity and permeability of system boundaries, thereby fostering a more dynamic view of system development in relation to other systems (Konrad et al., 2008). For instance, previous studies have investigated the system interactions between the electricity and waste management system (Raven, 2007), and between the electricity and electric transport systems (Haley, 2015). While the broader sustainability transitions literature has emphasized urban contexts as key nodes for socio-technical transitions, functioning as sites where multiple socio-technical systems converge, interact, and evolve (Frantzeskaki & Rok,

I. Mignon (✉) · K. Onufrey · L. Thomson
Chalmers University of Technology, Gothenburg, Sweden
e-mail: ingrid.mignon@chalmers.se


M. Steen et al. (eds.), *Sustainability Transitions and Industrial Transformation*,
https://doi.org/10.1007/978-981-95-4893-4_4

2018), most existing research on multi-system interactions has largely focused on the national level, leaving localized contexts comparatively underexplored. This represents a missed opportunity, as municipal-level analyses provide rich empirical settings to engage simultaneously with multiple socio-technical systems, offering explanations beyond those grounded in isolated systems (Grin et al., 2017).

Another notable limitation of existing multi-system analyses lies in their predominant focus on dual interactions, i.e., mapping and explaining interconnections between two systems. This narrow approach becomes increasingly problematic as transitions accelerate, triggering simultaneous changes across multiple systems. The resulting effects of emerging and evolving cross-system interactions go beyond dual relationships, spanning a broad range of socio-technical systems and their corresponding societal functions. Hence, recent studies have called for "a shift in research from focusing on single system transitions plus wider interactions towards investigating a 'patchwork' of interacting transitions" (Andersen & Geels, 2023, p. 10). In order to begin investigating such "patchworks" of interconnected transitions across multiple systems, the level of municipalities provides an appropriate context.

Accordingly, this chapter goes beyond the typical focus on dual interactions by considering mutual interconnections across three systems, i.e., the food, energy, and transport systems. A transition is urgently required in each one of these systems. The food system is widely recognized as environmentally and socially unsustainable (Hebinck et al., 2021; Meynard et al., 2017), scaling up the diffusion of clean energy production is necessary to limit climate change (Mignon & Bergek, 2016), and the transport system needs to shift from fossil fuels to renewable alternatives (ER, 2017). However, none of these systems exists in isolation and must be considered within their broader socio-technical context, including neighbouring systems with which they interact (Andersen & Geels, 2023; Andersen et al., 2023; Kanger et al., 2021; Sutherland et al., 2015). Earlier studies have highlighted the interconnections between food, energy, and transport systems at the national level (Huttunen et al., 2014; Sutherland et al., 2015). In this chapter, we chose to study the interactions linking these three systems in the urban context of Gothenburg municipality.

Specifically, we perform an in-depth review of policy documents to map the interactions between key actors, infrastructures/technologies, and institutions across these systems in Gothenburg municipality. In doing so, we identify the strengths and weaknesses of the structures (e.g., projects, plans, networks) put in place by municipal actors to foster interactions between these systems and formulate policy recommendations on how to navigate imbalances and asymmetries inherent to urban multi-system transitions.

Following, we present the analytical framework, focusing on the concept of inter-system couplings and connecting it to the urban context. We then elaborate on the methodology and present the study findings in accordance with the theoretical framework. Finally, we discuss the key insights and summarize the contributions, along with suggestions for further research.

## Analytical Framework

### Multi-System Interactions and Couplings

Interactions across systems result in the reshaping of system boundaries, affecting actor possibilities for achieving their goals (Rosenbloom, 2020). Therefore, previous studies have paid significant attention to explore sites of interactions (Rosenbloom, 2019), their types, intensity, and impact on the involved systems (Breitschopf et al., 2023). Two systems can be involved in different types of interactions simultaneously, ranging from competitive to symbiotic, and the type of interactions can change over time (Geels, 2007; Raven & Verbong, 2009). In this chapter we focus particularly on the interfaces, or couplings that link different systems and are key to achieving compatibility between them (Andersen et al., 2023).

Literature distinguishes between two primary types of couplings, structural and functional (Konrad et al., 2008). Structural couplings are defined as system elements such as technologies (including physical infrastructure), actors, and institutions that connect multiple systems (Konrad et al., 2008). For instance, the installation of harbour charging stations can be understood as a technology-related structural coupling that connects the maritime transport sector with the electricity system (Andersen et al., 2023). This coupling is fundamental as it provides the physical and organizational framework necessary for interaction. Technology-related couplings represent interactions between supply chains and upstream systems during technology development and diffusion, as well as complementary technologies in use contexts. Actor-related couplings emerge when business actors diversify into other systems or when users interact with and transfer learning and experiences across systems. Institution-related couplings emerge when actors address institutional mismatches or when common meta-rules develop across multiple systems (Andersen & Geels, 2023).

Functional couplings refer to the flow of resources between systems (Konrad et al., 2008), which are facilitated by the structural couplings (Andersen & Geels, 2023). For example, the electrical energy flowing through the charging stations at the harbour constitutes a functional coupling between the maritime transport and electricity systems, creating a functional interdependency (Andersen et al., 2023). Ongoing transitions can result in the emergence of new functional couplings across systems. For example, Huttunen et al. (2014) showed how biogas production can enhance new resource flows between transport, agri-food, and energy systems: municipal food waste, among other input materials, undergoes anaerobic digestion and hence serves as an input resource to produce energy or, after additional refining, to produce transport fuels; in turn, the rest products from anaerobic digestion can flow back into the agri-food system in the form of nutrients and fertilizers. However, the feasibility of functional couplings can be hampered by a lack of corresponding structural couplings, e.g., in terms of incoherent policies across different involved systems (Andersen & Geels, 2023; Huttunen et al., 2014).

Building on the foundational work of Konrad et al. (2008), Andersen et al. (2023) proposed a dynamic and processual nexus building perspective elaborating how inter-system couplings are created across the four dimensions of technology, institutions, actors, and resources. They show that the process of building couplings involves tensions, conflicts and can result in asymmetric adjustments where one system has to undergo larger changes than the other due to power imbalances between actors from different systems. Developing technical and institutional compatibility can be a politically fraught process with disagreements on which solutions are most desirable, sometimes leading to the implementation of sub-optimal workarounds (Andersen et al., 2023). Furthermore, as transitions develop from early phases towards acceleration, multi-system interactions become more intense and complex (Markard et al., 2020), which can be seen in a form of knock-on effects from diffusion of individual innovations on wider systems (McMeekin et al., 2019) or in a form of increasingly interconnected transitions in parallel systems, driven by the same factors and competing for the same resources. For example, Nykamp et al. (2023) showed how grid connection challenges were a common constraint in the electrification of both ferry transportation, construction sites, and ammonia production.

## Inter-System Couplings in the Urban Context

Multi-system interactions as well as transitions in general are spatially embedded (Ferloni, 2022). Hence, urban environments serve as important contexts where different types of multi-system interactions are created (Frantzeskaki et al., 2017). In terms of technology couplings, cities serve as test beds for developing, testing, and upscaling new technological solutions (Sharp et al., 2024). In one and the same place, different innovations are combined with different existing urban systems (Hodson et al., 2024). Such processes of experimentation and reconfiguration involve competitive as well as complementary interactions across technologies and systems (Hodson et al., 2017). During contemporary transitions where outcomes are still evolving and remain uncertain, urban contexts can support unexpected connections between previously separate elements and systems (Grin et al., 2017; Rohracher & Späth, 2017).

Regarding institutional couplings, in the localized contexts such as municipalities, understandings of sustainability transition goals are often contested. Actors from different systems are driven by different visions, leading to institutional environments that are sometimes contradictory and sometimes mutually reinforcing (Hodson et al., 2017). As multiple interpretations of visions and goals open for multiple development directions, governance of urban transitions requires integrative approaches across different systems, considering conflicts between different solutions and systems along with synergies (Grin et al., 2017; Rohracher & Späth, 2017).

As for the actor couplings, urban studies highlight the roles of various types of actors, including local governments, companies, networks, community groups, etc. (Huang & Broto, 2018). In particular, local governments can take a central role in driving and intermediating urban transitions, with support and engagement of other actors (Broto, 2017; Fudge et al., 2016; Holscher et al., 2019). Through this role, local governments can intermediate between and within different scales of governance (i.e., local, national, and international), hence linking actors, practices, and policies of different scales (Gustafsson & Mignon, 2020).

# Method

## Case Motivation

Earlier transition research has highlighted the interconnections between food, energy, and transport systems, despite these systems not typically being considered together (Huttunen et al., 2014). For instance, biogas from agricultural byproducts and the co-location of renewable energy technologies on agricultural land demonstrate the potential for systemic sustainability transitions across these sectors (Sutherland et al., 2015). While previous studies have addressed such interconnections at the national level, our study shifts the focus to the municipal level, examining how key actors, infrastructures/technologies, and institutions interact across the food, energy, and transport systems in Gothenburg.

Gothenburg, located on the southwest coast of Sweden at the mouth of the Göta River, provides a suitable setting for studying these multi-system interactions. As Sweden's second-largest city, with a population of 600,000 and over 1 million in the greater metropolitan area, Gothenburg experiences substantial resource flows across these critical systems, providing a rich environment for analysing their interconnections. From a transport system perspective, Gothenburg stands out as a major hub in Sweden, housing key infrastructure such as the municipally owned Port of Gothenburg—the largest in Scandinavia, an airport, extensive rail and road networks, and global vehicle manufacturers, including Volvo Cars and Volvo Trucks that are heavily involved in urban testbeds. The energy system in Gothenburg is similarly noteworthy, with the city owning its own energy company, Göteborg Energi, and operating a vast district heating system. It also has Sweden's largest installed solar power capacity, emphasizing its focus on energy resilience and sustainability. Gothenburg's influence on the food system is notable as well, with the city managing large-scale school meal programmes and 3000 hectares of agricultural land. This is particularly relevant in the context of Sweden's low national food self-sufficiency, which adds pressure on Gothenburg to enhance food system resilience. Overall, the populous urban context of Gothenburg, coupled with the specific characteristics of each system, makes it a promising case for studying multi-system interactions.

## Data Collection

This study was conducted through an in-depth review of policy documents from Gothenburg municipality, which serve as a crucial data source for understanding the municipality's role in shaping the interactions between the food, energy, and transport systems. Policy documents are particularly valuable as they capture the municipality's strategic intentions, plans, and priorities, offering insights into how these systems are managed and interconnected within the urban context.

The documents were purposively sampled based on their potential to illuminate inter-system interactions. This targeted approach helped to ensure that the most relevant and informative documents were included in the analysis. Table 4.1 offers an overview of the selected documents, highlighting the primary system for which they were selected.

Although not formally part of the data collection and analysis process, the researchers involved in this study bring with them prior experience and insights gained from their previous research on the food (e.g.,

**Table 4.1** Overview of documents selected for analysis

| Document | Type of Document | Reviewed by author focusing on | | |
|---|---|---|---|---|
| | | Food | Energy | Transport |
| Circular food system report | Policy recommendations | X | | |
| Energy Plan 2022–2030 | Action plan | | X | |
| Electrification Plan 2022–2030 | Action plan | | | X |
| Waste Management Plan 2021–2030 | Program | X | | |
| Environment and Climate Programme 2021–2030 | Program | X | | |
| Procurement guidelines | Guidelines | X | | |
| Business program | Program | | X | |
| Traffic Strategy 2035 | Strategy | | | X |
| Västra Götaland County food system transformation agenda | Strategy | X | | |

Thomson, 2022), energy (e.g., Mignon et al., 2023), and transport systems in Gothenburg (e.g., Werner & Onufrey, 2022; Werner et al., 2021). The researchers' background knowledge enriched the current analysis, allowing for a more informed and nuanced interpretation of the multi-system interactions within the urban context of Gothenburg.

## Data Analysis

The data analysis was conducted in two phases. In the first phase, each author independently analysed the empirical material related to their respective system (food, energy, or transport). This involved an iterative process of reading and re-reading the material, with a focus on identifying key statements that related to the theme of multi-system interactions. These statements were coded according to their relevance to structural or functional couplings and were further categorized based on the specific dimension of socio-technical systems they pertained to (actors, infrastructure/technologies, or institutions). Additionally, each author preliminarily identified the system or systems to which each coded section was related (food, energy, transport).

Following the initial independent coding, a collaborative discussion was conducted to consolidate and refine the findings. In this joint session, each author presented the coded segments from their analysis for collective review. The goal was to develop a cohesive understanding by comparing and integrating the independent analyses. This collaborative process allowed for a more comprehensive cross-system analysis, where structural and functional couplings identified in one system (e.g., food) were examined in the context of findings from the other systems (e.g., energy and transport). This integrative approach facilitated the identification of connections and interdependencies across the systems, thereby deepening the understanding of how these systems interact. By collectively leveraging the insights and perspectives of the researchers, this phase ensured a more thorough examination of the cross-system interactions.

## Findings and Analysis

### Technology Couplings

Analysis of the strategic planning documents from the municipality of Gothenburg reveals several key established, emerging and planned technology couplings across the energy, transport, and food systems (see Table 4.2). Among the established couplings, the connection of food waste with the energy and transport systems stands out. For instance, waste food is processed and incinerated at municipal co-generation plants (e.g., Sävenäs), providing electricity and district heating. Biogas production from waste slurry, transported to various biogas facilities in the region, reflects how the need for more local renewable energy can create new operational dependencies (e.g., on large fleets of trucks for waste transport). However, these couplings also highlight potential tensions. For example, the intensified use of transport to move slurry (involving around 30 trucks per week) potentially decreases the benefits of biogas. This tension underscores how multi-system interactions can generate unintended knock-on effects if the necessary supporting infrastructure (e.g., pipelines or local treatment plants) is underdeveloped.

Among the emerging technology couplings, notable examples include a pilot project for indoor food production and initiatives focused on alternative charging methods for electric vehicles. The indoor food production pilot, which links the energy, transport, and food systems, involves the development of a food production facility that utilizes excess heat from a data centre owned by the municipal energy company. Due to its urban location, the transportation required to deliver produce to customers is significantly reduced. The municipality's focus on alternative charging (e.g., wireless systems, high-powered stations) indicates the broader electrification drive in the region. Yet, these initiatives remain in pilot or early roll-out phases, highlighting the risk of fragmented implementation if city-wide coordination lags behind technological advances.

Finally, among the planned technology couplings are the ongoing evaluations of a future carbon capture and storage (CCS) and hydrogen production plant, which point to a strategy to transform local industry and transport networks. Documents indicate that such a plant would have the potential to capture emissions from the local industry (including refineries and other manufacturing industries), to transform and store it within and outside of the municipality's boundaries. However, the technical viability of this would require extensive inter-system coordination on how carbon is stored and transported.

**Table 4.2** Overview of couplings identified in the data (E stands for electricity, T for transport, F for food)

| Document | Structural couplings | | | Functional couplings |
|---|---|---|---|---|
| | *Technology couplings* | *Institutional couplings* | *Actor couplings* | |
| Energy documents | - electricity grid E+T+F<br>- district heating grid E+T+F<br>- waste-to-energy plants E+T+F<br>- hydrogen production plant E+T<br>- carbon capture and storage plant E+T+F<br>- smart control systems/batteries E+T<br>- charging stations for heavy transport E+T<br>- biocarbon production E+T+F | - action plan for energy E+T+F<br>- environment and climate strategy E+T+F<br>- arenas for dialogues among transport & energy actors E+T<br>- explicit goals to influence policies outside the boundaries of the cities E+T+F<br>- explicit goals to influence citizens and local organizations' practices. E+T+F | - Göteborg Energi AB<br>- Göteborg Energi Nät AB<br>- Vattenfall and Svenska Kraftnät<br>- Renova AB<br>- Gryaab AB<br>- Preem AB and St1 AB<br>- Göteborgs Hamn AB<br>- business region Göteborg<br>- Förvaltning AB Framtiden<br>- Higab<br>- Älvstranden Utveckling AB<br>- Göteborg municipality | - electricity production based on various inputs (e.g., solar, wind, (bio)gas, waste incineration) $\Rightarrow$ grid $\Rightarrow$ batteries, charging stations.<br>- electricity generation and waste incineration (including food) Þ district heating<br>- food (or other organic) residue $\Rightarrow$ electricity generation (and district heating) or slurry $\Rightarrow$ biogas production (and transport fuel)<br>- CO2 from various industries $\Rightarrow$ CCS plant $\Rightarrow$ storage areas<br>- food (or other organic) residue $\Rightarrow$ biocarbon (will be used locally or transported). |

(continued)

(continued)

| Document | Structural couplings | | | Functional couplings |
|---|---|---|---|---|
| | *Technology couplings* | *Institutional couplings* | *Actor couplings* | |
| Transport documents | - charging infrastructure for electric personal cars and heavy vehicles E+T<br>-wireless charging infrastructure for personal cars, via charging plates integrated into the asphalt E+T<br>- electric road systems for goods transport E+T<br>- charging infrastructure for hydrogen cars E+T<br>- energy services related to electric charging E+T<br>- technologies for electricity grid improvements: Capacity enhancements, load/power control, local electricity production, V2G, battery storage E+T+F | - electrification plan E+T+F<br>- transport strategy E+T<br>- strategic programme for business development E+T<br>- business region Göteborg coordinates the implementation of the electrification plan E+T+F<br>- explicit ambition for coordination at the system level to avoid possible goal conflicts/sub-optimizations E+T+F<br>- explicit ambition to have an overall strategy and plan for charging infrastructure and cross-sector electrified transport and energy supply E+T<br>- policy goal for availability of charging options matching societal needs E+T | - business region Gothenburg<br>- Volvo cars & Volvo AB<br>- Chalmers, RISE, Lindholmen Science Park<br>- Renova AB<br>- Göteborg Energi<br>- Vattenfall and Svenska Kraftnät<br>- Göteborg Hamn<br>- Göteborg municipality | - electricity ⇔ electric vehicles E+T<br>- City's purchased transport services include transportation of food, e.g., to schools. T+F<br>- within urban goods transport, food waste collection is one of the primary functions T+F |

(continued)

(continued)

| Document | Structural couplings | | | Functional couplings |
|---|---|---|---|---|
| | *Technology couplings* | *Institutional couplings* | *Actor couplings* | |
| Food documents | - water treatment plant E+F<br>- multi-functional greenhouse concept E+F<br>-waste-to-energy plant E+T+F<br>- Gothenburg harbour E+T+F | - City plan E+T+F<br>- waste management plan E+F<br>- Gbg City guidelines for procurement E+T+F<br>- Gbg City report on circular food production E+T+F<br>- public meals policy E+T+F | - Göteborg municipality<br>- Göteborg energi AB<br>- Renova AB<br>- St1 AB<br>- business region Göteborg<br>- Gryaab AB | - food waste Þ processing facility Þ slurry transported around the west coast of Sweden for biogas production E+T+F<br>- waste water and sewage ⇒ biogas and fertilizer ⇒ food production. E+F<br>- data Centre residual heat ⇒ indoor growing E+F<br>- food waste ⇒ waste-to-energy plant ⇒ electricity E+T+F<br>- Most of the food consumed in Gothenburg imported from abroad, much arrives through Gothenburg harbour, as well as the majority of food produced in Gothenburg being exported to the rest of the world T+F |

## Institutional Couplings

The data highlights both explicit and implicit institutional couplings that shape multi-system interactions. Explicit institutional couplings stem from formal policy documents, such as the Action Plan for Energy and the Electrification Plan, where couplings between systems are described, planned, and framed. For instance, the Action Plan for Energy, a document setting the agenda for the period 2022 to 2030, states that:

> "The *energy system in Gothenburg consists of several interconnected systems…It is necessary to see the connections between the systems to understand the whole and to avoid suboptimization when a single goal is to be*

> *reached. To focus on the connections between the subsystems will be necessary to create synergies and achieve the municipality's energy and climate goals".*

(Göteborg Stads Energiplan 2022–2030, p9).[1]

These documents explicitly establish goals, visions, responsibilities, budgets, and targets that often shape multiple systems. For instance, the action plan for energy sets objectives for electricity production that also support the electrification of the transport sector, defining responsibilities and mandates for actors operating at the intersection of the energy, food, and transport systems. Each document also specifies whether the policies are binding or non-binding and whether the objectives fall within the direct or indirect influence of the municipality. In some cases, such as the municipality's procurement guidelines, the policies are binding, while in others, such as the municipality's infrastructure planning document, the policies are non-binding and serve as long-term strategic guidance.

Implicit institutional couplings involve less formalized objectives and ambitions, where influence is exerted indirectly, often without quantified targets or direct enforcement. These concern institutional influences that the municipality intends to have on actors or decision-makers outside its geographic boundaries. For instance, the action plan for energy indicates that the municipality intends to influence the government and the European Commission's decisions and policies related to biogas. Likewise, in the report on circular food production, the municipality indicates that although its direct influence on sustainable consumption and production is limited (to its own organization), it aims to indirectly influence actors (such as citizens) in choosing more sustainable food alternatives. However, the existence of both explicit and implicit couplings exposes the potential for institutional tensions that can result in fragmented or uneven implementation where some goals are fully enforced (e.g., waste collection) while others remain advisory (e.g., circular consumption and production).

[1] Quotes were translated by the authors.

## Actor Couplings

Actor couplings emerge as a principal driver of multi-system interactions, with substantial overlaps visible between the focal system documents. Central actors linking the three systems include, for instance, Gothenburg municipality (encompassing various departments), which serves as a central actor overseeing the food, energy, and transport systems.[2] Gothenburg municipality is the main owner for a number of key companies, including the municipal energy company Göteborg Energi AB, the resource management and recycling company Renova AB, the port of Gothenburg Göteborgs Hamn AB, and the municipal economic development agency Business Region Göteborg. Each of these companies are local nodes between systems. For example, Göteborg Energy AB supplies electricity, district heating, and energy services, and has a growing role in electric vehicle infrastructure. Likewise, Renova AB, besides being a key player in the resource management of the municipality, also operates 70% of the city's own heavy vehicles, tying transport to energy through waste incineration and biogas.

In addition to these multi-system actors, there are also actors that have strong couplings between two systems. For example, Gryaab AB is the wastewater handling and treatment plant owned by Gothenburg municipality; it links the food and energy systems through the treatment of sludge (from food waste) which is used for both fertilizer and biogas production. Likewise, the municipal real estate companies Förvaltning AB Framtiden, Higab, Älvstranden Utveckling AB connect the energy and transport systems through their use of electricity and vehicle charging infrastructure, and in some cases, installation of solar cells on their buildings.

## Functional Couplings

The analysis identifies multiple functional couplings among energy, transport, and food. Electricity, produced from waste incineration and

[2] It should however be noted that the lack of a comprehensive food perspective has led to calls for an official collaboration function focusing on food system transformation.

biogas digestion, is fed into the transport system, powering electric vehicles. Conversely, discarded food and wastewater become inputs for energy production, illustrating a circular approach, but also revealing conflicts if food waste is valued primarily as an energy feedstock rather than being prevented or redistributed.

The transport system not only demands energy (electricity, biofuels) but enables resource flows (e.g., delivering imported food, collecting food residuals). Transport ensures that goods, such as food produced outside municipal boundaries, can reach their destination. Moreover, collection of discarded food, which is an important input for electricity production, is also possible only if it can be transported. In turn, transportation of urban waste is among the first electrified applications of urban goods transport in Gothenburg. Future plans for carbon capture systems and hydrogen production will also hinge on the ability of the transport system to support these resource flows.

Finally, the food system is often portrayed as a downstream user of energy and transport, with emphasis on waste disposal over production inputs. The municipal documents note an aspiration to reduce external dependencies (e.g., imports) but lack detail on how local production will scale. This imbalance suggests a power dynamic where energy and transport systems dominate municipal planning, leaving food somewhat peripheral except as a source of residuals for energy generation.

## Discussion

When drawing on the results of the mapping of interactions presented above, several notable patterns emerged that allow for the identification of interaction weaknesses and the drawing of empirical and theoretical implications for a sustainable transition in the urban context involving multiple systems.

Firstly, it is interesting to note that the municipality of Gothenburg explicitly recognizes the interdependence of transition processes and systems, advocating that they must be considered collectively rather than independently. Yet, despite this understanding, the mapping of couplings shows that dual system couplings between the energy and transport systems remain the most prevalent ones. While some couplings involving the food system do exist, they tend to be more implicit. Notably, among actor and functional couplings, key actors from the food system were absent. As a consequence, it is fair to stress that there is a mismatch between the ambitions of the municipality, and the structures put into place to facilitate and organize such interactions.

From a policy perspective, this means that awareness and ambition alone are not sufficient to foster multi-system interactions. Instead, it is important for governing bodies such as municipalities (1) to keep track of the interactions taking place and the structures put into place for these interactions (2) to identify weaknesses in the interactions and in the structures put in place, and (3) to adjust the structures to remedy for the lack of interactions. To exemplify, through the mapping of technological, institutional, actor, and functional couplings (Table 2), it appears that the food system is not explicitly mentioned in some strategic documents, plans, or policy goals, despite it being impacted by these policies. Likewise, several projects and dialogue arenas have been created, where representatives of one system are missing. The next step would then be for the municipality to adjust this lack of couplings to facilitate interactions.

From a theoretical perspective, the prevalence of dual interactions (i.e., rather than multi-system interactions) presented in the mapping reflects rather well the current state of the literature on multi-system interactions. It shows that while scholars and practitioners understand the need to consider multiple systems together, it is seldom done in practice. Urban environments, while being sites for multi-system interactions (Frantzeskaki et al., 2017), appear insufficient by themselves to foster meaningful attention to these interactions. This calls for future research to develop new methods to study multi-system interactions that can be used for both researchers and practitioners. A further development of coupling mapping, which was performed in this study, may be a way forward for increased awareness of multi-system interactions in policymaking. For instance, interviews with involved actors would be a way to reveal the underlying tensions and conflicts present in multi-system interactions.

Another significant pattern in the mapping is that actor couplings show the largest number of interlinkages between systems. As illustrated in Table 2, in a municipality such as Gothenburg, several actors play key roles across multiple systems. This pattern of actor couplings represents a potential strength allowing for broader multi-system interactions. Indeed, the studied actor couplings in the Gothenburg municipality are a starting point where additional couplings (technology, institutional) can be initiated and where synergies can be highlighted and exploited. As highlighted above, this emphasizes the critical need to map key actor couplings and identify missing actors and resolve the gaps by involving key actors. In our case, important actors from the food system, such as food producers and retailers, were missing, e.g., in dialogue groups. This stands in stark contrast to the energy and transport systems where private sector actors are highly involved and exert significant influence. This lack of representation of the food system could explain the limited number of technological, institutional, and functional couplings highlighted between the systems across the studied documents. Once identified, the lack of representation (and the uncovered potential of interactions with the food system) may rather simply be handled by involving relevant actors. Theoretically, while previous studies highlighted the importance of different types of couplings and positioned structural couplings as prerequisites for functional ones (Andersen & Geels, 2023; Konrad et al., 2008), our analysis suggests that actor couplings can serve as a foundation for building other types of structural couplings.

Apart from having more limited interactions with the other systems, the food system exhibits minimal structural similarity to the energy and transport systems, especially from the vantage point of a municipal actor within an urban context. While municipalities often exert significant control over the energy and transport sectors, the food system is markedly different. It is characterized by a decentralized, heterogeneous, and intricate mix of subsystems, lacking the centralized oversight typically seen in energy and transport systems. This decentralization results in a more complex and less uniform structure, posing unique challenges for urban governance and integration into broader socio-technical transitions. Such complexity may explain why food in our study was predominantly considered by the municipality from the perspective of disposal and the focus on earlier stages of the supply chain (i.e., food production) was entirely absent.

Finally, it is interesting to notice that in contrast with the energy and transport systems, where increased or stronger interactions with other systems are typically encouraged, the food system in our study showed signs of aiming to reduce its interactions with the transport system. For example, the pilot indoor growing project by virtue of being situated in the urban context, inherently reduces the need for extensive transportation, thereby lessening its interaction with the transport system.

From a policy perspective, this raises the question of how to organize multi-system interactions, when these interactions aim at being reduced. In such situations, e.g., in cases where dependencies with negative consequences on the environment, it may be necessary for a neutral, long-term thinking actor with no (or little) commercial interest to act as an orchestrator of multi-system interactions. Gothenburg municipality, as a public actor with societal and long-term interests (rather than commercial interest), and as the owner of several municipal companies within the energy, food, and transport systems, distinguishes itself as the primary governing actor of urban multi-system interactions and may therefore be suitable for playing such an orchestrating role. More concretely, this means that representatives of the municipality may lead dialogues and collaborations in projects where one of the aims is to lower the level of interactions between systems.

From a theoretical perspective, this finding adds nuance to prior research highlighting that multi-system interactions can have adverse effects, such as asymmetries or sub-optimal solutions (Andersen et al., 2023). In particular, our results suggest that multi-system interactions should not be a goal in and of themselves; in some cases, reducing or limiting interactions may prove more conducive to supporting sustainability transitions.

## Conclusion and Suggestions for Future Research

In this chapter, we went beyond the typical focus on dual interactions, i.e., mapping and explaining interconnections between two systems, which have until now mainly been the focus on previous empirical studies of multi-system interactions, and instead considered mutual interconnections across three systems, i.e., the food, energy, and transport systems. Specifically, we mapped the inter-system interactions between key actors, infrastructures/technologies, and institutions across the food, transport, and energy systems in Gothenburg municipality, through the analysis of municipal documents, including policy recommendations, guidelines, action plans, and strategies.

Mapping inter-system linkages revealed planned, emerging and established technology couplings, implicit and explicit institutional couplings, and multiple key actor couplings across the energy, food, and transport systems. Notably, while the municipality of Gothenburg explicitly recognizes the interdependence of transition processes and systems, consistent with urban transition studies (Grin et al., 2017; Hodson et al., 2024), the prevailing focus in the documents is on dual interactions between the energy and the transport systems, rather than broader multi-system interactions. This highlights the challenges of effectively integrating and prioritizing multi-system linkages and underscores the difficulties of operationalizing visions for multi-system transitions, even in urban environments, which are often seen as natural sites for such interactions (cf. Frantzeskaki & Rok, 2018).

The mapping performed in this study proved to be a useful tool for identifying strengths and weaknesses of existing structures put in place to foster multi-system interactions. Among the strengths, structural actor couplings showed the largest number of interlinkages across systems. From a policy perspective, this provides a potential way forward to drive multi-system transitions; when key multi-system actors are identified, additional interactions, synergies, and roadmaps can be developed. Actor couplings therefore represent interesting avenues for future research aimed at accelerating multi-system transitions. For instance, it would be interesting to consider the key multi-system actors from an intermediation perspective, considering their potential roles as system entanglers (Löhr & Chlebna, 2023).

At the same time, existing structures, exemplified by municipal plans, projects, and dialogue arenas, showed a notable weakness in terms of enabling sufficient and balanced (i.e., multi rather than dual) interactions across all three studied systems. In particular, the food system was underrepresented and, when present, considered from a limited perspective of the disposal stage. This could at least partially be explained by particularities of the food system, representing a decentralized, heterogeneous, and intricate mix of subsystems, lacking the centralized oversight typically seen in, e.g., energy and transport systems.

This study predominantly focused on the analysis of the policy documents, leading to some limitations. First, the emphasis was placed on structural preconditions rather than the processual dynamics of multi-system interactions. Second, the mapping reflected the perspective of municipal departments that were in charge of writing the policy documents. To address these gaps, we encourage further in-depth studies involving a wider range of organizations' sources of data. Such studies could better capture processual dynamics, including tensions and conflicts, and provide deeper insights into the roles of various actors in urban multi-system transitions.

## References

Andersen, A. D., & Geels, F. W. (2023). Multi-system dynamics and the speed of net-zero transitions: Identifying causal processes related to technologies, actors, and institutions. *Energy Research & Social Science, 102*, 103178. https://doi.org/10.1016/j.erss.2023.103178

Andersen, A. D., Geels, F. W., Steen, M., & Bugge, M. M. (2023). Building multi-system nexuses in low-carbon transitions: Conflicts and asymmetric adjustments in Norwegian ferry electrification. *Proceedings of the National Academy of Sciences of the United States of America, 120*(47), e2207746120. https://doi.org/10.1073/pnas.2207746120

Breitschopf, B., Grimm, A., Billerbeck, A., Wydra, S., & Köhler, J. (2023). Towards understanding interactions between socio-technical systems in sustainability transitions. *Energy Research & Social Science, 106*, 103323. https://doi.org/10.1016/j.erss.2023.103323

Broto, V. C. (2017). Urban governance and the politics of climate change. *World Development, 93*, 1–15.

ER. (2017). *Strategisk plan för omställning av transportsektorn till fossilfrihet (Strategic plan for fossil-free transtion of the transport sector).*

Ferloni, A. (2022). Transitions as a coevolutionary process: The urban emergence of electric vehicle inventions. *Environmental Innovation and Societal Transitions, 44*, 205–225. https://doi.org/10.1016/j.eist.2022.08.003

Frantzeskaki, N., Broto, V. C., Coenen, L., & Loorbach, D. (2017). *Urban sustainability transitions* (Vol. 5). Routledge.

Frantzeskaki, N., & Rok, A. (2018). Co-producing urban sustainability transitions knowledge with community, policy and science. *Environmental Innovation and Societal Transitions, 29*, 47–51. https://doi.org/10.1016/j.eist.2018.08.001

Fudge, S., Peters, M., & Woodman, B. (2016). Local authorities as niche actors: The case of energy governance in the UK. *Environmental Innovation and Societal Transitions, 18*, 1–17. https://doi.org/10.1016/j.eist.2015.06.004

Geels, F. W. (2007). Analysing the breakthrough of rock 'n' roll (1930–1970) multi-regime interaction and reconfiguration in the multi-level perspective. *Technological Forecasting and Social Change, 74*(8), 1411–1431. https://doi.org/10.1016/j.techfore.2006.07.008

Grin, J., Frantzeskaki, N., Broto, V. C., & Coenen, L. (2017). Sustainability transitions and the city: Linking to transition studies and looking forward. In *Urban sustainability transitions* (pp. 359–367). Routledge.

Gustafsson, S., & Mignon, I. (2020). Municipalities as intermediaries for the design and local implementation of climate visions. *European Planning Studies, 28*(6), 1161–1182. https://doi.org/10.1080/09654313.2019.1612327

Haley, B. (2015). Low-carbon innovation from a hydroelectric base: The case of electric vehicles in Quebec. *Environmental Innovation and Societal Transitions, 14*, 5–25. https://doi.org/10.1016/j.eist.2014.05.003

Hebinck, A., Klerkx, L., Elzen, B., Kok, K. P. W., König, B., Schiller, K., Tschersich, J., van Mierlo, B., & von Wirth, T. (2021). Beyond food for thought-directing sustainability transitions research to address fundamental change in Agri-food systems. *Environmental Innovation and Societal Transitions, 41*, 81–85. https://doi.org/10.1016/j.eist.2021.10.003

Hodson, M., Geels, F. W., & McMeekin, A. (2017). Reconfiguring urban sustainability transitions. *Analysing Multiplicity. Sustainability, 9*(2), 299. https://doi.org/10.3390/su9020299

Hodson, M., McMeekin, A., & Lockhart, A. (2024). Urban infrastructure reconfiguration and digital platforms: Who is in control? *Environmental Innovation and Societal Transitions, 50*, 100816. https://doi.org/10.1016/j.eist.2024.100816

Holscher, K., Frantzeskaki, N., McPhearson, T., & Loorbach, D. (2019). Tales of transforming cities: Transformative climate governance capacities in new York City, U.S. and Rotterdam. *Netherlands. J Environ Manage, 231*, 843–857. https://doi.org/10.1016/j.jenvman.2018.10.043

Huang, P., & Broto, V. C. (2018). Interdependence between urban processes and energy transitions: The dimensions of urban energy transitions (DUET) framework. *Environmental Innovation and Societal Transitions, 28*, 35–45. https://doi.org/10.1016/j.eist.2018.03.004

Huttunen, S., Kivimaa, P., & Virkamäki, V. (2014). The need for policy coherence to trigger a transition to biogas production. *Environmental Innovation and Societal Transitions, 12*, 14–30. https://doi.org/10.1016/j.eist.2014.04.002

Kanger, L., Schot, J., Sovacool, B. K., van der Vleuten, E., Ghosh, B., Keller, M., Kivimaa, P., Pahker, A.-K., & Steinmueller, W. E. (2021). Research frontiers for multi-system dynamics and deep transitions. *Environmental Innovation and Societal Transitions, 41*, 52–56. https://doi.org/10.1016/j.eist.2021.10.025

Konrad, K., Truffer, B., & Voss, J. P. (2008). Multi-regime dynamics in the analysis of sectoral transformation potentials: Evidence from German utility sectors. *Journal of Cleaner Production, 16*(11), 1190–1202. https://doi.org/10.1016/j.jclepro.2007.08.014

Löhr, M., & Chlebna, C. (2023). Multi-system interactions in hydrogen-based sector coupling projects: System entanglers as key actors. *Energy Research & Social Science, 105*, 103282. https://doi.org/10.1016/j.erss.2023.103282

Markard, J., Geels, F. W., & Raven, R. (2020). Challenges in the acceleration of sustainability transitions. *Environmental Research Letters, 15*(8), 081001. https://doi.org/10.1088/1748-9326/ab9468

McMeekin, A., Geels, F. W., & Hodson, M. (2019). Mapping the winds of whole system reconfiguration: Analysing low-carbon transformations across production, distribution and consumption in the UK electricity system (1990–2016). *Research Policy, 48*(5), 1216–1231. https://doi.org/10.1016/j.respol.2018.12.007

Meynard, J. M., Jeuffroy, M. H., Le Bail, M., Lefèvre, A., Magrini, M. B., & Michon, C. (2017). Designing coupled innovations for the sustainability transition of agrifood systems. *Agricultural Systems, 157*, 330–339. https://doi.org/10.1016/j.agsy.2016.08.002

Mignon, I., & Bergek, A. (2016). System- and actor-level challenges for the diffusion of renewable electricity technologies: An international comparison. *Journal of Cleaner Production, 128*, 105–115. https://doi.org/10.1016/j.jclepro.2015.09.048

Mignon, I., Lindahl, J., & Melander, L. (2023). *The impact of value logics on business model design—The case of solar parks in Sweden 2023 EURAM conference, June 2023*. Dublin.

Nykamp, H., Andersen, A. D., & Geels, F. W. (2023). Low-carbon electrification as a multi-system transition: A socio-technical analysis of Norwegian maritime transport, construction, and chemical sectors. *Environmental Research Letters, 18*(9), 094059. https://doi.org/10.1088/1748-9326/acf67a

Ohlendorf, N., Löhr, M., & Markard, J. (2023). Actors in multi-sector transitions - discourse analysis on hydrogen in Germany. *Environmental Innovation and Societal Transitions, 47*, 100692. https://doi.org/10.1016/j.eist.2023.100692

Raven, R. (2007). Co-evolution of waste and electricity regimes: Multi-regime dynamics in The Netherlands (1969–2003). *Energy Policy, 35*(4), 2197–2208. https://doi.org/10.1016/j.enpol.2006.07.005

Raven, R. P., & Verbong, G. P. (2009). Boundary crossing innovations: Case studies from the energy domain. *Technology in Society, 31*(1), 85–93. https://doi.org/10.1016/j.techsoc.2008.10.006

Rohracher, H., & Späth, P. (2017). Cities as arenas of low-carbon transitions: Friction zones in the negotiation of low-carbon futures. In *Urban sustainability transitions* (pp. 287–299). Routledge.

Rosenbloom, D. (2019). A clash of socio-technical systems: Exploring actor interactions around electrification and electricity trade in unfolding low-carbon pathways for Ontario. *Energy Research & Social Science, 49*, 219–232. https://doi.org/10.1016/j.erss.2018.10.015

Rosenbloom, D. (2020). Engaging with multi-system interactions in sustainability transitions: A comment on the transitions research agenda. *Environmental Innovation and Societal Transitions, 34*, 336–340. https://doi.org/10.1016/j.eist.2019.10.003

Sharp, D., Raven, R., & Farrelly, M. (2024). Pluralising place frames in urban transition management: Net-zero transitions at precinct scale. *Environmental*

*Innovation and Societal Transitions, 50*, 100803. https://doi.org/10.1016/j.eist.2023.100803
Sutherland, L.-A., Peter, S., & Zagata, L. (2015). Conceptualising multi-regime interactions: The role of the agriculture sector in renewable energy transitions. *Research Policy, 44*(8), 1543–1554. https://doi.org/10.1016/j.respol.2015.05.013
Thomson, L. (2022). Leveraging the value from digitalization: A business model exploration of new technology-based firms in vertical farming. *Journal of Manufacturing Technology Management, 33*(9), 88–107. https://doi.org/10.1108/Jmtm-10-2021-0422
Werner, V., & Onufrey, K. (2022). If electric trucks are the solution, what are the problems? A study of agenda-setting in demonstration projects. *Energy Research & Social Science, 91*, 102722. https://doi.org/10.1016/j.erss.2022.102722
Werner, V., Onufrey, K., & Magnusson, T. (2021). From protection to selective exposure: Commercial demonstrations as steppingstones for upscaled technology diffusion. *International Journal of Automotive Technology and Management, 21*(3), 250–272. https://doi.org/10.1504/IJATM.2021.116604

# 5

# The Multi-Sectoral Interactions of Electrification in Norwegian Coastal Shipping

Tuukka Mäkitie, Allan D. Andersen, and Markus Steen

## Introduction

Climate change mitigation requires sustainability transitions towards net-zero in all sectors, including mobility, energy supply, and manufacturing (Victor et al., 2019). Such transitions often imply structural changes in the wider economy, as novel innovations transform sectors in terms of key actors, institutions, and value creation opportunities (Andersen et al., 2020). Moreover, for innovations to diffuse widely, new couplings

T. Mäkitie (✉) · M. Steen
Department of Technology Management, SINTEF Digital, Trondheim, Norway
e-mail: tuukka.makitie@sintef.no

T. Mäkitie · A. D. Andersen
TIK Center for Technology, Innovation and Culture, University of Oslo, Oslo, Norway

A. D. Andersen
Department of Food and Resource Economics, University of Copenhagen, Copenhagen, Denmark


M. Steen et al. (eds.), *Sustainability Transitions and Industrial Transformation*,
https://doi.org/10.1007/978-981-95-4893-4_5

between technologies and sectors are often needed. The electrification of transport, for example, requires new couplings between transport (automobiles, trains, ships, etc.) and power sectors in the form of battery charging infrastructure. Such inter-sectoral dynamics in sustainability transitions are important to understand as they can be central to how parallel sectoral transformations towards sustainability may open possibilities for mobilizing cross-sectoral synergies that can accelerate change, or create bottlenecks (Andersen & Geels, 2023; Mäkitie, Hanson et al., 2022). Inter-sectoral dynamics may be less critical in the early phase of transition, but can become highly pronounced in the acceleration phase. Moreover, they are important in considering the broader effects of sustainability transitions on industrial transformations and thus on creation and losses of jobs in different sectors. The recent literature has advanced the conceptual and theoretical understanding of multi-system interactions in sustainability transitions (Kanger et al., 2021; Rosenbloom, 2020). However, this literature has seldom sought to achieve a holistic perspective of the different systems of provision associated with sustainability transitions, especially in terms of comprehensive empirical studies cutting across multiple sectors (Andersen et al., 2020).

The recently proposed 'multi-sectoral perspective' (MSP) (Andersen et al., 2026) aims to remedy conceptual gaps in the literature by outlining a heuristic framework to conceptualize the complex interactions between sectors that affect a sustainability transition in a focal socio-technical system, such as transport or food. Drawing on the multi-level perspective (MLP) (Geels, 2002) and the technological innovation system (TIS) (Bergek et al., 2008; Hekkert et al., 2007) approaches, MSP outlines three main analytical dimensions. First, the MLP literature has provided insights on interdependencies between production, distribution, and end-use sectors of various resources, such as energy and food (Holtz et al., 2008; McMeekin et al., 2019). In MSP, we conceptualize these as "horizontal" interdependencies across sectors (see Fig. 5.1). Second, the TIS literature has elaborated on the interdependencies of focal technologies with various component and material supply sectors, including batteries and materials in electric mobility (Geels, 2004; Stephan et al., 2017). In MSP, we refer to these as "vertical" interdependencies. Third, adjacent sectors may influence transition processes in a focal sector despite no direct inter-sectoral couplings. Examples of such effects include technology spillovers from adjacent sectors into a focal sector, or competition over scarce resources. All involved sectors have their own regime and niche dynamics that may affect transition processes.

Notably, earlier transition literature has discussed the implications of horizontal, vertical, and adjacent dimensions on system transitions separately, picking up on specific sectoral interactions, but not seeking to understand the full picture (Andersen et al., 2026). Therefore, building on MLP and TIS, the MSP proposes a *holistic* view of the structural changes and interaction patterns within and across sectors that are part-and-parcel of transition processes. Taken together, the ambition is a framework that can help us better understand how sectors change both structurally and functionally in transitions.

The MSP was iteratively developed through a systematic literature review synthesizing and integrating various conceptual and empirical

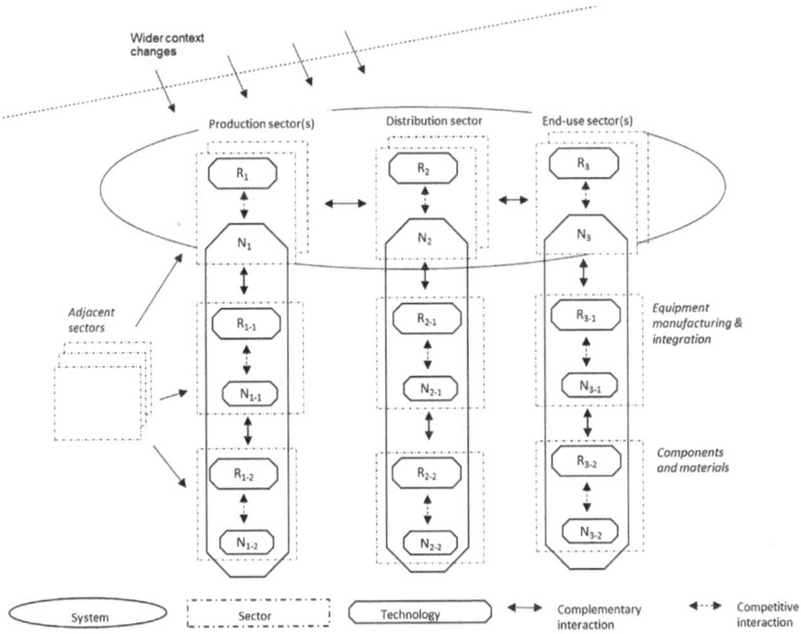


Fig. 5.1 Multi-sectoral perspective (Andersen et al., 2026)

insights (Andersen et al., 2026). The purpose of this chapter is to provide a first empirical application of the MSP, using the case of rapidly diffusing electrification in Norwegian coastal shipping to exemplify the multi-sectoral dynamics along the aforementioned vertical, horizontal, and adjacent sector dimensions. Our empirical analysis is guided by the following research question: *how have multi-sectoral interactions affected the electrification of the Norwegian maritime sector?*

Electrification of coastal shipping in Norway was chosen as it presents a case where multi-sectoral interactions have played a significant role in shaping the transition process, and because it has been subject to multiple previous studies, many of which the authors of this book chapter have been part of (Andersen et al., 2023; Bach et al., 2020; Bergek et al., 2021, 2023; Bjerkan & Ryghaug, 2021; Bjerkan et al., 2021; Bjerkan & Seter, 2021; Bugge et al., 2022; Damman & Steen, 2021; Mäkitie, Hanson et al., 2022; Njøs et al., 2020; Nykamp et al., 2023; Sjøtun, 2018; Steen et al., 2019; Steen, Bjerkan et al., 2024; Sæther & Moe, 2021). The analysis in this chapter is thus based on comprehensive primary and secondary data and prior analysis.[1] This book chapter further contributes to the study of this topic by providing a holistic view of the key drivers and barriers for electrification emanating from the interactions of the focal sector (Norwegian coastal shipping) with other sectors.

In the next section we present the MSP, the analytical approach and the case study. We then apply the analytical framework to the case study of maritime electrification. In the final section we discuss insights gained from this analysis and the implications for policymakers and managers, as well as methodological issues related to multi-sectoral analysis.

---

[1] We ourselves have studied electrification of the maritime sector in various research projects since 2017, including GREENFLEET, TRAZEPO, GONST, FME NTRANS, PATHWAYS, funded by the Research Council of Norway and Nordic Energy Research.

## Theoretical Framework and Methodology

### Multi-sectoral Perspective

As sustainability transitions accelerate, meaning, when niche innovations start to mature and diffuse widely, the impacts on the wider economy become of interest for researchers, policymakers, and other decision takers.

Evolutionary economics (Nelson, 2020; Nelson & Winter, 1977, 1982), a foundational building block for transition studies (Rip & Kemp, 1998), has extensively studied how major innovations can cause industrial transformations across multiple sectors (Dahmén, 1988; Perez, 2009) and changes in the sectoral composition of the economy (Carlsson, 2016; Saviotti & Pyka, 2004). Transition studies, however, have typically focused on single-sector or -technology studies (Andersen & Markard, 2020).

MSP provides a novel synthesis of two central transition studies frameworks, the MLP and TIS, to portray and make sense of recent transition phenomena involving industrial transformation across multiple, interlinked sectors. It discerns two main types of inter-sectoral relations. One concerns material resource throughput in "horizontal" value chains, which describes the transformation of resources into societal services (e.g., from coal to electricity over distribution to energy service). The sectors involved in the horizontal chain serve sub-functions of the entire chain (e.g., mobility, energy, or food) by deploying various technologies to transform resources into services. Each sector employs different technologies to serve its function (e.g., electricity supply uses hydro, wind, or gas power plant technology, and the distribution sector uses transformers, cables, and batteries) (McMeekin et al., 2019). The development and production of these technological artefacts takes place in multi-sectoral "vertical" technology value chains, which constitute the second type of inter-sector relation (Stephan et al., 2017). Operation and transformation of both horizontal and vertical value chains rely on alignment and coordination across actors and sectors. While extant frameworks and studies have accounted for the dynamics of vertical and horizontal value chains in isolation, the MSP proposes to analyse the inter-relationships between the two types of value chains. Besides material throughputs and technologies, adjacent sectors may provide indirect effects to transitions, such as offshore wind benefiting from knowledge spillovers from (offshore) oil and gas (Mäkitie et al., 2018).

MSP functions, firstly, as a mapping tool for identifying the sectoral configuration of socio-technical systems, i.e., the number and types of sectors involved in horizontal and vertical value chains that together condition transition processes. The view of sectoral configurations is useful for understanding whether and how change in one sector may require changes in other sectors, and for zooming in and out of different parts of the system during a transition. The broad framing implies that transitions do not necessarily involve change in all related sectors or at the same time, but rather sequentially through push and pull pressures caused by sectoral changes. Further on, we propose that such multi-sectoral interactions are inherent to most system transitions.

Beyond serving as a heuristic to map the multi-sectoral interactions of transitions, the MSP outlines a tentative set of multi-sectoral interaction processes that may drive or hinder a transition in a focal sector (see Table 5.1). These processes point to dynamics between two or more sectors conditioning the focal transition and can be divided into the structural dimensions of multi-sectoral interactions (Andersen et al., 2026).

In technological terms, the emergence of niche technologies in a focal sector, such as shipping, is, first, often conditioned by processes of creating new technological linkages with other sectors, as well as potential changes in the configurations of such sectors. For example, technological relatedness with an adjacent sector may boost technological development in a focal sector. Second, technology diffusion is dependent on expansion of value chains, which may require or lead to significant change processes in interconnected sectors. Such need for expansion may lead to bottlenecks, including difficulties in upscaling the supply of scarce materials and resources, such as sustainably sourced biomass. Third, beyond novel technologies, a purposeful decline in some sectors may be necessary for the acceleration of transitions. In sustainability transitions context, an example is the need to phase out the extraction and use of fossil energy resources.

**Table 5.1** Summary of multi-sectoral processes in the electrification of Norwegian coastal shipping

| | Multi-sectoral interaction processes | Multi-sectoral dynamic | Examples of multi-sectoral drivers and barriers in the electrification of Norwegian coastal shipping |
|---|---|---|---|
| Technology | 1. Emergence of electrification technologies | New technology requires linkages with other sectors or/and changes in them<br>Adjacent sector boosts/limits niche emergence | Lack of capacity in local grids creating hurdles for electrification in remote areas (H)<br>Grid bottlenecks leading to experimentation with novel solutions (V)<br>Local collaboration across the battery-electric value chain conducive for innovation (V)<br>Battery technology spillovers from land transport accelerating technology development in shipping (A) |
| | 2. Wider diffusion of electrification technologies | Diffusion requires, often rapid, upscaling of both horizontal and vertical chains, and sometimes induce transitions in those sectors | Norwegian renewable electricity system conducive for diffusion of electric vessels (H)<br>Frontrunner status conducive for establishment of battery module factories in Norway (V) |
| | 3. Technology decline | Dissolving inter-sectoral linkages in both horizontal and vertical chains | Public policies making fossil fuels less attractive (H&V) |

(continued)

**Table 5.1** (continued)

| | Multi-sectoral interaction processes | Multi-sectoral dynamic | Examples of multi-sectoral drivers and barriers in the electrification of Norwegian coastal shipping |
|---|---|---|---|
| Institutions | 4. Evolving formal institutions | Misaligned regulations and policies in interlinked sectors can hold back actors and technological change | Differing incentives and goals between electricity distribution and shipping hindering electrification (H) |
| | 5. Evolving informal institutions | The (differences in) legitimacy, culture, and practices of interlinked sectors shapes transitions | High consensus of electrification as a legitimate decarbonization solution (V)<br>Differing business cultures, worldviews, and practices between electricity distribution and shipping slowing down electrification (H) |
| Actors and networks | 6. Business integration | Uncertainty over ability/ willingness of input sectors to supply what is needed leads to multi-sector actors | Port authorities and electricity companies collaborate to found OPS companies (H) |
| | 7. Building cross-sectoral networks and coalitions | Coordinate/ orchestrate collective action across value chain sectors | Clusters and port authorities coordinating collaboration across the value chain (H&V) |

(continued)

**Table 5.1** (continued)

| Multi-sectoral interaction processes | Multi-sectoral dynamic | Examples of multi-sectoral drivers and barriers in the electrification of Norwegian coastal shipping |
|---|---|---|
| 8. Diversification from adjacent sectors | Diversifying actors realize resource flows with adjacent sectors | Established maritime and electrical supplier companies bringing in complementary assets, boosting battery-electric innovation (A) |
| 9. Institutional work | Actors work to build new cross-sectoral institutions or defend existing ones | Cross-sectoral collaborations to create technical standards (V)<br>Entrepreneurial work by state agencies to create legitimacy and collaboration across the value chain (H&V) |

H = Horizontal dimension (material/energy throughputs required by technology), V = Vertical dimension (components and materials of technological artefacts), A = Adjacent dimension (spillovers and competition for technology from other value chains). Adapted from Andersen et al. (2026)

Socio-technical transitions may require changes in the institutional set-up that conditions multi-sectoral interactions. The fourth process thus points to the evolvement of formal institutions governing multi-sectoral interactions, such as standards and regulations, and how institutions in different sectors may be misaligned and thus constrain transition processes. The same can be true for informal institutions, such as norms and expectations, which is recognized as the fifth process (Andersen et al., 2026).

Transitions are also dependent on novel couplings between actors in different sectors. The sixth process is the potential of actors in choosing to vertically integrate across sectors, for example to ensure necessary inputs. Seventh, transitions often also require the coordination or intermediation of collective action across sectors by actors. Eight, novel multi-sectoral interactions can be shaped by the diversification of actors into other (often related) sectors, which influences the flow of material and non-material resources. Ninth process, finally, points to conscious efforts by actors across sectors to build or shape missing or misaligned institutions (Andersen et al., 2026).

In summary, the MSP thus proposes a multifaceted set of processes that influence multi-sector interactions in transitions. In the following sections we employ this framework to analyse the multi-sectoral dynamics shaping the decarbonization of coastal shipping in Norway via electrification. Rather than unpacking these dynamics in full, which would be beyond the scope of this chapter, the empirical analysis is meant to illustrate the breadth of multi-sector interactions and to provide examples of how these interaction processes may foster or hinder the transition, i.e., the electrification.

## Analysis

Using our expert knowledge of the research that has examined this case study, we identified the key literature reporting (implicitly or explicitly) on the multi-sectoral features of the electrification of Norwegian coastal shipping. This resulted in a body of 21 scientific papers.[2] After this, we reviewed the papers in terms of insights regarding the effect of multi-sectoral interactions on maritime electrification, and vice versa. We used the two key dimensions of MSP to guide this review: (1) the types of multi-sectoral interaction (horizontal, vertical, and adjacent), and (2) the multi-sectoral interaction processes (see Table 5.1) that may drive or hinder transitions along the main structural dimensions of socio-technical system analysis (technology, institutions, and actors and networks). In the review, each identified multi-sectoral interaction was thus assessed in terms of its type of interaction (step 1) and process (step 2). These examples of interactions were recorded in a spreadsheet format together with text excerpts and references. The findings of this review are reported below. The empirical examples were checked and further contextual details were added based on our own experience and knowledge of the case.

---

[2] These papers are cited in Section "Multi-sectoral Interaction Processes in the Electrification of Norwegian Coastal Shipping".

## Case Description: Electrification in the Norwegian Coastal Shipping

Electrification of shipping in this chapter refers to both battery-electric solutions (fully electric and hybrid) providing energy for propulsion and other energy needs, and vessels connecting to onshore power supply (OPS) when docked, thus replacing (fossil fuel) combustion engines conventionally used to provide power for example lighting, lifting, heating, or cooling. In Norway, battery-electric solutions have mostly been applied in coastal shipping segments that have limited operating distances (e.g., ferries) and/or high peak energy demand or otherwise operational profiles amenable to (hybrid) electrification (e.g., offshore supply vessels) (Bergek et al., 2021). OPS, especially the low voltage variant, is applicable for many shipping segments and is frequently implemented in Norwegian ports (Steen, Bjerkan et al., 2024). High voltage OPS has also seen a relatively rapid diffusion, and especially serves cruise and other large passenger and offshore supply vessels that have significant power demands also when not sailing.

In terms of alternative energy solutions for propulsion, on-board batteries have been the most diffused alternative solution among Norwegian shipowners in recent years (Mäkitie, Steen, et al., 2022). The electrification of Norwegian shipping has progressed rapidly from the first fully electric vessel in 2015 (Sjøtun, 2018) to more than 200 battery-electric vessels (both hybrid and fully electric) in early 2023. To put this into perspective, there are over 1500 transport ships (both goods and passenger) sailing under a Norwegian flag, and for example nearly 5500 fishing vessels. This diffusion of battery-electric technology has been driven by procurement and tendering processes especially within the (public) car passenger ferry and (private) offshore supply vessel segments, setting low-carbon emissions as a requirement for the awarded vessels. This has created a demand for low- and zero-carbon solutions (Bergek et al., 2021; Bjerkan et al., 2019). These policies have been shaped by Norway's goal of halving the carbon emissions of domestic maritime by 2030. Many ports have sought to foster electrification notably through establishing charging infrastructure, and often in response to local climate policy targets (Bjerkan & Seter, 2021; Damman et al., 2019).

Multi-sectoral interactions of the electrification of shipping are illustrated in Fig. 5.2. In the vertical dimension, we observe a transformed propulsion and energy system value chain from combustion engines towards (hybrid) electric power, while novel value chains with OPS equipment, batteries, and energy system integration have emerged. In the horizontal dimension, there are new types of interactions between grid companies, ports and shipping companies which in some instances have proven challenging to satisfy. The case thus reveals 'whole-system'transition dynamics and associated multi-sectoral change, and how developments in different sectors have affected each other over time, and the electrification of coastal shipping in general.

## Multi-sectoral Interaction Processes in the Electrification of Norwegian Coastal Shipping

Prior literature has identified multiple ways in which multi-sectoral interactions have affected the electrification of coastal shipping in Norway. In this section, we use the MSP framework to analyse multi-sectoral interaction processes in the structural categories of technologies, institutions, and actors along the dimensions of vertical, horizontal, and adjacent sectoral interactions, and how they may have driven or hindered the electrification of Norwegian shipping. Table 5.1 summarizes these findings.

### Technologies

Multi-sectoral interaction processes in the technology dimension have been particularly visible in implementing fully electric ferries in peripheral and/or sparsely populated areas in Norway, where existing power grids are often weak and unable to provide enough electricity for

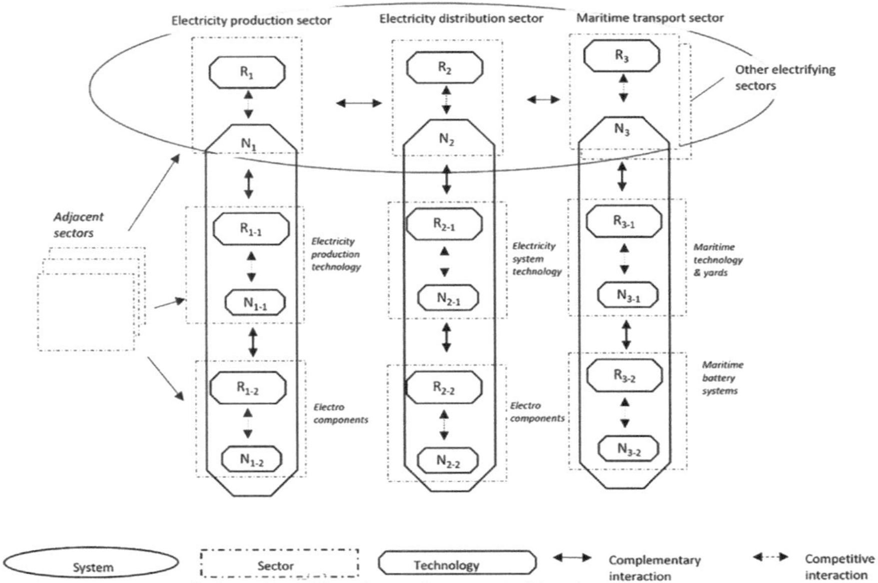


**Fig. 5.2** Multi-sectoral perspective of the electrification of coastal shipping

charging during the short ferry layover times, making continuous operation of battery-electric ferries challenging (Andersen et al., 2023). Expansion of grid capacity has often been a slow process, in part because of reluctance from distribution system operators (DSOs) due to conflicting incentives and high costs (Andersen et al., 2023). This inadequacy of the electricity distribution sector to provide power has slowed down electrification in remote areas. However, the lack of grid capacity has also led actors to experiment with work-around technological solutions, such as slow-charging onshore stationary batteries which are able to supplement or replace charging directly from the grid (Bergek et al., 2021). This problem-solving engagement of electronics and maritime technology sectors has been conducive for electrification in shipping.

Indeed, the development of various battery-electric solutions has benefited from the presence of a strong domestic maritime technology sector in Norway, including world-leading ship designers, yards, and manufacturers of maritime electronics and propulsion systems. In contrast to battery-electric solutions for automobiles, which are standardized and mass-produced, battery systems for ships to a greater extent require tailored design and overall system integration. This is because of the immense variety of different ship and vessel types, and the need to refurbish existing (fossil fuel powered) vessels with battery packs (Steen, 2018). In such instances, technological innovation benefits from close user-producer interactions between Norwegian shipowners and technology suppliers stimulated by their close geographical, institutional, and organizational proximities. Close collaboration between domestic actors in different sectors in the technology dimension (vertical) has driven the development and deployment of electric vessels in Norway (Bach et al., 2020; Steen et al., 2019). Such collaboration to develop innovative battery-electric technology for shipping has been particularly prevalent in Western Norway, where strong regional competences connected to ship design, power electronics, and automation have contributed to creation of a regional frontrunner market for electric vessels (Bugge et al., 2022; Njøs et al., 2020).

In addition to local competences, the decades-long global development of battery technologies for land-based transport has paved the way for the application of battery systems in vessels, creating valuable knowledge spillovers for maritime electrification (Mäkitie, Hanson et al., 2022). While the earlier battery technology development has largely been taking place outside of Norway, the development of a Norwegian frontrunner market in shipping stimulated the establishment of domestic maritime battery assembly factories (Bach et al., 2020), both by new entrants (Corvus) and established maritime suppliers (Siemens). This development of industrial capacity (Steen, Mäkitie et al., 2024) for manufacturing and servicing batteries is a precondition for the availability and cost reduction of maritime batteries, but also further fosters user-producer interactions between battery-suppliers, ship designers, yards, and shipping companies.

A key factor affecting diffusion of both electric propulsion and OPS has been the availability of renewable electricity (Steen, Bjerkan et al., 2024). Over the last decades, Norway has had a power system with a surplus of renewable electricity which has created good overall framework conditions for electrification. This has, for example, provided legitimacy for battery-electric technology as a scalable zero-carbon solution for shipping. In urban and more densely populated areas with existing strong grids, access to electricity has mostly not been a major issue (Mäkitie, Hanson et al., 2022). In more peripheral areas, however, insufficient capacity in the electricity distribution sector has constituted a significant barrier to electrification (Karlsen, 2024).

The domestic policy that has supported electrification in shipping has also included raising the cost of fossil fuel consumption, seeking to invoke decline in their attractiveness as energy solutions. For example, an annually increasing carbon fee is collected from $CO_2$ emitters. Moreover, many tenders to operate publicly governed ferry connections have included emission reduction requirements unattainable for fossil fuels (Bugge et al., 2022). Such deteriorating conditions for fossil fuels have been conducive for electrification in shipping as it encourages shipowners to explore alternatives to conventional fuels (Bach et al., 2021). However, in the short term, fossil fuels remain economically attractive alternatives, especially in situations of high electricity prices.

## Institutions

In the case of maritime electrification in Norway, a key reported institutional mismatch has been connected to supplying the necessary charging capacity from local electricity grids in remote locations. A key issue has been the mismatches between guiding incentive systems and revenue models of electricity (grid) distribution system operators (DSOs) and shipping companies. The revenue structures of DSOs favour serving as many customers as possible with the lowest possible capital cost. Building expensive extra grid capacity in remote locations to serve a single electric ferry is at odds with this incentive system. Although DSOs are legally obliged to provide grid access, several felt penalized and engaged only reluctantly with shipping actors. DSOs do not have a mandate to favour "green" projects. Rather, all electricity needs from various sectors regardless of use and purpose are treated identically, without favouring, for example, electricity use that replaces fossil fuel use. This is not fully aligned with the goal of urgently reducing emissions in the domestic transport sector and the goal of several shipping companies to implement zero-carbon technologies such as battery-electric vessels. The mismatch between electricity distribution sector and shipping is thus both techno-economical and institutional, given that actors in the different sectors have differing incentive structures and goals (Andersen et al., 2023; Mäkitie, Hanson et al., 2022; Nykamp et al., 2023). This lack of shared goals and logics across sectors (horizontally) leads to delays for electrification in shipping (Andersen et al., 2023).

In the vertical dimension, upstream maritime technology suppliers, yards, and designers have been relatively well-aligned with shipowners. As mentioned, Norway has a strong maritime technology sector where several technology suppliers with electricity competences (such as ABB, Siemens, and many others) have been active in developing and offering technological solutions for battery-electric vessels. Moreover, Norwegian shipowners have been relatively favourable to decarbonization efforts, and early adopters of especially battery-electric solutions (Mäkitie, Steen et al., 2022). Meanwhile, the Norwegian Shipowners' Association, the interest organization of shipowners in Norway, has pledged that only vessels with zero-emission technologies would be ordered from

2030 onwards. Such shared goals and expectations between technology suppliers and shipowners help to align actors and sectors in the vertical technology value chain, which is thus conducive for electrification.

Indeed, Sæther and Moe (2021) report that there has been a general consensus among actors in different sectors that electrification is beneficial and feasible, especially for the (car passenger) ferry segment. Early on, there was an absence of lobbying to delay or slow down electrification in the ferry segment. This paved the way for using public procurement as a key mechanism to promote electrification in the ferry segment (Sæther & Moe, 2021). Similarly, Bjerkan and colleagues noted that when implementing OPS in the ports of Oslo and Kristiansand, electrification in the passenger vessel segment had high legitimacy as a decarbonization solution among the key actors, including shipowners, port owners and operators, and technology and energy suppliers (both vertical and horizontal). This high and shared legitimacy across sectors of electrification as means to decarbonize shipping was thus found to be conducive for electrification (Bjerkan & Ryghaug, 2021; Bjerkan et al., 2021).

This high legitimacy is due to several reasons, some of which point to multi-sectoral dynamics. The electricity system in Norway is almost entirely based on renewables, and was until recently perceived to be able to cater to the increasing electricity demand from shipping. On the contrary, other shipping decarbonization solutions, such as biofuels, have suffered from the perceived lack of scalability in production capacity due to resource constraints, such as sustainably sourced biomass, thus lowering their legitimacy and perceived feasibility (Mäkitie, Hanson et al., 2022). Finally, due to the presence of domestic companies across the entire maritime battery-electric value chain (vertical), electrification has been seen to have a potential for domestic value creation in different sectors, which has increased the legitimacy of domestic policies seeking to foster electrification (Bugge et al., 2022). In sum, the generally positive perceptions across key sectors have provided legitimacy to electrification. However, such legitimacy may also change over time, for example if electricity becomes substantially more expensive or scarce.

## Actors and Networks

The multi-sectoral processes in the actor dimension have been visible in the actions of Norwegian port authorities and electricity suppliers when seeking to develop OPS infrastructure. In the absence of private companies which would construct, operate, and maintain such infrastructure, the Port of Oslo joined forces with a local energy provider to create a joint enterprise that would carry out this mediation task between electricity supply and shipping (Bjerkan & Ryghaug, 2021). Similarly, the Port of Bergen and the regional utility company BKK (later called Eviny) created a dedicated OPS company, Plug AS, which has established itself as a dedicated provider and operator of OPS equipment in several domestic ports (Andersen et al., 2023). In other words, business integration between sectors, i.e., electricity utilities and port authorities adopting the role of a charging infrastructure developer, was necessary for electrification as there were not yet other actors that would provide this service.

Regional cluster organizations and research centres have also been important in fostering cross-sectoral collaboration. A notable example is the NCE Maritime CleanTech cluster, which comprises actors from the maritime, energy, and other (supplier) sectors especially in the region between Bergen and Stavanger on the west coast, and which has worked to establish interactions across the whole battery-electric value chain (Bugge et al., 2022; Sjøtun & Njøs, 2019). Moreover, large companies in the electronics sector, such as Siemens and ABB, have been instrumental in developing electric propulsion systems and associated control systems and automation (Bergek et al., 2021). Also suppliers of conventional propulsion systems in the maritime technology sector, such as Wärtsilä and Rolls-Royce, have been fast to diversify to develop and provide battery-electric systems (Bugge et al., 2022). Diversification of actors from adjacent technology value chains has thus been driving electrification in shipping.

Meanwhile, development contracts by the Norwegian Public Road Administration (NPRA) for new technologies have created a platform for different actors to collaborate in the development of electric vessels, such as Ampere vessel, the first fully electric ferry (Andersen et al., 2023).

In another form of institutional work, the NPRA has been described as instrumental in acting "entrepreneurially" in the electrification of the ferry segment, building legitimacy for battery-electric solutions, and establishing high ambitions for emission reductions (Bugge et al., 2022; Sæther & Moe, 2021).

Another example of cross-sectoral institution building is from the early 2010s when the Port of Oslo together with a shipowner, technology, and R&D providers established a collaboration to develop technical designs and practices for OPS, which at the time were immature. This collaboration resulted in a technical solution which now can be considered as an industrial standard (Bjerkan et al., 2021). In this way the Port of Oslo used its existing networks to coordinate OPS-related discussions involving actors from different sectors, facilitating discussions to make OPS usable and more acceptable across various types of actors (Bjerkan & Ryghaug, 2021).

## Discussion and Conclusion

In this chapter we employed a novel multi-sectoral perspective (MSP) (Andersen et al., 2026) to analyse how sustainability transitions often involve complex interaction processes within and across different sectors. Our application of MSP to the case of electrifying coastal shipping in Norway has shown that the analysis of such multi-sectoral processes can help to reveal and explain some of the key drivers and barriers for this electrification. We thus propose that MSP framework can be useful as an overarching and systematized analytical perspective to understand the dynamics and implications of a diverse set of multi-sectoral interactions for sustainability transitions (Andersen et al., 2020). It thus contributes to the literature on sustainability transition studies with a holistic perspective on multi-system dynamics at the level of sectors.

The experience from applying MSP to an empirical study in this present book chapter has revealed some of its strengths and weaknesses as an analytical framework. In terms of strengths, MSP was found to support the analysis of the effect of multi-sectoral interactions on transitions at least in two ways.

First, it enables a basic typology of differentiating between different types of multi-sectoral interactions, namely "horizontal" throughput interactions, "vertical" technology component interactions, and "adjacent" interactions with other sectors. This allows for more detailed understanding of the types of interdependencies that a transition in a focal sector (in our case coastal shipping) may have with interconnected sectors, and the implications of specific types of sectoral interactions for the transition. For example, in the case of electrification in coastal shipping, it could be seen that while horizontal interactions had resulted in bottlenecks for electrification (e.g., limited grid capacity in remote locations), vertical and adjacent interactions were often conducive for this transition (e.g., technology providers eagerly supplying new technological solutions, and spillovers regarding battery technologies from land-based transport). Second, the MSP allows for a more systematic analysis and assessment of multi-sectoral interaction processes and whether or not these have been conducive for the transition. In turn, this supports the identification of measures to dissolve bottlenecks or to reinforce strengths both within the focal sector as well as in other sectors. MSP may thus be useful, for example, in providing insights for policy analyses regarding multi-sectoral drivers and barriers for transitions.

In terms of weaknesses, MSP is analytically complex due to its ambition to conceptualize a "full picture" multi-sectoral view of transitions. While its application was highly useful in making sense of and categorizing interactions in a previously well-researched transition, namely the electrification of coastal shipping in Norway, its full application would be a major task in guiding the analysis of a yet little-studied transition. Moreover, MSP is potentially best applied in situations where the transition has already entered the acceleration phase, rather than still being in the early formative phase, due to the higher presence and visibility of multi-sectoral interactions and their prevailing dynamics. Moreover, if the analysis had included also other emerging decarbonization solutions for shipping, such as hydrogen and biofuels, and the interactions between such emerging decarbonization solutions, the analytical task would have become even more daunting. The same applies to accounting also for geographical differences and how multi-sectoral interactions are influenced by dynamics at different geographical scales (local, national, international). These issues highlight the analytical challenge of seeking to understand various temporally unfolding interactions between sectors in a single analysis, calling for further methodological and theoretical development. Alternatively, MSP framework could be applied as an overall heuristic for analysts that may want to "zoom into" specific sectoral interactions and problem areas, while treating other interactions as the context for such specific interactions.

We conclude by proposing MSP as a framework that may help analysts and practitioners to analyse and make sense of the multi-sectoral complexities of real-world transition processes. Thus, it may provide an analytical framework in and of itself, or an overall "map" or "guide" of multi-sectoral interactions to better understand and appreciate the full intricacy of multi-sectoral dynamics, and how they may shape transitions.

## References

Andersen, A. D., & Geels, F. W. (2023). Multi-system dynamics and the speed of net-zero transitions: Identifying causal processes related to technologies, actors, and institutions. *Energy Research & Social Science, 102*, 103178. https://doi.org/10.1016/j.erss.2023.103178

Andersen, A. D., Geels, F. W., Steen, M., & Bugge, M. M. (2023). Building multi-system nexuses in low-carbon transitions: Conflicts and asymmetric adjustments in Norwegian ferry electrification. *Proceedings of the National Academy of Sciences, 120*(47), e2207746120. https://doi.org/10.1073/pnas.2207746120

Andersen, A. D., Mäkitie, T., Steen, M., & Wanzenböck, I. (2026). Conceptualizing the relationship between industrial transformation and accelerating sustainability transitions: A multi-sectoral perspective. *Energy Research & Social Science 131*, 104509. https://doi.org/10.1016/j.erss.2025.104509

Andersen, A. D., & Markard, J. (2020). Multi-technology interaction in socio-technical transitions: How recent dynamics in HVDC technology can inform transition theories. *Technological Forecasting and Social Change, 151*, 119802. https://doi.org/10.1016/j.techfore.2019.119802

Andersen, A. D., Steen, M., Mäkitie, T., Hanson, J., Thune, T. M., & Soppe, B. (2020). The role of inter-sectoral dynamics in sustainability transitions: A comment on the transitions research agenda. *Environmental Innovation and Societal Transitions, 34*, 348–351. https://doi.org/10.1016/j.eist.2019.11.009

Bach, H., Bergek, A., Bjørgum, Ø., Hansen, T., Kenzhegaliyeva, A., & Steen, M. (2020). Implementing maritime battery-electric and hydrogen solutions: A technological innovation systems analysis. *Transportation Research Part d: Transport and Environment, 87*, 102492. https://doi.org/10.1016/j.trd.2020.102492

Bach, H., Mäkitie, T., Hansen, T., & Steen, M. (2021). Blending new and old in sustainability transitions: Technological alignment between fossil fuels and biofuels in Norwegian coastal shipping. *Energy Research & Social Science, 74*, 101957. https://doi.org/10.1016/j.erss.2021.101957

Bergek, A., Bjørgum, Ø., Hansen, T., Hanson, J., & Steen, M. (2021). Sustainability transitions in coastal shipping: The role of regime segmentation. *Transportation Research Interdisciplinary Perspectives, 12*, 100497. https://doi.org/10.1016/j.trip.2021.100497

Bergek, A., Hansen, T., Hanson, J., Mäkitie, T., & Steen, M. (2023). Complexity challenges for transition policy: Lessons from coastal shipping in Norway. *Environmental Innovation and Societal Transitions, 46*, 100687. https://doi.org/10.1016/j.eist.2022.100687

Bergek, A., Jacobsson, S., Carlsson, B., Lindmark, S., & Rickne, A. (2008). Analyzing the functional dynamics of technological innovation systems: A scheme of analysis. *Research Policy, 37*(3), 407–429. https://doi.org/10.1016/j.respol.2007.12.003

Bjerkan, K. Y., Karlsson, H., Snefuglli Sondell, R., Damman, S., & Meland, S. (2019). Governance in maritime passenger transport: Green public procurement of ferry services. *World Electric Vehicle Journal, 10*(4), 74. https://www.mdpi.com/2032-6653/10/4/74

Bjerkan, K. Y., & Ryghaug, M. (2021). Diverging pathways to port sustainability: How social processes shape and direct transition work. *Technological Forecasting and Social Change, 166*, 120595. https://doi.org/10.1016/j.techfore.2021.120595

Bjerkan, K. Y., Ryghaug, M., & Skjølsvold, T. M. (2021). Actors in energy transitions. Transformative potentials at the intersection between Norwegian port and transport systems. *Energy Research & Social Science, 72*, 101868. https://doi.org/10.1016/j.erss.2020.101868

Bjerkan, K. Y., & Seter, H. (2021). Policy and politics in energy transitions. A case study on shore power in Oslo. *Energy Policy, 153*, 112259. https://doi.org/10.1016/j.enpol.2021.112259

Bugge, M. M., Andersen, A. D., & Steen, M. (2022). The role of regional innovation systems in mission-oriented innovation policy: Exploring the problem-solution space in electrification of maritime transport. *European Planning Studies, 30*(11), 2312–2333. https://doi.org/10.1080/09654313.2021.1988907

Carlsson, B. (2016). Industrial dynamics: A review of the literature 1990–2009. *Industry and Innovation, 23*(1), 1–61. https://doi.org/10.1080/13662716.2015.1120658

Dahmén, E. (1988). 'Development blocks' in industrial economics. *Scandinavian Economic History Review, 36*(1), 3–14. https://doi.org/10.1080/03585522.1988.10408102

Damman, S., Kenzhegaliyeva, A., Bjerkan, K. Y., & Steen, M. (2019). *Mot nullutslippshavner i 2030? En studie av handlingsrom med fokus på havnene i Oslo, Narvik og Kristiansand* (2019:01304) (SINTEF report).

Damman, S., & Steen, M. (2021). A socio-technical perspective on the scope for ports to enable energy transition. *Transportation Research Part d: Transport and Environment, 91*, 102691. https://doi.org/10.1016/j.trd.2020.102691

Geels, F. W. (2002). Technological transitions as evolutionary reconfiguration processes: A multi-level perspective and a case-study. *Research Policy, 31*(8), 1257–1274. https://doi.org/10.1016/S0048-7333(02)00062-8

Geels, F. W. (2004). From sectoral systems of innovation to socio-technical systems: Insights about dynamics and change from sociology and institutional theory. *Research Policy, 33*(6), 897–920. https://doi.org/10.1016/j.respol.2004.01.015

Hekkert, M. P., Suurs, R. A. A., Negro, S. O., Kuhlmann, S., & Smits, R. E. H. M. (2007). Functions of innovation systems: A new approach for analysing technological change. *Technological Forecasting & Social Change, 74*(4), 413–432. https://doi.org/10.1016/j.techfore.2006.03.002

Holtz, G., Brugnach, M., & Pahl-Wostl, C. (2008). Specifying "regime"—A framework for defining and describing regimes in transition research. *Technological Forecasting and Social Change, 75*(5), 623–643. https://doi.org/10.1016/j.techfore.2007.02.010

Kanger, L., Schot, J., Sovacool, B. K., van der Vleuten, E., Ghosh, B., Keller, M., Kivimaa, P., Pahker, A.-K., & Steinmueller, W. E. (2021). Research frontiers for multi-system dynamics and deep transitions. *Environmental*

*Innovation and Societal Transitions, 41*, 52–56. https://doi.org/10.1016/j.eist.2021.10.025

Karlsen, J. (2024). *Å seile mot strømmen: Utfordringer ved elektrifisering av ferjesamband i Nordland*. Norwegian University of Science and Technology. https://ntnuopen.ntnu.no/ntnu-xmlui/handle/11250/3138380

McMeekin, A., Geels, F. W., & Hodson, M. (2019). Mapping the winds of whole system reconfiguration: Analysing low-carbon transformations across production, distribution and consumption in the UK electricity system (1990–2016). *Research Policy, 48*(5), 1216–1231. https://doi.org/10.1016/j.respol.2018.12.007

Mäkitie, T., Andersen, A. D., Hanson, J., Normann, H. E., & Thune, T. M. (2018). Established sectors expediting clean technology industries? The Norwegian oil and gas sector's influence on offshore wind power. *Journal of Cleaner Production, 177*, 813–823. https://doi.org/10.1016/j.jclepro.2017.12.209

Mäkitie, T., Hanson, J., Steen, M., Hansen, T., & Andersen, A. D. (2022). Complementarity formation mechanisms in technology value chains. *Research Policy, 51*(7), 104559. https://doi.org/10.1016/j.respol.2022.104559

Mäkitie, T., Steen, M., Saether, E. A., Bjørgum, Ø., & Poulsen, R. T. (2022). Norwegian ship-owners' adoption of alternative fuels. *Energy Policy, 163*, 112869. https://doi.org/10.1016/j.enpol.2022.112869

Nelson, R. R. (2020). A perspective on the evolution of evolutionary economics. *Industrial and Corporate Change, 29*(5), 1101–1118. https://doi.org/10.1093/icc/dtaa045

Nelson, R. R., & Winter, S. G. (1977). In search of useful theory of innovation. *Research Policy, 6*(1), 36–76. https://doi.org/10.1016/0048-7333(77)90029-4

Nelson, R. R., & Winter, S. G. (1982). *An evolutionary theory of economic change*. Belknap Press.

Njøs, R., Sjøtun, S. G., Jakobsen, S.-E., & Fløysand, A. (2020). Expanding analyses of path creation: Interconnections between territory and technology. *Economic Geography, 96*(3), 266–288. https://doi.org/10.1080/00130095.2020.1756768

Nykamp, H., Andersen, A. D., & Geels, F. W. (2023). Low-carbon electrification as a multi-system transition: A socio-technical analysis of Norwegian maritime transport, construction, and chemical sectors. *Environmental Research Letters, 18*(9), 94059. https://doi.org/10.1088/1748-9326/acf67a

Perez, C. (2009). Technological revolutions and techno-economic paradigms. *Cambridge Journal of Economics, 34*(1), 185–202. https://doi.org/10.1093/cje/bep051

Rip, A., & Kemp, R. (1998). Technological change. In S. Rayner & E. L. Malone (Eds.), *Human choice and climate change. Resources and technology.* (pp. 327–399). Battelle Press.

Rosenbloom, D. (2020). Engaging with multi-system interactions in sustainability transitions: A comment on the transitions research agenda. *Environmental Innovation and Societal Transitions, 34*, 336–340. https://doi.org/10.1016/j.eist.2019.10.003

Saviotti, P. P., & Pyka, A. (2004). Economic development by the creation of new sectors. *Journal of Evolutionary Economics, 14*(1), 1–35. https://doi.org/10.1007/s00191-003-0179-3

Sjøtun, S. G. (2018). A ferry making waves: A demonstration project 'doing' institutional work in a greening maritime industry. *Norsk Geografisk Tidsskrift—Norwegian Journal of Geography, 73*(1), 16–28. https://doi.org/10.1080/00291951.2018.1526208

Sjøtun, S. G., & Njøs, R. (2019). Green reorientation of clusters and the role of policy: 'The normative' and 'the neutral' route. *European Planning Studies, 27*(12), 2411–2430. https://doi.org/10.1080/09654313.2019.1630370

Steen, M. (2018). Et grønt maritimt skifte? Muligheter of utfordringer for en miljøvennlig skipsfart. In G. Rusten & H. Haarstad (Eds.), *Grønn omstilling - norske veivalg* (pp. 45–62). Universitetsforlaget.

Steen, M., Bach, H., Bjørgum, Ø., Hansen, T., & Kenzhegaliyeva, A. (2019). *Greening the fleet: A technological innovation system (TIS) analysis of hydrogen, battery electric, liquefied biogas, and biodiesel in the maritime sector*. SINTEF Digital. https://ntnuopen.ntnu.no/ntnu-xmlui/handle/11250/2613837

Steen, M., Bjerkan, K. Y., Hansen, L., & Seter, H. (2024). Implementing decarbonisation measures in Norwegian ports. *Transportation Research Interdisciplinary Perspectives, 23*, 100993. https://doi.org/10.1016/j.trip.2023.100993

Steen, M., Mäkitie, T., Hanson, J., & Normann, H. E. (2024). Developing the industrial capacity for energy transitions: Resource formation for offshore wind in Europe. *Environmental Innovation and Societal Transitions, 53*, 100925. https://doi.org/10.1016/j.eist.2024.100925

Stephan, A., Schmidt, T. S., Bening, C. R., & Hoffmann, V. H. (2017). The sectoral configuration of technological innovation systems: Patterns of knowledge development and diffusion in the lithium-ion battery technology in Japan. *Research Policy, 46*(4), 709–723. https://doi.org/10.1016/j.respol.2017.01.009

Sæther, S. R., & Moe, E. (2021). A green maritime shift: Lessons from the electrification of ferries in Norway. *Energy Research & Social Science, 81*, 102282. https://doi.org/10.1016/j.erss.2021.102282

Victor, D. G., Geels, F. W., & Sharpe, S. (2019). *Accelerating the low carbon transition. The case for stronger, more trageted and coordinated international action.* http://www.energy-transitions.org/content/accelerating-low-carbon-transition

# 6

# Conceptualizing Multi-System Configurations: The Case of Hydrogen in Germany

Meike Löhr

## Introduction

Achieving net-zero emissions by 2050 requires extensive decarbonisation efforts that extend beyond individual sectors and existing technologies (IEA, 2021). This entails phasing out of fossil fuels, deploying low-carbon technologies such as hydrogen, and increased electrification in all sectors. The latter is facilitated through sector coupling, the connection of two or more sectors (Fridgen et al., 2020; Löhr & Chlebna, 2023).

From a theoretical perspective, the emerging literature on multi-system interactions (MSI)[1] addresses these connections as structural and functional couplings (Andersen & Geels, 2023; Konrad et al., 2008).

[1] There are several terms used for the emerging literature, including multi-system interactions (MSI) and, more general, multi-system dynamics (MSD).

M. Löhr (✉)
Department of Environmental Social Sciences, Eawag: Swiss Federal Institute of Aquatic Science and Technology, Dübendorf, Switzerland
e-mail: Meike.loehr@eawag.ch


M. Steen et al. (eds.), *Sustainability Transitions and Industrial Transformation*,
https://doi.org/10.1007/978-981-95-4893-4_6

The MSI literature conceptualizes the nature and type of system interactions (Papachristos et al., 2013; Raven & Verbong, 2007), defines inter-sectoral dynamics (Andersen et al., 2020; Rosenbloom, 2020), bridges policy and transition studies to conceptualize system emergence in multi-system settings (Löhr et al., 2024), and advances the concept of system entanglers (Kanger et al., 2021; Löhr & Chlebna, 2023; van der Vleuten, 2019).

Core elements such as structural and functional couplings, types of interaction, system boundaries, overlaps or sites of interaction, and connecting actors such as system entanglers emerge from this literature. Building on these, this contribution proposes a conceptual framework for "multi-system configurations" that brings together core elements at different stages of interaction. Multi-system configurations describe the interaction of two or more systems at different stages of interconnection, such as emerging connections, interaction, or integration. It covers system[2] level changes, including system boundaries, and accounts for the activities of actors working towards connecting systems in different configurations. Typically, research depicts one point in time and shows how two systems couple, e.g., by overlapping and thus starting to connect. However, system interaction is a dynamic process that comprises many different phases, stages, and configurations. To stimulate more complex and at the same time more precise analyses about the actual system configurations observed, this contribution proposes a conceptualization of five different multi-system configurations: (1) unconnected systems, (2) emerging connections, (3) interconnected systems, (4) integrated systems, and (5) new system emergence. Some of them may build on each other but they do not necessarily describe a linear development or normative "best" configuration. The conceptualization comprises interactions among actors, institutions, and technologies (structural couplings) at the site of interaction. It includes strategies of system entanglers, a specific type of actor who connects different systems by "fostering between-system links" or applying "cross-sectoral

[2] A system is here understood as a socio-technical system at the level of the sector, consisting of different actors, technologies, and institutions, that, together fulfil a function (Geels 2004). The term system and sector are used interchangeably in this contribution. System rather has a theoretical, sector a more empirical connotation.

competencies and learning" (Löhr & Chlebna, 2023). While their core activities have been defined, it has not been conceptualized how the activities of system entanglers vary in different system configurations.

To empirically illustrate the conceptualization, this contribution draws on a qualitative content analysis, based on document analysis and supplementary expert interviews, which examines three cases of system interaction in Germany involving industry, mobility, and heating, all connected through hydrogen with the electricity sector (Löhr & Chlebna, 2023).

This research contributes to MSI by advancing the conceptual understanding of distinct system configurations. Such a refined comprehension facilitates identifying, e.g., competence gaps in real-life projects and can inform efforts to accelerate the transition process, e.g., in terms of sector coupling and electrification.

The remainder of the article is organized as follows. The next section presents the state of the art and theoretical background, focusing on multi-system interactions and the role of actors therein. The following section conceptualizes five multi-system configurations. The conceptualization is then empirically illustrated. The final section discusses and summarizes the analysis.

## State of the Art and Theoretical Background

### Multi-System Interactions

Reaching net-zero by 2050 comes not only with empirical challenges but also with theoretical ones. The single system perspective that prevailed within transition studies (Köhler et al., 2019) is being extended to a multi-system perspective which includes the interplay of several systems.

Early contributions analyse 'multi-regime dynamics' and 'multi-regime interaction' (Konrad et al., 2008; Raven, 2007; Raven & Verbong, 2007). The latter extends the multi-level perspective and proposes the nature of interactions across regime boundaries as competitive, symbiotic, integrative, or spillover (Raven & Verbong, 2007). Sandén and Hillman (2011) propose six interaction modes of technology interaction: competition, symbiosis, neutralism, parasitism, commensalism, and amensalism and the idea of structural overlaps in socio-technical systems. Systems may overlap in one or more dimensions. Another typology of system interactions between different levels includes niche transfer, niche interference, niche autonomy, and niche emergence. The authors also analyse the nature of interactions, e.g., reinforcing or disruptive (Papachristos et al., 2013). Recently, inter-system interactions in socio-technical systems were summarized as interfaces between systems, relationships between systems, impacts, and intensity (Breitschopf et al., 2023).

A core idea of MSI describes structural and functional couplings (Andersen & Geels, 2023; Konrad et al., 2008). The latter describe input–output relationships. Structural couplings refer to a shared use of system elements like actors, institutions, technologies, or infrastructures. This idea has been adopted for system interactions between a TIS (Technological Innovation System) and its context (Bergek et al., 2015). Structural couplings were also used for different subsystems in the global innovation systems literature (Binz & Truffer, 2017).

While the former contributions take a system or technology perspective, few include a temporal one. Exceptions are Andersen et al. (2023) who suggest four analytical dimensions, actors, technologies, institutions, and resources to analyse system connections in four phases: experimentation, stabilization, mass diffusion, and anchoring and reconfiguration of a wider system.

To better understand how actors create institutional couplings, Käsbohrer et al. (2024) analyse multi-system interactions through institutional work. They propose four options for the reconfiguration of system boundaries: failing institutional work, successful institutional work, and overlap of both systems; dominance of actors coupling successfully, and couplings driven by interface actors. They do not show the evolution of system integration or the work of system entanglers.

To sum up, earlier papers analyse system interactions between niches and regimes. They suggest types of system (or technology) interactions (and analytical patterns) distinguishing, e.g., cooperation and competition. Many of the early contributions focus on changes within one system (often in the sense of a sector), later contributions add a multi-system perspective that covers mostly two systems. While the concept of couplings has been revived lately, there still is a need of better understanding how systems are connected, by whom, through which activities, and in which configurations. Some contributions add a temporal perspective, e.g., transition phases, but this mostly does not address changes of (and within) the interacting systems, e.g., shifting system boundaries or activities at the site of interaction. To address these gaps, I suggest conceptualizing different multi-system configurations and show how couplings vary across different configurations. I focus on the role of actors such as system entanglers therein.

## Actors in Multi-System Interactions: System Entanglers

Many actors, including incumbents and challengers from industry, associations, politics, civil society, and research, are developing low-carbon technologies, negotiating policies, and shaping transition processes. Actors who bridge interests, technologies, organizations, or systems play a key role in this context. Among the bridging actors intermediaries are described as the "key catalysts that speed up change" (Kivimaa et al., 2019, p. 1063). They intermediate between multiple actors and interests, align interests, and facilitate change (Kivimaa et al., 2019; Klerkx & Leeuwis, 2009). From a political science and collective agency perspective, policy brokers act as bridging actors for advocacy coalitions. They negotiate agreements between coalitions that may disagree on beliefs. Their role is crucial because coalitions based on shared beliefs can accelerate transitions (Hess, 2019; Löhr, 2020; Löhr et al., 2024; Markard et al., 2016; Wesche et al., 2024). While these actor concepts have usually been applied within a single system perspective, a specific type of actor, the system entangler, has been conceptualized for a multi-system perspective, operating across several systems (Löhr & Chlebna, 2023). Initially defined in the deep transitions literature on a macro level as 'connecting transitions' by interweaving socio-technical systems, system entanglers cross multiple systems and work to connect them (Schot & Kanger, 2018; van der Vleuten, 2019). This idea has been further developed building on the heuristic of 'transition work' (Löhr et al.,

2022) to describe actors who actively work to connect different systems (Löhr & Chlebna, 2023). The activities and characteristics of system entanglers are defined as follows. Especially the two latter distinguish system entanglers from other actors:

- Innovating and implementing
- Envision systemic change
- Network and coalition building across systems
- Foster between system links
- Cross-sectoral competencies & learning.

These activities enhance our understanding of how actors connect systems. The processual perspective of transition work and its core idea of creating, maintaining, and disrupting institutions,[3] actors, and technologies may help to understand how actors establish connections between systems in different configurations.

While many actors can contribute to further connecting systems, e.g., by working on a policy that addresses several sectors, by being part of a cross-sectoral association or by installing a technology that can be used in two sectors, only some actors pursue system entanglement (connecting different systems) as an active goal. This differentiates system entanglers from actors (unintentionally) contributing to system entanglement.

A system entangler may be positioned within a system, at the site of interaction between systems or, working together with another system entangler, in both (or more) systems of interest (Löhr & Chlebna, 2023).[4] A system entangler acts at the site of interaction which can be understood as constructed "space" between systems. The site of interaction represents the space in which actors working to connect different systems on a technological, institutional, and organizational

[3] The heuristic has been inspired by institutional work which focuses on institutions Lawrence Thomas, B., & Suddaby, R. (2006). Institutions and institutional work. *Handbook of Organizations Studies.*, 215–254.

[4] In this regard, it differs from intermediaries who act, e.g., between networks, entities, or institutions (Kivimaa et al., 2019). The deep transitions literature links system entanglers to intermediaries stating that intermediaries shape single system transitions, while the deep transitions perspective calls for 'inter-systemic intermediaries' or 'transnational system entanglers' to create connections between systems (Kanger et al., 2021).

level interact. In reality, it is unlikely to "see" a space in-between systems. It does not represent a physical gap between systems but a constructed one. An example for a site of interaction could be a (demonstration) project which provides regulatory freedom, funding, etc. for the implementing actors to create connections between two sectors. Other possible sites of interactions could be created by actors uniting in an alliance with the purpose to work on a cross-sectoral issue, within the realm of an (international) company or situated across value chains.

## Conceptualization of Multi-System Configurations

To better grasp different stages, constellations, and couplings of MSI, I propose conceptualizing five distinct multi-system configurations and the corresponding activities for actors within each: (1) unconnected systems, (2) emerging connections between systems, (3) interconnected systems, (4) integrated systems, and (5) new system emergence. Some of the proposed configurations suggest a linear development and a closer integration of (ideal type) systems. However, this does not necessarily reflect reality. Some systems may never integrate or reverse and even disentangle. Furthermore, this does not imply that system functioning can only be assured through full integration or that one configuration is superior to others.

A system is here understood as a socio-technical system at the level of a sector—the conceptual idea may also be applied to other (levels or types of) systems.[5] A socio-technical system consists of different technologies (innovations), institutions (formal and informal institutions such as policies, standards, norms, etc.), and actors (individual and collective actors such as policy actors, firms, associations, etc.) that interact to fulfil a specific societal function (Geels, 2004). Examples are the electricity, mobility, or heating sectors. I will focus on five analytical

[5] The degree of aggregation of socio-technical systems differs in analyses. A socio-technical system may be larger and encompass several sectors. In this case, the sector would be a subsystem of the socio-technical system. This perspective is, however, not pursued in this paper.

dimensions typically applied in transition studies—systems, technologies, institutions, and actors, and add a specific actor type, the system entangler.

The conceptualization builds on Löhr and Chlebna (2023). I re-analysed the hydrogen case study to identify references to different stages, phases, or configurations of system interaction mentioned by the interviewees. The segments coded in MAXQDA on multi-system interactions were central. For the conceptualization, I kept institutions, actors, and technologies as key elements of a system (Geels, 2004; Löhr et al., 2022) and added the system entangler and system category to describe system configurations in their multi-dimensionality. This new perspective on the case study served as the basis for deriving a conceptualization of multi-system configurations (Eisenhardt, 1989). The case study material allowed to conceptualize five different configurations. While the first three can be fully derived from the data, the two latter are based on expectations expressed by the actors, observations drawn from the cases and the broader literature. Therefore, I only empirically illustrate the first three configurations (unconnected systems, emerging connections, and interconnected systems).

## Unconnected Systems

Systems: Unconnected systems function mainly independently of each other without a site of interaction in place and separating system boundaries. Each system focuses on its own objectives, technologies, institutions, and actors without considering potential synergies with each other (see Fig. 6.1). This independence can lead to inefficiencies and missed opportunities for cross-system problem-solving and innovation, but it could also ensure more resilient and diverse systems. One example for the latter is critical infrastructure in systems like nuclear energy where independence from systems like the internet can enhance safety and resilience.

Technologies: Technologies within unconnected systems do not facilitate cross-system connectivity. The technologies within each system may be distinct, reflecting rather isolated technological silos or similar but not

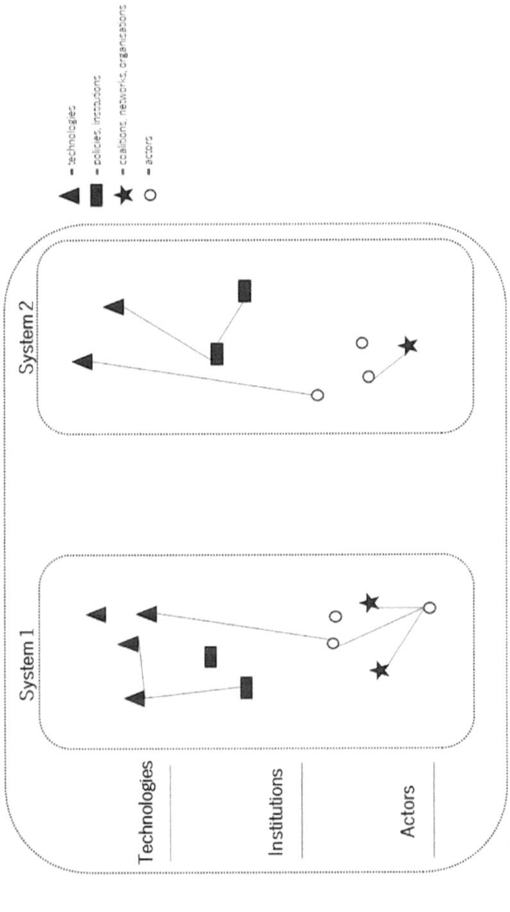


**Fig. 6.1** Unconnected systems

functionally coupled. There are no interoperable platforms or tools that enable information exchange and cooperation across systems.

Institutions: Institutions are designed to address issues within individual systems, lacking mechanisms for cross-system engagement. Policies, regulations, and norms are focused on the functioning within one system.

Actors: Actors within unconnected systems have little to no interaction with those in other systems. Each actor operates within the confines of its own system, is competent within the system boundaries but lacks opportunities for cooperation, knowledge exchange, and learning across systems.

System entanglers: In unconnected systems, system entanglers—if they exist—begin to envision systemic change by identifying opportunities for cross-system interactions. They lay the foundation for interconnecting different systems and establishing cross-system linkages.

## Emerging Connections Between Systems

Systems: Connections between systems emerge, with the site of interaction taking shape and system boundaries becoming more permeable. The site of interaction is depicted as space in-between systems to illustrate the observed gap it bridges (see Fig. 6.2). Initial connections form between previously independent systems, allowing for preliminary collaboration and coordination. These interactions mark the beginning of a shift from an isolated system functioning towards a more interconnected one. This becomes visible in the example of cars and artificial intelligence, with connections starting to emerge through self-driving pilot tests.

Technologies: Innovation and new technology development (and implementation) determine this configuration, while, at the same time, established technologies start to phase out. New collaboration tools (e.g., platforms, hubs) are adopted to facilitate these emerging technological interactions, enabling more effective communication and coordination between different systems (for technology development in a TIS life cycle as well as connections to the context see Markard (2020)).

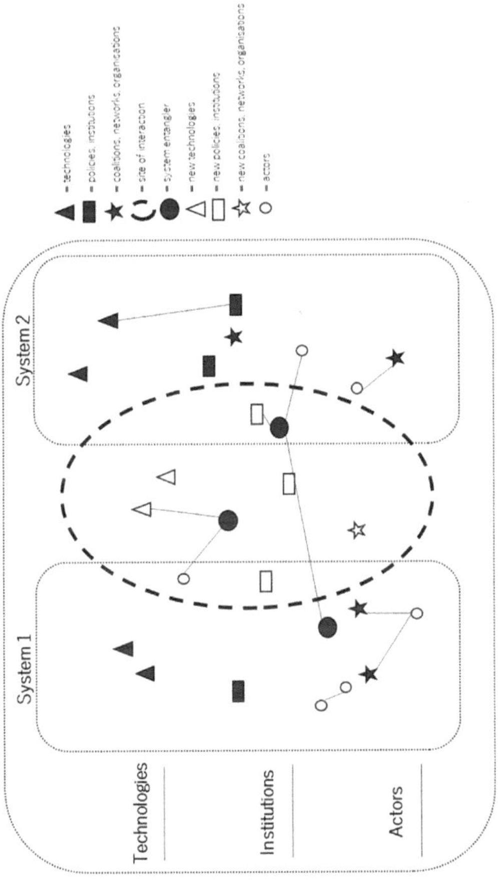


**Fig. 6.2** Emerging connections between systems

Institutions: Policies, institutions, and rules are created or adapted to support the emerging connections between systems. This may involve removing or modifying existing institutions to better align the systems, ensuring they can support cross-system activities. If institutions do not align in this configuration, this might create tensions and bottlenecks later.

Actors: Actors start to interact across systems, initiating cross-system collaborations and exchanges (Raven & Verbong, 2007). These interactions include the sharing of knowledge, resources, and best practices. Actors might have different interests in connecting systems. Some might refrain from increasing connections as they would weaken their businesses, purposes, or responsibilities. Tensions, e.g., about resources, may emerge in this configuration.

System entanglers: System entanglers play a crucial role in this configuration by creating new networks and coalitions and dissolving (or destabilizing) existing ones (e.g., those whose purpose is no longer given). They foster links between systems and develop and employ (their) cross-sectoral competencies. Positioned within or between systems, they facilitate the emerging connections and promote the necessary skills and knowledge for effective collaboration. System entanglers work both towards further connecting systems by entangling technologies, institutions, and actors as well as by disentangling those elements that hinder a further system integration (system disentangler) or would lead to system failures.[6]

## Interconnected Systems

Systems: Systems are well interconnected, exhibiting a high degree of interdependence while maintaining some level of independence. The connections between systems are characterized by rather loose couplings, allowing for flexibility and adaptability while ensuring effective collaboration and coordination (Dubois & Gadde, 2002; Weick, 1976). The site of interaction is well-established bridging the two systems (see Fig. 6.3).

[6] The activities of system entanglers correspond to creation and disruption as described in transition work (see Löhr et al., 2022).

The interconnected system covers parts of the prior existing systems, integrating some of their technologies, institutions, and actors while also containing new ones. The system boundaries are rather permeable. An example for interconnected systems is the battery sector and the power sector, where batteries for the grid act as loose couplings between two systems that remain largely independent.

Technologies: Technologies are in a growth phase and diffusing. They are well-developed and support the interconnected nature of the systems. These technologies ensure reliable and relatively efficient physical links, facilitating interactions between the different systems.

Institutions: Institutions are being embedded and practices routinised to support interconnected systems. Actors establish standard protocols, processes, and norms that facilitate ongoing collaboration and coordination across systems, ensuring interdependence. Actors from different sectors cooperate to solve a shared goal (see also policy integration (Metz et al., 2020; Tosun & Lang, 2017)). They set up new policies, possibly as part of a cross-system policy mix, that address the characteristics of different systems to allow for a smooth interaction (e.g., in the form of cross-ministerial groups or cross-sectoral strategies). Cross-system policies address broader, common goals.

Actors: Actors interact regularly across systems, engaging in continuous collaboration and exchange. These interactions become more routinised, facilitating cooperation, and reaching shared objectives.

System entanglers: System entanglers focus on expanding and deepening networks. Positioned within or in-between systems, they play a key role in maintaining and enhancing the interconnected systems, further fostering links between systems, and ensuring that the network ties remain strong. System entanglers increasingly act to maintain newly established technologies, institutions, and actors (networks) across systems (see transition work (Löhr et al., 2022)). Tension may rise and system entanglers can take over a buffering function smoothening possible frictions between the different systems. They support the connections being stable. Part of their characteristics might become embedded in the system structure, notably in the established connections, creating a self-supporting cross-sectoral structure. But as the

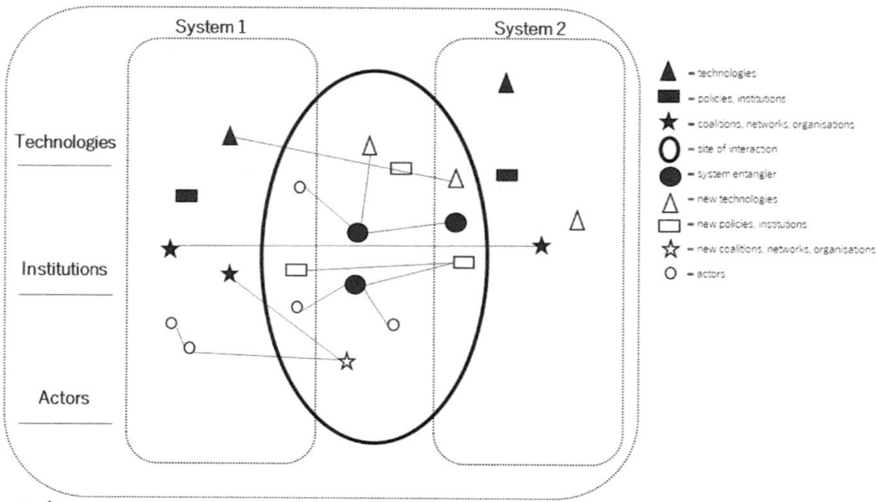


**Fig. 6.3** Interconnected systems

systems are still separate—apart from the overlaps at the site of interaction—system entanglers are needed to adjust if misalignments evolve and reinforce and strengthen well-working connections.

## Integrated Systems

Systems: In this configuration, the parts of the systems that overlapped, i.e., subsystems, are (fully) integrated (see also the integration of regimes—and their disintegration through regime split ups in Raven and Verbong (2007)). Within this subsystem, the former system boundaries (between systems 1 and 2) vanish (become seamless), the site of interaction loses its importance (see Fig. 6.4). The couplings between components are rather tight, which may lead to inflexibilities and increase risks (Dubois & Gadde, 2002; Weick, 1976). In countries with high diffusion rates like Norway, electromobility is an example for an integrated system between the transport and electricity sectors.

Technologies: The new technologies are well-established and fully integrated into the system. They support seamless operations and interactions, enabling the unified system to function efficiently and effectively. The focus is on maintaining and optimizing these mature technologies.

Institutions: Institutions encompass elements from both (prior) systems, are combined into new cohesive ones, stabilized, and deeply embedded within the integrated system. Institutional practices become routinised and standardized. The policy mix (Oltmer & Löhr, 2025; Rogge & Reichardt, 2016), which addresses cross-system issues, contributes to reducing system boundaries.

Actors: Actors interact within the integrated system as the new status quo. Collaboration and coordination are inherent in the actors' roles and activities.

System entanglers: System entanglers focus on preserving and strengthening the integrated system. Their activities and competencies are embedded in the system structure; their role as system entanglers may become (partly) superfluous. Positioned within the integrated system,

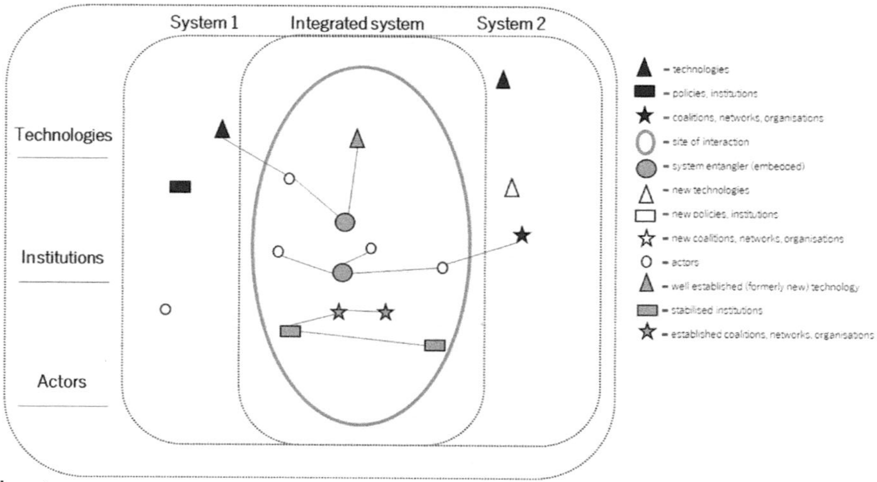


**Fig. 6.4** Integrated systems

they ensure that the integrated nature of the system is maintained. Alternatively, without system entanglers, the status of integrated institutions may deteriorate.

## New System Emergence

Systems: At the site of interaction, a new (sub)system 3 emerges that partly overlaps with the adjacent systems 1 and 2 but develops a largely independent structure driven by new technologies and actors working with them (see Fig. 6.5). The system elements do not originate mainly from the prior existing systems but are newly created and distinct. The new system operates independently but may interact with the original systems and incorporate some of its elements (also see the policy literature on nascent subsystems for system emergence (Ingold et al., 2017), niche emergence in MSI literature (Papachristos et al., 2013), or the formative phase of a nascent TIS (Markard, 2020). System boundaries are permeable but continue to take shape. An example of a new system emerging is the hydrogen policy subsystem which evolves at the intersection of existing (sub)systems (electricity, transport, heating, industry) but is sufficiently different in terms of technologies, regulation, and actors involved (Löhr et al., 2024). Another example is the solar photovoltaic sector that emerged between the semiconductor and power sectors.

Technologies: New technology development drives the emergence of a new, distinct system. This new technology was not part of the existing systems. New technologies, tools, and platforms are being created to support the new system. Some of these will allow connections to the original (and other adjacent) systems.

Institutions: In the new system institutions, policies, and rules are being created which may be influenced by the adjacent systems. These new policies, regulatory frameworks, and standards are distinct from the original systems. At the beginning, there are few institutions, and the institutional setting is unclear. Institutions might emerge as strategies, white papers or (informal) agreements, etc. before they stabilize (and are followed by laws, regulations, standards, etc.).

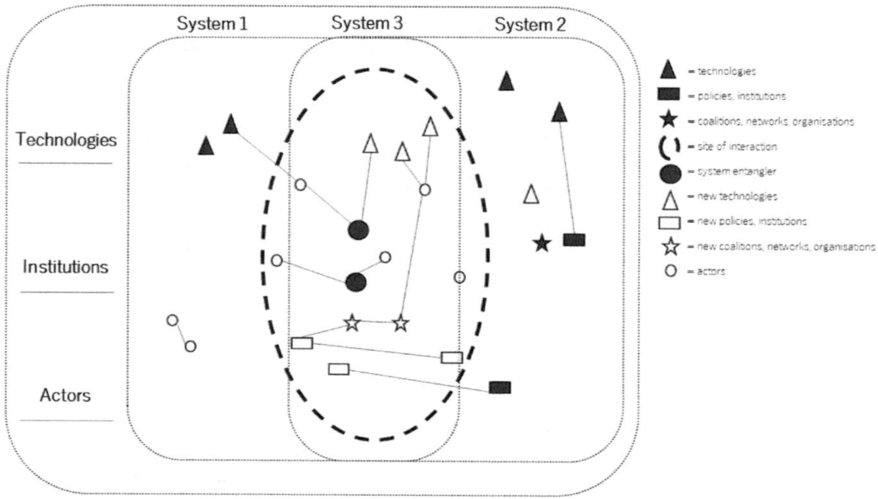


**Fig. 6.5** New system emergence

Actors: Actors from the existing systems and newly emerging actors (e.g., manufacturers, new interest groups) join the emerging system. They develop positions, ties and start to build coalitions (Ingold et al., 2017; Löhr et al., 2024). Actors work on creating new technologies, and institutions, and build ties to the original systems. Some actors may maintain their roles in the original system.

System entanglers: System entanglers may have less of a role in creating the new system but rather in bridging it to existing systems to transfer knowledge. For connections within one system, it does not necessarily need system entanglers as other actors may take this role.

Within the presented system configurations, actors including system entanglers pursue different roles and activities, institutions, and technologies change, influencing system boundaries and the (shape of the) site of interaction. The following section serves to illustrate some of these differences for three system configurations.

## Empirical Illustration

The empirical illustration of this contribution is based on re-assessing the cases in Löhr and Chlebna (2023). The qualitative comparative case study investigates hydrogen development in Germany. It analyses how actors couple the electricity with the mobility, heating, and industry sectors in three green hydrogen pilot projects. In all cases actors install electrolyzers and respective hydrogen technologies (e.g., fuel cell buses) to connect the electricity sector to one of the three mentioned sectors. In the mobility case, fuel cell buses run on hydrogen. In heating, hydrogen is blended into the gas grid. In the industry case, hydrogen shall substitute natural gas in the production of steel. A main difference between the three cases is the implementation progress. While the mobility and heating projects are already implemented, the steel industry case is less advanced. This serves to illustrate different system configurations.

Hydrogen in Germany is a timely and relevant case, considering the ambitions of decarbonising its energy-intensive and industry heavy economy (Belova et al., 2023; Ohlendorf et al., 2023). Hydrogen gained traction as transportable, storable, and climate neutral (if produced from renewable electricity) energy carrier to decarbonise hard-to-decarbonise industries such as steel, chemicals, and shipping (Kern et al., 2023; Öhman et al., 2022). The German hydrogen strategy, published in 2020 (BMWi, 2020), has been updated in 2023 (BMWK, 2023), aiming at a minimum electrolysis capacity of 10 GW until 2030 (Löhr et al., 2024). Numerous pilot projects and state co-funded projects are set up to roll out the hydrogen economy.

The analysis is based on a triangulation of data sources: press releases about the pilot projects, newspaper articles, project websites, and complementary expert interviews (see Table 6.1 in the appendix). The qualitative content analysis was done partly deductively (based on forms of transition work as defined in Löhr et al. (2022) to identify activities of actors in the three dimensions technologies, institutions, and organizations), and partly inductively. A more detailed description of the methods applied can be found in Löhr and Chlebna (2023). I draw on this paper to empirically illustrate three configurations of system interactions in the hydrogen sector.

## Unconnected Systems: Industry Case

In this case, technologies from the two systems are hardly coupled. New technologies connecting the two systems have not been installed yet. In the industry case, this means that neither the electrolyzer nor the wind park (200 turbines) delivering the renewable electricity is built, while steel production is still fuelled by natural gas (technology). The steel company expects change to happen through external factors such as legislation or high energy prices, not through visionary actors. The regulatory frameworks for traditional steel production are in place (institutions). The company mainly engages in sector-based associations such as the steel association. If so, it cooperates within its sector but mainly pursues client-supplier relationships to receive necessary services. In the future, subcontractors will deliver the hydrogen needed, making the steel company a hydrogen bulk buyer as it cannot produce the required amounts itself (I11-industry, B43-industry) (actors). The steel company primarily focuses on its own sector. Cooperation with other steel producers or service providers is limited due to fears of knowledge leakage, which hampers cross-sectoral learning (I-10-industry). The industry case also shows a lack of cross-sectoral competencies: "In my opinion all of this needs to be considered together, and in the world of steel you will have to *develop some competency* for the electricity market" (I-10-industry) (system entangler). In sum, activities mainly address the steel sector. System entanglement is hindered, which has no major consequences in this configuration but would complicate stronger system interactions if that were the objective.

## Emerging Connections Between Systems: Heating Case

In this case, two project partners set up the technologies and tools connecting two systems, here illustrated by the heating case blending hydrogen into the gas grid. The wind park and the gas grid are in place. The two project developers tested and installed the electrolyzer and a superordinate control system connecting the two sectors. "The superordinate control system collects and processes signals from the involved partners and plants, which are needed to determine how (and when) the electrolyser runs" (I-6-heating) (technology). Partners from both sectors cooperate closely and develop shared standards and standardized communication channels (I-6-heating). "That's what makes it such a challenge because there are no standardised solutions, it's still very individual and everyone is trying things out, but also making it open and opening a space for discussion. How can we implement this well together? (…) I had the feeling that although they (project partners) naturally looked at their own systems a lot, they were always open to the system as a whole" (I-6-heating). They form informal institutions and influence formal ones that address both sectors: "(…) we can use it (this project) to have a great effect in the political and social debate, i.e., it is a perfect illustration of what is currently missing in the energy transition and (it) shows what is actually possible and where the regulatory system still needs to improve" (I-6-heating) (institutions). The actors engage not only in associations within their sector but also merge existing associations. Creation and disruption are at play, reducing parallel and superfluous structures and leading to the emergence of a new cross-sectoral association. "We have the renewable energy associations. Mr X introduced the idea to couple it there as well. (…) They thereby created the platform for sector coupling" (I7-heating) (actors).

The actors work closely together, share their knowledge, drive cross-sectoral cooperation, and by implementing the hydrogen sector coupling project foster links between the systems. Having experience in both, the electricity and the heating sector, the project partners bring and apply their cross-sectoral competencies (system entangler). System entanglers drive the emerging connections between the two systems, which is central in this configuration.

## Interconnected Systems: Mobility Case

In this configuration, the connections between systems are more established. The mobility case in which the project developer couples the electricity sector with the mobility sector through electrolyzers, high pressure tanks (to transport the hydrogen), hydrogen refuelling stations, and fuel cell buses and cars, serves as an example. The technologies are set up and working. A common control system monitors the electricity and hydrogen production and consumption in the electricity and mobility sector. Initially, this was done partly manually. It will be automated in future, coordinating several projects, and bringing together supply and demand in an interconnected system (I-2-mobility). Here, upscaling the modular built components and replicating them is key. The successful implementation of the pilot projects stokes demand in other municipalities and leads to upscaling and diffusion. "The rocket is being fired now" illustrates the project developers' ambition for upscaling (B24-mobility). To further connect the sectors, the company engages in developing and selling hydrogen-based vehicles (technology).

The company engages in policymaking at the regional, national, and increasingly also on the European level, pursuing the objective to change existing institutions and to strengthen more integrative sector coupling policies (I-3-mobility) (institutions). Sector coupling is understood as a multi-dimensional process which "needs to take place not just in technological but also in administrative terms" (B24-mobility).

The company implementing this project has experience in both the electricity and the mobility sector. It characterizes itself as "one contact for generation, distribution, and use across all three sectors" and as "sector coupling company" (I-3-mobility). It started off with the idea of "*creating* a hydrogen-ecosystem with green hydrogen. (…) To *instigate a market,* we also buy fuel cell buses ourselves" (I-1-mobility). The company acts as a system entangler which is (still) key in this configuration. It started early on to create the links between the systems in which system boundaries become more permeable. It now focuses on diffusion and strengthening the links (system entangler).

## Discussion and Conclusion

This chapter presented five multi-system configurations and the respective technological, organizational, and institutional couplings within them. Some configurations may build on each other, aligning with evolutionary perspectives such as transition phases (Rotmans et al., 2001) or the technological innovation system (TIS) life cycle (Markard, 2020). Others, such as new system emergence—conceptualized, e.g., in political science as nascent subsystems (Ingold et al., 2017; Löhr et al., 2024) and in transition studies as niche emergence (Papachristos et al., 2013)—may emerge from any of the described configurations.

In the following, I will discuss core concepts of this contribution—the site of interaction, system boundaries, couplings, and system entanglers—and how they relate to prior research which focused on overlaps, couplings, and modes of interaction.

Site of interaction and overlaps: Existing literature has captured this space as 'overlap', (Käsbohrer et al., 2024; Sandén & Hillman, 2011) or 'interface' (Andersen & Geels, 2023; Andersen et al., 2023; Breitschopf et al., 2023) or also 'site of interaction' (Rosenbloom, 2020). As an overlap, this space is mostly depicted as a shared space between two systems at one point in time, often during interaction. This conceptualization extends the understanding of the site of interaction. It captures the changing space between two systems which may be with (3, 4, 5) or without system overlap (1), situated in-between systems (2,3), within the overlapping systems (4) or crossing the system boundaries of both systems (3,5, numbers refer to the five configurations). The site of interaction involves shifting system boundaries (e.g., varying permeability) (see also Käsbohrer et al. (2024)). It is a dynamic space that varies across system configurations and includes, e.g., lacking, emerging, or strong interconnections (couplings).

In this contribution changes at the site of interaction were discussed as occurring across all dimensions simultaneously. However, changes may affect only one or two dimensions, resulting in more complex system configurations. Sandén and Hillman (2011) suggest that 'overlaps' may occur in one dimension while remaining absent in others, yet this has not been integrated into the specific configurations. Further, the 'overlap' between systems may encompass a 'larger' part of one system, resulting in more actors, technologies, and institutions from the dominant system shaping the interaction. In such cases, the interests of actors from the dominant system determine the interaction (Käsbohrer et al., 2024), potentially increasing conflicts and resistance (Andersen et al., 2023). In such situations, system entanglers might need to navigate power dynamics and manage conflicts to ensure effective system interaction. Further, one system could interact with multiple other systems and consequently have several sites of interaction. Future research could develop more configurations to address these scenarios. An open question is whether one or more system entanglers would connect systems when more than two systems are involved.

Couplings: Prior research has analysed activities across system boundaries, e.g., as couplings (Konrad et al., 2008; Markard, 2020; Raven & Verbong, 2007) and identified the 'mode' or nature of system or technology interaction, e.g., as competitive or symbiotic (Raven & Verbong, 2007; Sandén & Hillman, 2011). The modes of interaction could be added to the interactions described here. While functional couplings (resource flows of, e.g., hydrogen and electricity) were less in focus, this contribution analyses structural couplings (technological, organizational, and institutional) and system-level changes and shows how they vary across five system configurations. It further demonstrates how these variations influence the (reconfiguration of the) site of interaction and the system boundaries. By doing so, it advances the concept of the site of interaction as a dynamic space where systems connect and interact.

System entangler: The empirical illustration showed graduations of system interaction in which the role and activities of system entanglers differ. While fostering links between systems and cross-sectoral competencies are key for the emerging connections and the interconnected systems configurations, the former is more characterized by creation and disruption, the latter by maintenance (Löhr & Chlebna, 2023; Löhr et al., 2022). This involves creating cross-sectoral networks, testing and implementing technologies, and formulating standards. In contrast, system entanglers in interconnected systems focus on diffusing technologies, stabilizing cross-sectoral network structures and advancing policy integration. The illustration showed lacking system entanglement in the unconnected system configuration and the potential challenges faced if further system integration, e.g., to achieve an encompassing transition, was the objective. However, it did not address trade-offs for system entanglers who invest significant time and resources in system entanglement at the expense of their activities within their original sector.

This chapter contributes to the MSI literature by conceptualizing multi-system configurations, including changes at the site of interaction, shifting system boundaries, couplings, and the role of system entanglers in each configuration. It adds a conceptual framework for analysing multi-system interactions taking into account different stages and constellations of system interaction. A valuable future contribution would be to apply and validate the conceptualization to the integrated systems and new system emergence configurations.

**Acknowledgements** This chapter builds on work carried out with Camilla Chlebna in the Emmy Noether-project "Regional energy transition as a social process" (REENEA) which was funded by the German Research Foundation and conducted at the University of Oldenburg (grant no. 316848319). I further acknowledge funding from Scene. Scene is a Joint Initiative co-financed by the ETH Board. I kindly thank Bernhard Truffer and Alejandro Núñez Jiménez as

Table 6.1 List of cited data per case

| Abbreviation | Case | Data type |
|---|---|---|
| I-2 | Mobility | Interview with company |
| I-3 | Mobility | Interview with company |
| B24 | Mobility | Press release, media article |
| I-6 | Heating | Interview with company |
| I-7 | Heating | Interview with company |
| I-10 | Industry | Interview with company |
| I-11 | Industry | Interview with company |
| B43 | Industry | Press release, media article |

well as two reviewers for their feedback and valuable insights on earlier versions of this chapter.

## Appendix

## References

Andersson, J., Hellsmark, H., & Johansson, E. (2026). Varieties of disagreement in transformative policy missions: A Q study on the decarbonization of Swedish industry. *Environmental Innovation and Societal Transitions*, 58, 101069.

Andersen, A. D., & Geels, F. W. (2023). Multi-system dynamics and the speed of net-zero transitions: Identifying causal processes related to technologies, actors, and institutions. *Energy Research & Social Science, 102*, Article 103178. https://doi.org/10.1016/j.erss.2023.103178

Andersen, A. D., Geels, F. W., Steen, M., & Bugge, M. M. (2023). Building multi-system nexuses in low-carbon transitions: Conflicts and asymmetric adjustments in Norwegian ferry electrification. *Proceedings of the National Academy of Sciences, 120*(47), Article e2207746120. https://doi.org/10.1073/pnas.2207746120

Andersen, A. D., Steen, M., Mäkitie, T., Hanson, J., Thune, T. M., & Soppe, B. (2020). The role of inter-sectoral dynamics in sustainability transitions: A comment on the transitions research agenda. *Environmental Innovation and Societal Transitions, 34*, 348–351. https://doi.org/10.1016/j.eist.2019.11.009

Belova, A., Quittkat, C., Lehotský, L., Knodt, M., Osička, J., & Kemmerzell, J. (2023). The more the merrier? Actors and ideas in the evolution of German hydrogen policy discourse. *Energy Research & Social Science, 97*, Article 102965. https://doi.org/10.1016/j.erss.2023.102965

Bergek, A et al. (2015). Technological innovation systems in contexts: Conceptualizing contextual structures and interaction dynamics. *Environmental Innovation and Societal Transitions, 16*, 51–64. https://www.sciencedirect.com/science/article/pii/S221042241530006X?casa_token=DNNogDplRQcAAAAA:dLjSVMfyVSTDEURUW4T-SA0ouqbxR-1oji7rx8_7IZfQ35pJi-w6h5bF4H0-y2blFvSOfELuOi0

Binz, C., & Truffer, B. (2017). Global Innovation Systems—A conceptual framework for innovation dynamics in transnational contexts. *Research Policy, 46*(7), 1284–1298. https://doi.org/10.1016/j.respol.2017.05.012

BMWi. (2020). *The national hydrogen strategy*. Berlin.

BMWK. (2023). *Fortschreibung der Nationalen Wasserstoffstrategie: NWS 2023*. Berlin.

Breitschopf, B., Grimm, A., Billerbeck, A., Wydra, S., & Köhler, J. (2023). Towards understanding interactions between socio-technical systems in sustainability transitions. *Energy Research & Social Science, 106*, Article 103323. https://doi.org/10.1016/j.erss.2023.103323

Dubois, A., & Gadde, L.-E. (2002). The construction industry as a loosely coupled system: Implications for productivity and innovation. *Construction Management & Economics, 20*(7), 621–631. https://doi.org/10.1080/01446190210163543

Eisenhardt, K. M. (1989). Building theories from case study research. *Academy of Management Review, 14*(4), 532–550. https://doi.org/10.2307/258557

Fridgen, G., Keller, R., Körner, M.-F., & Schöpf, M. (2020). A holistic view on sector coupling. *Energy Policy, 147*, Article 111913. https://doi.org/10.1016/j.enpol.2020.111913

Geels, F. W. (2004). From sectoral systems of innovation to socio-technical systems: Insights about dynamics and change from sociology and institutional theory. *Research Policy, 33*(6–7), 897–920. https://doi.org/10.1016/j.respol.2004.01.015

Geels, F. W. (2018). Disruption and low-carbon system transformation: Progress and new challenges in socio-technical transitions research and the Multi-Level Perspective. *Energy Research & Social Science*, 37, 224–231.

Hess, D. J. (2019). Cooler coalitions for a warmer planet: A review of political strategies for accelerating energy transitions. *Energy Research & Social Science, 57*, Article 101246. https://doi.org/10.1016/j.erss.2019.101246

IEA. (2021). *Net Zero by 2050: A roadmap for the global energy sector.*

Ingold, K., Fischer, M., & Cairney, P. (2017). Drivers for policy agreement in nascent subsystems: An application of the advocacy coalition framework to fracking policy in Switzerland and the UK. *Policy Studies Journal, 45*(3), 442–463. https://doi.org/10.1111/psj.12173

Kanger, L., Schot, J., Sovacool, B. K., van der Vleuten, E., Ghosh, B., Keller, M., Kivimaa, P., Pahker, A.-K., & Steinmueller, W. E. (2021). Research frontiers for multi-system dynamics and deep transitions. *Environmental Innovation and Societal Transitions, 41*, 52–56. https://doi.org/10.1016/j.eist.2021.10.025

Kern, F., Schmelzle, F., & Hummel, M. (2023). Hydrogen as a panacea for decarbonising everything? Exploring contested hydrogen pathways in Germany. *Environmental Research Letters, 18*(11), Article 114017. https://doi.org/10.1088/1748-9326/acff0e

Kivimaa, P., Boon, W., Hyysalo, S., & Klerkx, L. (2019). Towards a typology of intermediaries in sustainability transitions: A systematic review and a research agenda. *Research Policy, 48*(4), 1062–1075. https://doi.org/10.1016/j.respol.2018.10.006

Klerkx, L., & Leeuwis, C. (2009). Establishment and embedding of innovation brokers at different innovation system levels: Insights from the Dutch agricultural sector. *Technological Forecasting and Social Change, 76*(6), 849–860. https://doi.org/10.1016/j.techfore.2008.10.001

Konrad, K., Truffer, B., & Voß, J. -P. (2008). Multi-regime dynamics in the analysis of sectoral transformation potentials: Evidence from German utility sectors. *Journal of Cleaner Production, 16*(11), 1190–1202. https://doi.org/10.1016/j.jclepro.2007.08.014

Käsbohrer, A., Hansen, T., & Zademach, H.-M. (2024). Multi-system interactions and institutional work: Actor interactions at the interface of residential storage systems and electric vehicles in Germany. *Environmental Innovation and Societal Transitions, 51*, Article 100844. https://doi.org/10.1016/j.eist.2024.100844

Köhler, J., Geels, F. W., Kern, F., Markard, J., Onsongo, E., Wieczorek, A., Alkemade, F., Avelino, F., Bergek, A., & Boons, F. (2019). An agenda for sustainability transitions research: State of the art and future directions. *Environmental Innovation and Societal Transitions, 31*, 1–32. https://doi.org/10.1016/j.eist.2019.01.004

Lawrence Thomas, B., & Suddaby, R. (2006). Institutions and institutional work. *Handbook of Organizations Studies*, 215–254. https://ssrn.com/abstract=3197577

Löhr, M. (2020). *Energietransitionen: Eine Analyse der Phasen und Akteurskoalitionen in Dänemark*. Springer-Verlag.
Löhr, M., & Chlebna, C. (2023). Multi-system interactions in hydrogen-based sector coupling projects: System entanglers as key actors. *Energy Research & Social Science, 105*, Article 103282. https://doi.org/10.1016/j.erss.2023.103282
Löhr, M., Chlebna, C., & Mattes, J. (2022). From institutional work to transition work: Actors creating, maintaining and disrupting transition processes. *Environmental Innovation and Societal Transitions, 42*, 251–267. https://doi.org/10.1016/j.eist.2021.12.005
Löhr, M., Markard, J., & Ohlendorf, N. (2024). (Un) usual advocacy coalitions in a multi-system setting: The case of hydrogen in Germany. *Policy Sciences*, 1–31. https://doi.org/10.1007/s11077-024-09536-7
Markard, J. (2020). The life cycle of technological innovation systems. *Technological Forecasting and Social Change, 153*, Article 119407. https://doi.org/10.1016/j.techfore.2018.07.045
Markard, J., Suter, M., & Ingold, K. (2016). Socio-technical transitions and policy change–Advocacy coalitions in Swiss energy policy. *Environmental Innovation and Societal Transitions, 18*, 215–237. https://doi.org/10.1016/j.eist.2015.05.003
Metz, F., Angst, M., & Fischer, M. (2020). Policy integration: Do laws or actors integrate issues relevant to flood risk management in Switzerland? *Global Environmental Change, 61*, Article 101945. https://doi.org/10.1016/j.gloenvcha.2019.101945
Ohlendorf, N., Löhr, M., & Markard, J. (2023). Actors in multi-sector transitions-discourse analysis on hydrogen in Germany. *Environmental Innovation and Societal Transitions, 47*, Article 100692. https://doi.org/10.1016/j.eist.2023.100692
Öhman, A., Karakaya, E., & Urban, F. (2022). Enabling the transition to a fossil-free steel sector: The conditions for technology transfer for hydrogen-based steelmaking in Europe. *Energy Research & Social Science, 84*, Article 102384. https://doi.org/10.1016/j.erss.2021.102384
Oltmer, M., & Löhr, M. (2025). Tracing actors in policy mixes for transitions: A systematic literature review and insights from policy process theories. *Environmental Innovation and Societal Transitions, 56*, Article 100989. https://doi.org/10.1016/j.eist.2025.100989

Papachristos, G., Sofianos, A., & Adamides, E. (2013). System interactions in socio-technical transitions: Extending the multi-level perspective. *Environmental Innovation and Societal Transitions, 7*, 53–69. https://doi.org/10.1016/j.eist.2013.03.002

Raven, R. (2007). Co-evolution of waste and electricity regimes: Multi-regime dynamics in the Netherlands (1969–2003). *Energy Policy, 35*(4), 2197–2208. https://doi.org/10.1016/j.enpol.2006.07.005

Raven, R., & Verbong, G. (2007). Multi-regime interactions in the dutch energy sector: the case of combined heat and power technologies in the Netherlands 1970–2000. *Technology Analysis & Strategic Management, 19*(4), 491–507. https://doi.org/10.1080/09537320701403441

Rogge, K. S., & Reichardt, K. (2016). Policy mixes for sustainability transitions: An extended concept and framework for analysis. *Research Policy, 45*(8), 1620–1635. https://doi.org/10.1016/j.respol.2016.04.004

Rosenbloom, D. (2020). Engaging with multi-system interactions in sustainability transitions: A comment on the transitions research agenda. *Environmental Innovation and Societal Transitions, 34*, 336–340. https://doi.org/10.1016/j.eist.2019.10.003

Rotmans, J., Kemp, R., & van Asselt, M. (2001). More evolution than revolution: Transition management in public policy. *Foresight, 3*(1), 15–31. https://doi.org/10.1108/14636680110803003

Sandén, B. A., & Hillman, K. M. (2011). A framework for analysis of multi-mode interaction among technologies with examples from the history of alternative transport fuels in Sweden. *Research Policy, 40*(3), 403–414. https://doi.org/10.1016/j.respol.2010.12.005

Schot, J., & Kanger, L. (2018). Deep transitions: Emergence, acceleration, stabilization and directionality. *Research Policy, 47*(6), 1045–1059. https://doi.org/10.1016/j.respol.2018.03.009

Tosun, J., & Lang, A. (2017). Policy integration: Mapping the different concepts. *Policy Studies, 38*(6), 553–570. https://doi.org/10.1080/01442872.2017.1339239

van der Vleuten, E. (2019). Radical change and deep transitions: Lessons from Europe's infrastructure transition 1815–2015. *Environmental Innovation and Societal Transitions, 32*, 22–32. https://doi.org/10.1016/j.eist.2017.12.004

Weick, K. E. (1976). Educational organizations as loosely coupled systems. *Administrative Science Quarterly*, 1–19. https://doi.org/10.2307/2391875

Wesche, J., Löhr, M., Hess, D. J. (2024). Only changing policy? The role of coalitions in sustainability transitions. In J. Wesche, A. Hendriks (Eds.),

*Introduction to sustainability transitions research*. Cambridge University Press. https://doi.org/10.33774/coe-2024-tdwdx-v2

# 7

# Pulling the Brake: Worldwide Institutional Reactions to Uber's Entry

Nikolaos Koutsimpogiorgos, Koen Frenken, Andrea M. Herrmann, and Matthijs B. Punt

## Introduction

Digital platforms have become key intermediaries of economic transactions in many industries worldwide, shaping the way people interact, shop, travel, and socialize (Van Dijck et al., 2018). Amazon, Airbnb, Facebook, and Uber are among prime examples. As digital platforms scale their operations across countries, compliance with government regulations plays a secondary role for platforms as it would slow down their quest to gain market share. Instead, government regulation is generally dealt with by platforms at a later stage, a strategy known as "don't ask

N. Koutsimpogiorgos · K. Frenken (✉) · M. B. Punt
Copernicus Institute of Sustainable Development, Utrecht University, Utrecht, The Netherlands
e-mail: k.frenken@uu.nl

A. M. Herrmann
Nijmegen School of Management, Radboud University, Nijmegen, The Netherlands


M. Steen et al. (eds.), *Sustainability Transitions and Industrial Transformation*,
https://doi.org/10.1007/978-981-95-4893-4_7

permission, ask forgiveness" (Kenney & Zysman, 2016, p. 67). Governments and other stakeholders, on the other hand, tend to increasingly criticize the purposeful neglect of regulations by platforms.

One class of platforms are gig platforms, which connect freelance workers with requesters to carry out a one-time, short-term task such as a taxi ride (e.g., Uber), a delivery (e.g., Deliveroo), or a cleaning job (e.g., Helpling). Gig platforms prompted many regulatory conflicts worldwide (for reasons explained below), but the way such institutional frictions play out, may be quite different. Indeed, as countries have distinct national systems of institutions, an interesting question holds what can explain that digital platforms meet different responses from national authorities in different countries? More specially, our study will explore the question *what national institutions affect the likelihood of negative institutional reactions towards digital platforms?*

Empirically, we focus on Uber, as the most prominent example of a new platform service, also known as 'ride-hailing'. After the introduction of the Uber app, anyone could become a taxi chauffeur even including those without a taxi licence just by downloading the app and accepting ride requests, thus fostering strong competition in taxi markets around the world. Initially, the Uber service held a promise that, with taxis becoming much cheaper, many residents would do without a car of their own, leading to sustainability gains. However, empirical studies suggest that such benefits were not realized (Schaller, 2021; Zhang et al., 2022).

Uber rapidly expanded its service around the world, but encountered various types of regulatory friction (Thelen, 2018; Uzunca et al., 2018). In particular Uber's disregard for taxi licence regulations has resulted in severe protests from licensed taxi drivers and prompted regulatory action in many countries. Furthermore, Uber was seen as disruptive to existing employment regulations as Uber considers drivers to be independent contractors while exercising algorithmic power over them, as well as to national social security systems as independent contractors may lack social security and protection against dismissal by Uber.

While the institutional responses of governments and unions to ride-hailing platforms have been well documented for a large number of countries in case study research (Davis & Sinha, 2021; Heeks et al., 2021; Pelzer et al., 2019; Sanchez Vargaset al., 2022; Seidl, 2022; Thelen,

2018; Uzunca et al., 2018; Weber et al., 2021; Yuana et al., 2020), a global analysis of such responses is still lacking. Our study provides a first global analysis of this kind. For all 81 countries in which Uber entered after it was founded, we analyse what institutional factors can explain the variation in negative institutions reactions that Uber encountered. Our main finding holds that the clashes were most prominent in countries with a well-functioning justice system.

The entry of Uber into the taxi sector can further been understood as a process of convergence of two socio-technical systems—the ICT sector and the mobility sector—fostering a transition. As such, Uber's app can be understood as a disruptive innovation stemming from the tech sector and applied in the mobility sector from a multi-system transition perspective (Andersen & Geels, 2023; Rosenbloom, 2020). Uber's new service prompted 'multi-system conflicts' between digital platform companies adhering to institutions in the digital regime and incumbent business operating under institutions in a specific sectoral regime at hand, here the taxi regime (Andersen et al., 2021). Our study thus also contributes more broadly to the understanding of the role of national institutions in disruptive innovation processes in multi-system transition processes.

## Theoretical Framework

Uber was one of the first, and is today, one of the best-known platforms whose app proved very popular among drivers and clients alike. Uber introduced a digital service that allows drivers, using their own car and without a taxi licence, to find passengers in their vicinity who would pay electronically via the app and leave a rating afterwards. The app substantially lowered the costs for drivers to enter the taxi market, while clients valued the rapid response time, low prices and high quality of service as supported by the user rating system. However, tax offices expressed concerns about whether Uber drivers would declare their income at the tax office, with Uber refusing to share data on personal earnings to the tax authorities. Incumbent taxi companies were equally critical to Uber, as Uber—seen as illegal drivers as they do not adhere to regulations—was able to offer cheaper services, quickly taking away market share from incumbent taxi companies. And at a later stage, legal scholars pointed out that the power that Uber holds over the drivers—in terms of allocating rides, surveilling the quality of service and at times banning drives from the platform after poor performance—may qualify Uber as an employer of drivers, even if the latter formally agreed to work with Uber as an independent contractor and not as an employee (De Stefano & Aloisi, 2018). Many unions and legal scholars argue that platforms should be classified as employers and their workers as employees (represented by unions), giving some voice and control to workers (De Stefano & Aloisi, 2018; Ilsøe et al., 2020).

While Uber initially used essentially the same service—matching unlicensed drivers to clients via its app—around the world, different points of friction and institutional responses occurred in different countries. Davis and Sinha (2021) speak of 'Varieties of Uberisation' between those countries where Uber was introduced, as diverse pre-existing institutional arrangements resulted in diverse responses to this new service. For example, in Nigeria, where prospective drivers have limited access to formal finance or insurance, Uber filled this institutional void by developing its own finance and insurance services. In Indonesia, by contrast, local transportation markets were controlled by driver guilds, and Uber would open local offices to facilitate communication with drivers and leaders. Such striking differences were also found in the early study by Thelen (2018) on Germany, Sweden, and the US. She concludes that Uber's entry triggered the mobilization of different social actors and the formation of different coalitions, resulting in different institutional responses ranging from accommodations to outright bans. Other contributions to the analysis of institutional tensions prompted by taxi gig platforms include studies on China, Colombia, Egypt, India, Indonesia, China, Mexico, Nigeria, Panama, South Africa, The Netherlands, and UK (Davis & Sinha, 2021; Heeks et al., 2021; Pelzer et al., 2019; Sanchez Vargas et al., 2022; Uzunca et al., 2018; Weber et al., 2021; Yuana et al., 2020). Similar institutional tensions have also been studied in other gig platform contexts, including food delivery platforms (such as Deliveroo) and domestic cleaning platforms (such as Helpling) (Ilsøe et al., 2020; Koutsimpogiorgos et al., 2023; Tassinari & Maccarrone, 2020).

What these case studies on Uber have in common is the finding that negative institutional reactions to gig platforms expose a lack of legitimacy among their various stakeholders. Critically though, normative evaluations on legitimacy do not happen automatically or on their own. Rather, they are generated by social institutions (e.g., court of law, public media) that assess and evaluate other actors, resulting in normative judgements over their broader legitimacy (Bitektine & Haack, 2015). Thus, to understand how a specific firm creates (il-)legitimacy across countries, we should examine its interaction with the institutions that define its environment, including the prevailing national institutions.

More specifically, one can distinguish between three types of national institutions that may affect the legitimization of gig platforms (Punt et al., 2023): institutions relating to the rule of law, institutions supporting innovation, and institutions relating to labour. Importantly, previous studies have investigated which of these institutions affect the entry time of a platform in a country, assuming that platforms prefer first entering countries with favourable institutions to avoid negative institutional responses (Kim & Suh, 2021; Punt et al., 2023). In the present study, by contrast, we look at the post-entry effects of these institutions by investigating the actual institutional responses after the platform enters a country.

## Rule of Law

Theories on Multinational Corporations (MNCs) often emphasize that the rule of law—understood as enforcement of contracts, property rights, and provision of security (Haggard et al., 2008)—is important for foreign companies to operate properly. Institutions strengthening the rule of law can affect both MNCs decisions to enter a particular country and its post-entry operations. The effect on entry is best reflected in the location choices that MNCs make regarding future expansion. When selecting countries for expansion, MNCs prefer to avoid countries characterized by weak political institutions, as institutional volatility increases the risk of failure and could raise investment costs (García-Canal & Guillén, 2008).

The rule of law is particularly important as it can substantially hamper entrepreneurship whenever firms struggle to enforce contracts or secure patents (Contractor et al., 2020). Furthermore, corruption can be a significant impediment to business as firms have to deal with unpredictable costs and burdens (Bahoo et al., 2020). Interestingly, though, the opposite may be the case for firms whose business model is built on the disruption of traditional industries. As traditional firms are already embedded in national institutions, disrupting firms actively need to find their place within the existing regulatory framework, which can limit their growth prospects.

Considering that regulations are always embedded within a broader institutional framework, and given that their enforcement is conditioned upon a robust rule of law, we assume that disrupting gig platforms, like Uber, will face more institutional friction and reactions in countries with a strong rule of law. Furthermore, this provides more, and more efficient, legal avenues for other societal actors to turn against the platform. We thus expect that:

**Hypothesis 1:** Gig platforms are more likely to face negative reactions in countries with a stronger rule of law.

## Openness to Innovation

Another set of institutions that may affect the institutional responses to platforms' disruptive operations is the degree of an economy's openness to innovation and the quality of market regulation. The existence of innovation-friendly attitudes in the society, as well as pro-innovation institutions and policies, signal a more accommodating environment for innovative businesses in general and platforms in particular (Punt et al., 2023). And, reversely, the existence of barriers to entry and restrictions on economic activity act as a disincentive for MNCs (Djankov et al., 2006). Even though platforms would provoke different conflicts in different countries, the reference to gig platforms as innovative was common across various jurisdictions (Thelen, 2018). In fact, part of Uber's communication strategy was using innovative aspects of its business model as an argument in its attempt to create legitimacy (Pelzer et al., 2019). We thus assume that the existence of pro-innovation institutions will help gig platforms to minimize regulatory friction. Accordingly, we expect that:

**Hypothesis 2:** Gig platforms are more likely to face negative reactions in countries that are less open to innovation.

## Labour Protection

Labour market institutions form the third set of institutions which can stimulate negative institutional reactions towards gig platforms. The issue of employment classification of gig workers has become a core issue in the debate surrounding gig platforms—and a recurring point of conflict with formal institutions (Heeks et al., 2021; Koutsimpogiorgos et al., 2020). Particularly in the case of ride-hailing and food delivery, platforms exercise various forms of control over drivers, for example regarding the allocation of rides, price setting and the possibility of banning workers from the app (Heeks et al., 2021). This is the reason why some stakeholders argued that platform workers should not be classified as independent contractors but as employees of the gig platform (De Stefano & Aloisi, 2018; Frenken & Fuenfschilling, 2020; Ilsøe et al., 2020). Trade unions, in particular, have been vocal in their opposition to gig platforms, which they regard as subversive to established employment rights (Tassinari & Maccarrone, 2020). We, therefore, expect that gig platforms will experience more negative institutional reactions in countries with stricter labour market regulations, more social protection and higher unionization rates:

**Hypothesis 3:** Gig platforms are more likely to face negative reactions in countries with more stringent labour market regulations.

### Informality

Gig platforms tend to be successful in sectors where (semi-)informal labour is prevalent, including the taxi, delivery, cleaning, and handyman sector. Research shows that, in countries with low degrees of institutionalization of the labour market, platforms can bridge the gap by introducing formal rules, training, and some degree of security (Frey, 2020; Surie, 2017). In the case of Egypt, for example, Uber provided a new source of employment for young drivers, while it improved transportation safety for women by enforcing rules against sexual harassment (Uzunca et al., 2018). Those actions increased the platform's legitimacy among the public, thereby stifling opposition from entrenched interests. By providing operational assistance and reliable infrastructures, platforms actively build trust with gig workers and clients, effectively substituting formal institutions (Heeks et al., 2021; Jack, 2020; Weber et al., 2021). Consequently, one can assume that gig platforms may face less resistance in places with high degrees of informality in employment as disruption is more limited. We thus expect that:

**Hypothesis 4:** Gig platforms are more likely to face negative reactions in countries with lower degrees of labour informality.

## Data and Methods

### Dependent Variable

Our study focuses on Uber as one of the largest gig platforms operating in five continents. Uber was launched in 2012 in the US as an app to replace traditional ways of booking a taxi through call centres or street hailing. In its most common form, named UberPop across Europe and UberX in most other countries, anyone could work as a driver via the Uber app with their private-owned vehicle. Drivers were operating as independent contractors, tied with the company through a 'partnership' agreement. Drivers were free to decide when to work and for how many hours. Prices, however, were set by the Uber app, including a fee for the use of the app.

While Uber's introduction in new countries follows a pattern of city-by-city expansion, case studies showed that institutional reactions predominantly came from national actors, such as regulatory agencies, unions, and national politicians (Davis & Sinha, 2021; Pelzer et al., 2019; Thelen, 2018; Uzunca et al., 2018). We, therefore, took the country-level as the level of analysis, sampling all those countries where Uber has introduced its controversial UberX service with unlicensed drivers.

We used two different dependent variables to measure the likelihood of negative institutional reactions, namely volume and time-to-event. First, we counted the volume of negative institutional reactions as the sum of all negative institutional reactions for each country within the period under examination. If negative institutional reactions are more likely to occur, the count of such events will be higher depending on the time that Uber is active in a country. Second, for each country, we counted the time (in months) from UberX's introduction until the first negative institutional reaction. In this case, a higher likelihood means that the first negative reaction will occur sooner after Uber has entered a country.

Our data collection builds on Punt et al. (2023), who documented the global expansion of UberX from the US to other countries, by collecting data on countries where Uber entered up to February 2017.[1] Using their data as the basis, and updating it to include another nine countries where Uber entered after February 2017, resulted in a sample of 81 countries across five continents.[2]

We then used a set of keywords in Google search to identify all institutional reactions related to Uber's presence in these 81 countries. For all countries, the search period for which information on negative institutional reactions was collected is the period between the day of Uber's

[1] Puerto Rico and Andorra were excluded because data on independent variables were lacking. Further note that the US were not included as we analyse the entry of Uber in host countries, thus excluding the home country.

[2] The countries not covered by Punt et al. (2023) are Cote d'Ivoire, Ecuador, El Salvador, Finland, Honduras, Jamaica, Paraguay, and Andorra.

market entry in a country and the day we executed the search queries (25 November 2021).

Our search strategy includes terms for both country and city names where Uber entered, taken from Punt et al. (2023). More precisely, we used four search strings: {'Uber banned in' AND [country-name]}, {'Uber banned in' AND [city-name]}, {'Uber' AND '(labor OR labour)' AND [country-name]}, and {'Uber' AND '(labor OR labour)' AND [city-name]}. For every set of results, the first five pages were searched, resulting in 50 results per hit. Each of these results was checked for relevance before being included in our data. Negative reactions are those that (potentially) hinder Uber's operations, such as restrictive regulations, police prosecution of Uber drivers, and court cases lost by Uber. Further research was done in cases where a reaction was coded within another article, using tailor-made keywords. We explicitly excluded the (many) reactions by trade unions as these are beyond the scope of our research unless their reaction was combined with another institutional reaction, for example, when the union filed a lawsuit. Overall, we identified 264 negative institutional reactions to Uber in 71 of the 81 countries on our list. All coded reactions were then grouped into six categories:

- Ban: Every administrative action that results or intends to result in the halt of Uber's operations.
- Police Enforcement: The use of police enforcement measures, such as fines and impounding of vehicles, often targeting Uber drivers.
- Administrative Enforcement: The use of administrative measures aiming at the enforcement of regulations on Uber.
- Court Case: Every court case/lawsuit ruling with reference to Uber's activities.
- Uber Regulation: New regulations/laws or amendments to existing ones which affect Uber.
- Denial of Permit: The denial of Uber to operate legally within a jurisdiction.

## Independent Variables

Data were taken from the Institutional Profiles Database (IPD) 2016,[3] which contains 130 indicators of institutional characteristics for 143 countries. The IPD 2016 dataset is well suited for our research as it provides information on institutions that coincides with the time of Uber's entry in many countries. Given that national institutions hardly change in the short run (North, 2005), the IPD indicators are suitable to measure the independent variables. The dataset is composed of country-level indicators, measured on a continuous scale from 0 to 4.

The first set of three indicators focuses on political institutions, using those which are connected to the rule of law, namely, "function of the justice system", "level of corruption", and "influence of economic stakeholders". These indicators are relevant to our study as Uber's entry almost universally raises several legal and regulatory questions which are usually resolved through legal avenues. For the first indicator, higher scores indicate a well-functioning justice system, while the latter two are coded in reverse: with higher scores indicating lower levels of corruption and less influence of economic stakeholders on the formation of economic policy. Regarding the second indicator, lower levels of corruption reflect a stronger rule of law. The third indicator measures the lobby influence of various stakeholders, such as foreign companies and sectoral groups in the formation of economic policy within a state. Lower levels of lobbying result in a stronger rule of law.

To test the hypothesis on the effect of openness to innovation, we used a second set of three indicators from IPD, namely, "public support for innovation", "competition regulation", and "barriers to market entry". For the first two, higher values indicate increased state support for innovation and the existence of efficient market regulation. The latter indicator is reversely coded, with higher values indicating a lack of barriers for new competitors to enter markets for goods and services. Again, the choice of indicators for formal economic institutions reflects the specific issues surrounding Uber's entry into a new market. With Uber framing itself as an 'innovation company', which attempts to draw

[3] http://www.cepii.fr/institutions/EN/ipd.asp.

legitimacy from 'breaking up monopolies' and challenging established competition law, all three indicators are essential to this study.

Finally, we used a set of four indicators capturing employment regulations to explore the implications of labour institutions on Uber's entry Uber. These indicators include "trade unions freedoms", "compliance with employment law (formal sector)", "employment contract protection", and "effectiveness of social dialogue". For all four indicators, higher values indicate a stronger presence of the respective institutions. More specifically, the first indicator measures the ability of trade unions to freely exercise their operations and rights; the second variable measures the degree of respect for minimum wages in the formal sector; the third indicator refers to the share of permanent contracts and the protection against dismissal; and the fourth variable refers to the degree of effectiveness of the social dialogue among companies, industries, and countries.

The final indicator taken from IPD is the "significance of informal work". It is reversely coded, with lower values expressing a higher degree of informal labour. We used this variable to test Hypothesis 4

## Control Variables

We also included several control variables. The first is whether "English" is an official language spoken in the respective country. We thereby controlled for possible bias in the data collection of our dependent variable, given that our Google search only used keywords in English to identify institutional reactions to Uber. As a result, we may have omitted cases expressed in other languages and there is a higher chance including cases from countries where English is an official language. To control for this effect, we created a dummy language variable using data from the CIA Factbook.[4]

Second, we computed the variable "Time active" as the number of months that Uber was active in a country. This variable controls for the fact that longer presence would increase the probability of institutional

[4] https://www.cia.gov/the-world-factbook/field/languages/

reactions taking place. We computed the variable by taking the number of months between Uber's entry date and November 2021, the end of our study. For countries where Uber exited before this date, we take the months between Uber's entry and exit. Exit dates were identified using a similar process to that of the collection of the dependent variable, namely Google search with keywords {'Uber' AND 'exit' AND 'country name'}.

Third, we created a dummy variable, "Competing platform", to account for the presence of competing platforms before Uber entered a country. The presence of other platforms could affect the number of institutional reactions, as competing platforms may have already raised similar regulatory issues before Uber entered. That was mostly the case in non-Western countries where local players managed to establish a foothold before Uber. To identify the presence of competing platforms, we read the news articles from our dataset to locate instances that included Uber and another competing platform in the same text. Furthermore, we drew this information from general articles documenting competing platforms for Uber.[5]

Fourth, we added a control variable reporting each country's "GDP per capita". We computed this variable by using data from the CIA Factbook[iv] and log-transformed it.

Fifth, we used a final control variable "Regional spillovers" to account for the spillover effects of negative reactions within a continent, resulting from policy diffusion between countries sharing political, social, and cultural similarities (Bennett, 1991). Especially when faced with novel challenges, such as the gig economy, regulators may rely on copying best practice solutions that other countries have already enforced. The regional spillover variable measures the cumulative number of negative reactions that have occurred in countries within the same continent at time *t*.

[5] https://www.businessofapps.com/data/taxi-app-market/, https://www.marketingtutor.net/uber-competitors/, https://www.bstrategyinsights.com/top-10-uber-competitors-and-alternatives/, https://asia.nikkei.com/Business/Companies/Uber-s-Indonesia-exit-sets-off-driver-hunt-by-Grab-and-Go-Jek

### Regression Models

To test our hypotheses, we performed Negative Binomial (NegBin) regression to explain the count of negative reactions, and Cox regressions to explain the time to the first negative reaction occurring, both in R. We used a NegBin regression to account for the nature of the dependent variable which shows overdispersion with the variance exceeding the mean. Cox models are suitable for studying the impact of time-varying covariates on the risk of a specific event occurring, in this case, the first instance of a negative reaction to Uber in a country. For our analyses, Cox regressions are further useful as they account for the right-censored cases in our data, namely those countries which exhibited no reaction to Uber's presence during the timeframe covered, but might still do so later. Time is measured in months. We use extended Cox models to include a time-varying variable ("Regional spillover") in our analysis, avoiding violation of the proportional hazards assumption of a regular Cox model.[6] This time-varying variable is included in the extended Cox model, but cannot be included in the NegBin model.

## Results

### Institutional Reactions

Our research revealed a total of 250 negative institutional reactions to Uber's introduction. The average number of negative reactions per country is 3.4, the median is 3, whereas the maximum is 26 for the case of Canada.

In Fig. 7.1, we show the number of reactions of different types. In our dataset, the most prevalent type was *Court Cases* with 57 observed instances. Uber faced a wide range of legal complaints across different countries, which broadly reflect the range of regulatory questions created

---

[6] In order to estimate covariates associated with the risk of negative reaction, extended Cox models use exponential hazard functions that represent the risk that, if at time $t$ a country has not experienced any reaction, this country will experience one at some stage later on.

by the introduction of the platform. They were launched either by institutional actors in attempts to punish the platform for what they considered violations of existing laws, or by individuals in an effort to force the company to conform to the legal framework and/or seek damages suffered from Uber's operations.

The second most common reaction to Uber was the introduction of new *Uber Regulations* by national authorities, targeting the platform and other taxi apps. Overall, we recorded 53 new regulations that negatively impacted the platform's operations by prohibiting parts of its original business model and thus increasing operating costs. Examples include installing taxi metres in Uber cabs, mandatory licensing of drivers, and the obligation to return to base following every trip. In these ways, regulators attempt to create a level playing field between Uber and the incumbent taxi industry, which almost universally protested against what they saw as unfair competition from a multinational circumventing existing taxi regulations.

*Administrative Enforcement* constitutes a third category of reactions, which appeared 44 times in our database. More specifically, authorities took various measures to contain the platform, for example by issuing formal orders to comply, imposing fines on Uber, issuing rulings against

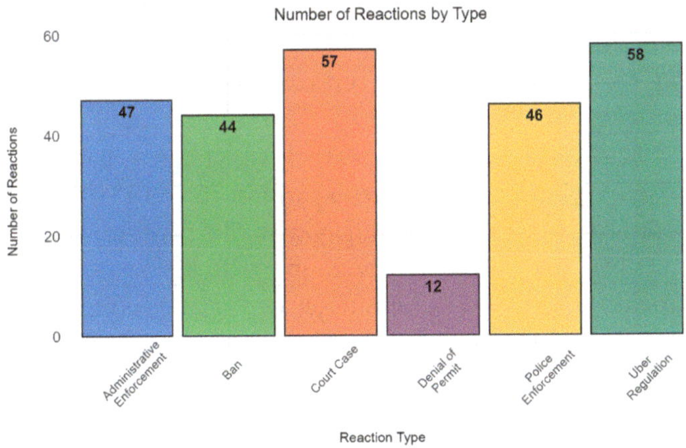


**Fig. 7.1** Typology of institutional reactions

the legality of its activities, or even raiding the company offices as part of an investigation.

A fourth type of reaction constitutes *Police Enforcement*, which occurred in 43 instances. These reactions were primarily based on Uber's characterization as an illegal transportation company, leading to police measures curbing Uber's activities on the road. Subsequently, Uber drivers became the prime targets, with the police issuing fines for illegal work, confiscation of vehicles and occasionally further criminal charges against drivers.

As a fifth category, we identified direct *Bans* of Uber's operations with 42 instances in total. Bans are issued by local, regional, or national administrations or transportation authorities. They aim to impose an immediate stop to Uber's activities within their respective jurisdiction. As Uber rarely asked for formal permission or approval before entering a new jurisdiction, several authorities responded by issuing cease-and-desist orders. Interestingly though, a direct ban rarely had the anticipated effect as Uber either challenged those measures judicially or ignored them for a prolonged period of time.

As the last and smallest category of institutional reactions, we observed the *Denial of Permit* to operate, of which we counted 11 instances. Most of these cases occurred in the UK, with the most prominent being the denial of the London transportation authority to grant a licence to Uber in 2017 and 2018.

Turning to the temporality of reactions, Fig. 7.2 juxtaposes the number of countries that experienced the introduction of Uber with the number of negative reactions per year. This highlights that the platform went through a phase of aggressive expansion in 2014, when it entered 39 new countries. In the following years, the platform continued its expansion, albeit at a much lower rate, up to 2018. The number of negative reactions closely follows Uber's trajectory of international expansion, with a short time-lag.

To further elaborate on the trajectories of different types of reactions, we plot the number of negative reactions per year for each category (Fig. 7.3). Direct bans of Uber's operations and police enforcement stand out as the most common initial reactions to Uber's entry. In later years, administrative enforcement and new regulations become increasingly

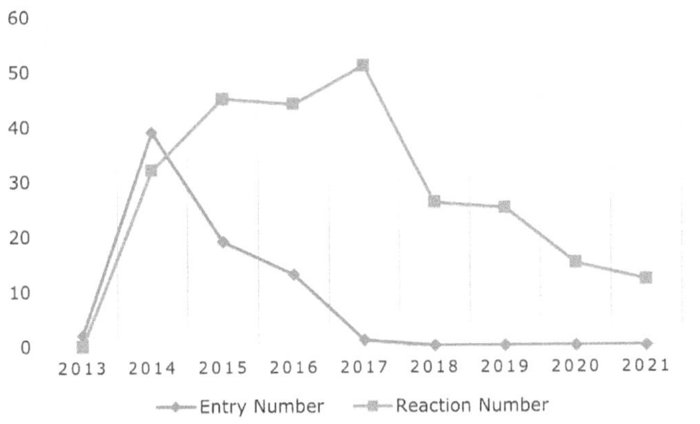


**Fig. 7.2** Number of Uber entries and negative reactions per year

important, indicating the heightened engagement of formal institutions with the platform.[7]

## NegBin Regression

Table 7.1 shows the descriptive statistics for the variables included in the regression models.[8] Figure 7.4 presents the dependent variable on a global map.

Table 7.2 presents five models to test our hypotheses on the effect of institutions on the negative reactions to Uber. Our first model includes only the control variables, which all show the expected signs. English as an official language, longer Uber activity and a country's GDP per capita all have a positive effect on the number of institutional reactions.

---

[7] Looking into the reactions, we find few reactions directly connected to the issue of employment status, with only a slight increase over time. This suggests that the primary concern for most stakeholders after Uber entered a country related to the illegality of the UberX service soliciting unlicensed drivers. The issue of whether drivers should be regarded as employees of Uber, providing them with minimum wage and social protection, became a more substantial concern only in later years.

[8] Correlation table is available upon request by email.

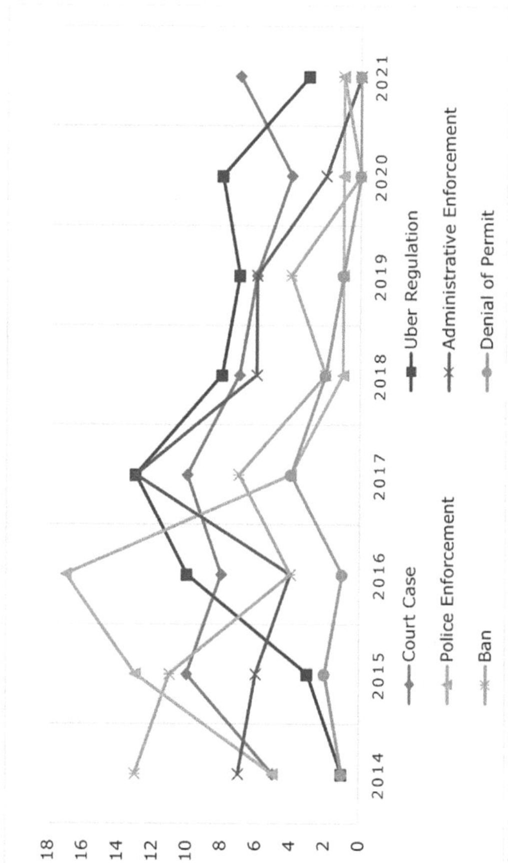


**Fig. 7.3** Number of institutional reactions per year per type

**Table 7.1** Descriptive statistics

| | N | Min | Max | Mean | Std. Dev |
|---|---|---|---|---|---|
| (1) Negative institutional reactions | 81 | 0 | 26 | 3.26 | 3.96 |
| (2) English official language (yes/no) | 81 | 0 | 1 | 0.28 | |
| (3) Competing platform (yes/no) | 81 | 0 | 1 | 0.42 | |
| (4) Time active | 81 | 5 | 100 | 66.52 | 24.55 |
| (5) GDP per capita (in 10,000 s) | 81 | 0.08 | 8.61 | 1.86 | 2.14 |
| (6) Level of corruption (lack of) | 79 | 0 | 4 | 1.77 | 1.24 |
| (7) Functioning of the justice system | 81 | 1 | 4 | 2.59 | 0.77 |
| (8) Influence of economic stakeholders | 81 | 0 | 3 | 1.52 | 0.64 |
| (9) Public support for innovation | 81 | 0 | 4 | 1.88 | 0.99 |
| (10) Barriers to market entry (lack of) | 81 | 0 | 4 | 1.68 | 0.94 |
| (11) Competition regulation | 79 | 0 | 4 | 2.34 | 0.86 |
| (12) Trade union freedoms | 77 | 0 | 4 | 2.92 | 0.99 |
| (13) Compliance with employment law | 75 | 1 | 4 | 3.18 | 0.77 |
| (14) Employment contract protection | 81 | 0.33 | 4 | 2.55 | 0.76 |
| (15) Effectiveness of social dialogue | 81 | 0 | 4 | 1.17 | 1.10 |
| (16) Significance of informal work (lack of) | 81 | 0 | 4 | 2.17 | 1.07 |

The existence of a competing platform shows the expected negative sign without being statistically significant.

The subsequent four models include the institutional variables used to test our four hypotheses. Model 2 includes all the variables capturing institutions relevant to the rule of law. Among them, the functioning of the justice system presents a strong and significant—and as expected positive—effect, with a positive coefficient of +2.206. This means that for each one-unit increase in the functioning of the justice system variable, the expected number of negative institutional reactions increases by a factor of 9.08 ($e^{2.206}$). The corruption variable is also significant, but, unexpectedly, shows a negative coefficient of –0.735. This indicates that countries with lower levels of corruption (i.e., stronger rule of law) display fewer negative reactions towards Uber. This runs counter to Hypothesis 1. Finally, the influence of economic stakeholders is not statistically significant. Overall, one variable supports and one variable contradicts the hypothesis, showing inconsistent results on the relationship between a strong rule of law and the volume of negative reactions to Uber.

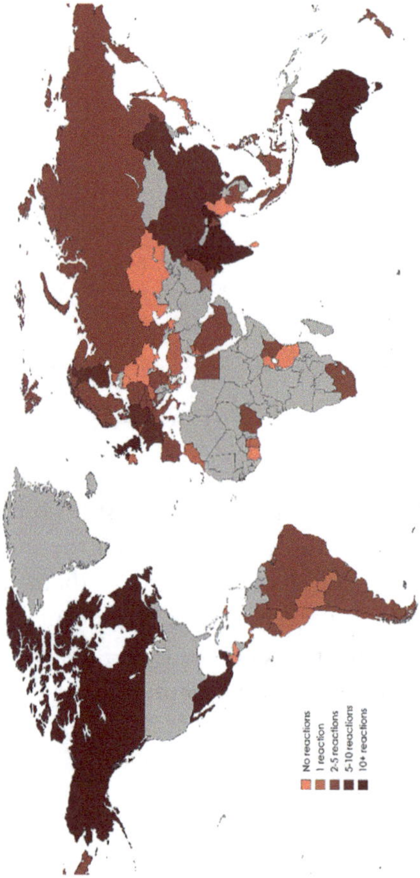


**Fig. 7.4** Global distributions of all negative institutional reactions to Uber

**Table 7.2** NegBin regression model predicting the total number of negative institutional reactions

| Model | (1) | (2) | (3) | (4) | (5) |
|---|---|---|---|---|---|
| English | **0.414**** | **0.423**** | **0.340*** | **0.461**** | **0.406**** |
| | (0.192) | (0.198) | (0.198) | (0.208) | (0.193) |
| Time active (log) | **0.542***** | **0.429**** | **0.479*** | **0.556***** | **0.532***** |
| | (0.180) | (0.175) | (0.197) | (0.195) | (0.185) |
| Competing platform | −0.165 | −0.043 | −0.089 | 0.069 | −0.134 |
| | (0.185) | (0.204) | (0.216) | (0.216) | (0.207) |
| GDP per capita (log) | **0.270***** | **0.262***** | **0.251***** | **0.296***** | **0.266***** |
| | (0.044) | (0.046) | (0.048) | (0.047) | (0.045) |
| Level of corruption (lack of) | | **-0.735**** | | | |
| | | (0.302) | | | |
| Functioning of the justice system | | **2.206***** | | | |
| | | (0.710) | | | |
| Influence of economic stakeholders | | 0.069 | | | |
| | | (0.342) | | | |
| Public support for innovation | | | 0.047 | | |
| | | | (0.266) | | |
| Barriers to market entry (lack of) | | | −0.194 | | |
| | | | (0.266) | | |
| Competition regulation | | | 0.576 | | |
| | | | (0.426) | | |
| Trade union freedoms | | | | 0.247 | |
| | | | | (0.475) | |
| Compliance with employment law | | | | 0.798 | |
| | | | | (0.637) | |

(continued)

**Table 7.2** (continued)

| Model | (1) | (2) | (3) | (4) | (5) |
|---|---|---|---|---|---|
| Employment contract protection | | | | −0.446 | |
| | | | | (0.461) | |
| Effectiveness of social dialogue | | | | −0.252 | |
| | | | | (0.332) | |
| Significance of informal work (lack of) | | | | | 0.061 |
| | | | | | (0.200) |
| Constant | -0.097 | **-1.931*** | −0.469 | −0.752 | −0.125 |
| | (0.795) | (0.995) | (0.847) | (1.002) | (0.795) |
| Observations | 81 | 79 | 79 | 72 | 81 |
| Log Likelihood | −164.149 | −155.474 | −160.958 | −144.371 | −164.102 |
| Likelihood-ratio test | 45.91*** | 17.350*** | 6.381* | 85.473*** | 0.093 |
| *Compared to* | Null model | Model 1 | Model 1 | Null Model | Model 1 |

*Note* *$p < 0.1$; **$p < 0.05$; ***$p < 0.01$

Model 3 shows that all three variables capturing pro-innovation institutions do not have a significant effect on the dependent variable. Therefore, we reject Hypothesis 2 on the relationship between limited openness to innovation and opposition to Uber.

In Model 4, we include the variables regarding labour protection, testing the hypothesis that more stringent labour protection leads to more negative institutional reactions. None of the four labour-related variables yields a statistically significant effect. Neither trade union freedom and the effectiveness of social dialogue, nor employment contract protection and compliance with employment law are effective predictors of the volume of reactions to Uber. Hypothesis 3 is therefore rejected. This insignificant result can be linked to the descriptive analysis of institutional reactions, where we found that negative reactions related to the issue of employment status are relatively few during the whole observation period, indicating that the fundamental issue of whether drivers should be regarded as employees of Uber, became a substantial concern only recently.

Model 5 tests the hypothesis that high levels of informal work might lead to less negative institutional reactions towards Uber. The variable regarding informal work yields a coefficient that is not significant. This suggests that, contrary to the findings of previous qualitative studies, Uber was not perceived as more legitimate in countries where informal labour is more prominent compared to countries where informal labour is less prominent. Accordingly, we reject Hypothesis 4.

Due to the limited number of cases (81 countries), we avoid including all 14 variables simultaneously in a single model to prevent overfitting, given the insufficient number of observations for comprehensive analysis.

## Time-to-Event Analysis

We complement our findings by performing a Cox regression. Our dependent variable in this case is the first instance of a negative institutional reaction in each country. We use the same independent variables as in the negative binomial regression except for the variable "Time active" as the time dimension is considered in the dependent variable in a time-to-event analysis. We do include a time-varying variable to capture institutional spillovers at the continent level ("Regional spillovers"). Table 7.3 shows the results.

Model 6 includes only the control variables. Regional spillovers and GDP per capita show significant positive effects, with only the latter remaining significant in subsequent models. Model 7 indicates results similar to those of the negative binomial regression. For the variable functioning of the justice system, the model produces a significant hazard ratio of +1.610, meaning that for each one-unit increase in the justice system variable, the hazard of a negative institutional reaction occurring increases by 61.0%, holding all other variables constant. This implies that in countries with a better-functioning justice system, negative institutional reactions happen more quickly after Uber's entry, partially confirming Hypothesis 1. In Model 8 on the effect of pro-innovation institutions, we find a significant hazard ratio of +1.396 for competition regulation, indicating that a one-unit increase in this variable is associated with a 39.6% increase in the hazard of negative institutional reactions. This means that countries with more efficient competition regulation are more likely to respond quickly to Uber's entry, finding some support for Hypothesis 2. Model 9 on labour market institutions and Model 10 on the informal sector show no significant effect. As in the NegBin regression analysis, the Cox regression analysis provides no evidence in support of Hypotheses 3 and 4.[9]

## Robustness Checks

We carried out a number of robustness checks. The results of these checks are available upon request. As a first robustness check, we include a

---

[9] We also conducted NegBin regressions that distinguish between severe reactions (the sum of national bans and court cases) and milder reactions (the sum of administrative enforcement and regulatory changes affecting Uber). The regression results show no differences in the support for our hypotheses across various types of institutional reactions, with the functioning of the justice system being the only variable to show a significant effect. Interestingly, when comparing the strength of the coefficients, this variable shows a higher coefficient for severe negative institutional reactions (1.917) compared to milder reactions (1.527), consistent with the theory. Results are available upon request by email.

**Table 7.3** Regression model predicting time to first negative institutional reaction

| | (6) | (7) | (8) | (9) | (10) |
|---|---|---|---|---|---|
| English | 0.848 | 0.650 | 0.714 | 0.773 | 0.815 |
| | (0.281) | (0.323) | (0.298) | (0.327) | (0.295) |
| Competing platform | 1.100 | 1.402 | 1.250 | 1.495 | 1.162 |
| | (0.271) | (0.304) | (0.304) | (0.313) | (0.296) |
| Regional spillovers | **1.039*** | 1.033 | 1.033 | 1.027 | 1.037 |
| | (0.023) | (0.025) | (0.026) | (0.027) | (0.024) |
| GDP per capita (log) | **1.318***** | **1.296***** | **1.303***** | **1.302***** | **1.306***** |
| | (0.063) | (0.066) | (0.072) | (0.068) | (0.066) |
| Level of corruption (lack of) | | 0.969 | | | |
| | | (0.172) | | | |
| Functioning of the justice system | | **1.610*** | | | |
| | | (0.283) | | | |
| Influence of economic stakeholders | | 0.948 | | | |
| | | (0.229) | | | |
| Public support for innovation | | | 0.956 | | |
| | | | (0.136) | | |
| Barriers to market entry (lack of) | | | 0.919 | | |
| | | | (0.165) | | |
| Competition regulation | | | **1.396*** | | |
| | | | (0.189) | | |
| Trade union freedoms | | | | 1.119 | |
| | | | | (0.231) | |
| Compliance with employment law | | | | 1.188 | |

(continued)

**Table 7.3** (continued)

| | | | | | |
|---|---|---|---|---|---|
| | | | | (0.261) | |
| Employment contract protection | | | | 1.181 | |
| | | | | (0.264) | |
| Effectiveness of social dialogue | | | | 0.935 | |
| | | | | (0.181) | |
| Significance of informal work (lack of) | | | | | 1.060 |
| | | | | | (0.130) |
| Observations | 1,431 | 1,419 | 1,370 | 1,234 | 1.431 |
| Events | 70 | 68 | 69 | 62 | 70 |
| Log Likelihood | −243.074 | −231.472 | −236.193 | −205.822 | −242.975 |
| Likelihood-ratio test | 20.196*** | 23.204*** | 13.763*** | 94.700*** | 0.197 |
| *Compared to* | Null model | Model 1 | Model 1 | Null model | Model 1 |

*Note* $*p < 0.1$; $**p < 0.05$; $***p < 0.01$

control variable for common law countries which often have a higher propensity for legal disputes viz. negative institutional reactions. If we run the regression including the common law dummy, this variable shows a positive and significant sign, while size and significance of other variables remain robust. As a second robustness check, we introduce a population variable to mitigate potential bias from relying solely on English keywords, as one may expect that smaller countries receive less coverage in the international press. The regression results indicate that, although the population variable consistently shows a significant positive effect in each model, our other results remain robust, with no substantial changes in the significance or direction of our coefficients. Finally, we split the dependent variable into serve reactions (the sum of national bans and court cases) and milder reactions (the sum of administrative enforcement and regulatory changes affecting Uber). The results for these regressions show no differences with the earlier results for all types of reactions except that the (anyway unexpected) negative coefficient for the corruption variable is no longer significant.

## Conclusion and Discussion

Our study looked at the role of national institutions in explaining the number of negative reactions that Uber received when introducing its service in 81 countries worldwide. This has been done by collecting data on events worldwide using English-language sources. While we tried to correct for the English-language bias in the regression analysis, the reliance of English-language sources remains a limitation of our study.

Our findings tend to show that Uber faced more negative reactions in countries with a strong rule of law. In particular, as our analysis suggests, Uber's entry created stronger institutional reactions in countries with well-functioning justice systems. This suggests that a robust legal system increases the number of available legal means which can be mobilized against disruptive platforms. Furthermore, it may indicate a greater familiarity with those means and possible lower enforcement costs in using such means.

We further find that pro-innovation institutions did not support Uber, despite the common observation that Uber typically framed their service as innovative in the context of the mobility regime, in improving customer service, ease of reservation and payment, low waiting times, and job opportunities for the unemployed (Davis & Sinha, 2021; Pelzer et al., 2019; Thelen, 2018; Uzunca et al., 2018). Apparently, this framing did not resonate much with stakeholders as reflected in the absence of any differences in the number of negative institutional reactions in countries with and without pro-innovation institutions.

Our quantitative approach was motivated by the possibility of covering all countries where Uber entered. Our approach comes at the price of using only general macro-institutional indictors and relying on media reports in English. What is more, we could not quantify specific differences in the taxi regimes across countries (which are anyway internally differentiated across cities within the same country). Future research could aim to collect more specific data on mobility institutions as well as on adverse reactions reported in local languages.

We hope to contribute to multi-system transition studies, with an interest in probing the role of institutional conditions in handling institutional conflicts across systems, here the tech and mobility sectors. While rapid scaling across countries is the typical strategy of platform companies, the institutional frictions arising from their disruptive business model may also hinder the very process of internationalization. Indeed, in the case of Uber, its aggressive expansion strategy backfired in various ways, forcing it to change its business model in Europe (Pelzer et al., 2019; Thelen, 2018) or to withdraw altogether from a number of key markets including China, Indonesia, Russia, the Philippines, Thailand, and Vietnam (Russell, 2018). Thus, while geographically sensitive internationalization strategy may hinder a rapid global expansion in the short run, but support a more viable business in the long run.

Furthermore, future studies could investigate the co-evolution of institutions and platform businesses in other multi-system contexts, including the intersection of digital platforms and domestic cleaning, healthcare, and food delivery (Koutsimpogiorgos et al., 2023; Margherita et al., 2024; Sanchez Vargas et al., 2022; Tassinari & Maccarrone, 2020; Yuana et al., 2020). In this way, future studies can further unpack the specific clashes between institutions making up the digital platform regime and the sectoral regime at hand, in addition to the country-level institutions that we have addressed in our analysis.

**Acknowledgements** Nikos Koutsimpogiorgos and Andrea M. Herrmann acknowledge the support by the Dutch Research Council (NWO) under Vidi grant number 452-17-017. Koen Frenken acknowledges support from the INTRANSIT centre funded by the Research Council of Norway.

## References

Andersen, A. D., Frenken, K., Galaz, V., Kern, F., Klerkx, L., Mouthaan, M., Piscicelli, L., Schor, J. B., & Vaskelainen, T. (2021). On digitalization and sustainability transitions. *Environmental Innovation and Societal Transitions, 41*, 96–98.

Andersen, A. D., & Geels, F. W. (2023). Multi-system dynamics and the speed of net-zero transitions: Identifying causal processes related to technologies, actors, and institutions. *Energy Research & Social Science, 102*, Article 103178.

Bitektine, A., & Haack, P. (2015). The 'macro' and the 'micro' of legitimacy: Toward a multilevel theory of the legitimacy process. *Academy of Management Review, 40*(1), 49–75.

Bahoo, S., Alon, I., & Paltrinieri, A. (2020). Corruption in international business: A review and research agenda. *International Business Review, 29*(4), Article 101660.

Bennett, C. J. (1991). What is policy convergence and what causes it ? *British Journal of Political Science, 21*(2), 215–233.

Bourgey-Gonse, T. (2024, February 16). Member states slam door shut on gig work directive. *Euractiv*. https://www.euractiv.com/section/gig-economy/news/member-states-slam-door-shut-on-gig-work-directive

Contractor, F. J., Dangol, R., Nuruzzaman, N., & Raghunath, S. (2020). How do country regulations and business environment impact foreign direct investment (FDI) inflows? *International Business Review, 29*(2), Article 101640.

Davis, G. F., & Sinha, A. (2021). Varieties of Uberization: How technology and institutions change the organization(s) of late capitalism. *Organization Theory, 2*(1), 263178772199519.
De Stefano, V., & Aloisi, A. (2018). European Legal Framework for "Digital Labour Platforms." *European Commission.* https://doi.org/10.2760/78590
Djankov, S., McLiesh, C., & Ramalho, R. (2006). Regulation and growth. *Economics Letters, 92*(3), 395–401.
Frenken, K., & Fünfschilling, L. (2020). The rise of online platforms and the triump of the corporation. *Sociologica, 14*(3), 101–113.
Frey, B. (2020). Platform labor and in/formality: Organization among motorcycle taxi drivers in Bandung, Indonesia. *Anthropology of Work Review, 41*(1), 36–49.
García-Canal, E., & Guillén, M. (2008). Risk and the strategy of foreign location choice in regulated industries. *Strategic Management Journal, 29*(10), 1097–1115.
Graham, M. (2020). Regulate, replicate, and resist – the conjunctural geographies of platform urbanism. *Urban Geography, 41*(3), 453–457.
Haggard, S., MacIntyre, A., & Tiede, L. (2008). The rule of law and economic development. *Annual Review of Political Science, 11*, 205–234.
Heeks, R., Gomez-Morantes, J. E., Graham, M., Howson, K., Mungai, P., Nicholson, B., & Van Belle, J. P. (2021). Digital platforms and institutional voids in developing countries: The case of ride-hailing markets. *World Development, 145*, Article 105528.
Ilsøe, A., Jesnes, K., & Hotvedt, M. (2020). Social partner responses in the Nordic platform economy. In K. Jesnes & S. Oppegaard (Eds.), *Platform Work in the Nordic Models: Issues, cases and responses.* Nordic Council of Ministers, TemaNord, no. 513, 68–78.
Jack, M. (2020). The socio-spatial installed base: Ride-hailing applications, parking associations, and precarity in tuk tuk driving in Phnom Penh. *Cambodia the Information Society, 36*(5), 252–265.
Kenney, M., & Zysman, J. (2016). The rise of the platform economy. *Issues in Science and Technology, 32*(3), 61–69.
Kim, H. J., & Suh, C. S. (2021, March). Spreading the sharing economy: Institutional conditions for the international diffusion of Uber, 2010–2017. *PLoS ONE, 16*(3), 2010–2017.
Koutsimpogiorgos, N., Frenken, K., & Herrmann, A. M. (2023). Platform adaptation to regulation: The case of domestic cleaning in Europe. *Journal of Industrial Relations, 65*(2), 156–184.

Koutsimpogiorgos, N., van Slageren, J., Herrmann, A. M., & Frenken, K. (2020). Conceptualizing the gig economy and its regulatory problems. *Policy and Internet, 12*(4), 525–545.
Margherita, G., Caffieri, A., & De Maio, N. (2024). When the algorithm 'matches' us: The experiences of early-career psychologists on online platform services for psychotherapy and counselling in Italy. *Counselling and Psychotherapy Research, 24*, 666–680.
North, D. C. (2005). *Understanding the Process of Economic Change*. Princeton University Press.
Pelzer, P., Frenken, K., & Boon, W. (2019). Institutional entrepreneurship in the platform economy : How Uber tried (and failed) to change the Dutch taxi law. *Environmental Innovation and Societal Transitions, 33*, 1–12.
Punt, M. B., van Kollem, J., Hoekman, J., & Frenken, K. (2023). Your Uber is arriving now: An analysis of platform location decisions through an institutional lens. *Strategic Organization, 21*(3), 501–536.
Rosenbloom, D. (2020). Engaging with multi-system interactions in sustainability transitions: A comment on the transitions research agenda. *Environmental Innovation and Societal Transitions, 34*, 336–340.
Russell, J. (2018, March 25). *It's official: Uber sells Southeast Asia business to Grab, TechCrunch.* https://techcrunch.com/2018/03/25/gruber-official/
Sanchez Vargas, D. Y., Maldonado Castañeda, O. J., & Hernández, M. R. (2022). Technolegal expulsions: Platform food delivery workers and work regulations in Colombia. *Journal of Labor and Society, 25*(1), 33–59.
Schaller, B. (2021). Can sharing a ride make for less traffic? Evidence from Uber and Lyft and implications for cities. *Transport Policy, 102*, 1–10.
Seidl, T. (2022). The politics of platform capitalism: A case study on the regulation of Uber in New York. *Regulation and Governance, 16*(2), 357–374.
Surie, A. (2017). Tech in work organising informal work in India. *Economic and Political Weekly, 52*(20), 12–15.
Tassinari, A., & Maccarrone, V. (2020). Riders on the storm: Workplace solidarity among gig economy couriers in Italy and the UK. *Work, Employment and Society, 34*(1), 35–54.
Thelen, K. (2018). Regulating uber: The politics of the platform economy in Europe and the United States. *Perspectives on Politics, 16*(4), 938–953.
Uzunca, B., Rigtering, C., & Ozcan, P. (2018). Sharing and shapinh: A cross-country comparison of how sharing economy firms shape their institutional environment to gain legitimacy. *Academy of Management Discoveries, 4*(3), 248–272.

Van Dijck, J., Poell, T., & De Waal, M. (2018). *The Platform Society: Public Values in a Connective World*. Oxford University Press.

Weber, C. E., Okraku, M., Mair, J., & Maurer, I. (2021). Steering the transition from informal to formal service provision: Labor platforms in emerging-market countries. *Socio-Economic Review, 19*(4), 1315–1344.

Yuana, S. L., Sengers, F., Boon, W., Hajer, M. A., & Raven, R. (2020). A dramaturgy of critical moments in transition: Understanding the dynamics of conflict in socio-political change. *Environmental Innovation and Societal Transitions, 37*, 156–170.

Zhang, Z., Zhai, G., Xie, K., & Xiao, F. (2022). Exploring the nonlinear effects of ridesharing on public transit usage: A case study of San Diego. *Journal of Transport Geography, 104*, Article 103449.

# 8

# (Re)interpreting Circularity? Understanding the Contested Directionalities of the Swedish Heavy-Duty Vehicle Sector Towards the Circular Economy

Sophie-Marie Ertelt

## Introduction

Recent sustainability transition research has witnessed a shift in focus from analysing change processes and innovation dynamics in singular socio-technical systems (hereafter systems) to an increased interest in the transformation dynamics spanning multiple systems (Andersen & Geels, 2023; Kanger et al., 2021). A noteworthy contribution to this stream of research is the Deep Transition (DT) framework (Kanger, 2022; Kanger & Schot, 2019; Schot & Kanger, 2018), which synthesizes insights from the Multi-Level Perspective (Geels, 2002, 2004) and Techno-Economic Paradigm framework (Freeman & Louca, 2001; Perez, 2002), to theorize a "series of connected and sustained fundamental transformations of a wide range of socio-technical systems in a similar

S.-M. Ertelt (✉)
Center for Sustainable Business, Örebro University School of Business, Örebro, Sweden
e-mail: sophie-marie.ertelt@oru.se

Institute of Social Studies, University of Tartu, Tartu, Estonia



M. Steen et al. (eds.), *Sustainability Transitions and Industrial Transformation*,
https://doi.org/10.1007/978-981-95-4893-4_8

direction" (Schot & Kanger, 2018, p. 1045) as a DT. The framework utilizes a rule-based logic to argue that radical changes within and between systems can be traced back to the emergence, diffusion, and alignment of rules, which are defined as "humanly devised constraints that structure human action, leading to a regular pattern of practice in systems" (Schot & Kanger, 2018, p. 1053).

Their seminal works propose that a limited number of so-called meta-rules, a single rule present in multiple systems (for example, the use of fossil fuels), have guided the co-evolution of systems into a shared directionality for the past two centuries (Kanger & Schot, 2019; Schot & Kanger, 2018). However, this shared directionality of historical development, referred to as the *First Deep Transition* towards industrial modernity (Kanger et al., 2022; Pahker et al., 2024), has contributed to a double challenge: the amplification of social inequalities and environmental degradation (Kanger & Schot, 2019). The increasing recognition of this double challenge over the last few decades has prompted attempts to deal with the problems of the first DT, potentially stimulating a *Second Deep Transition*, the emergence of new meta-rules that can alter the directionality of the evolution of a broad range of systems towards more sustainable and just practices (Kanger, 2022; Kanger & Schot, 2019; Pahker et al., 2024). The emerging meta-rules[1] around the CE, aimed at transforming prevailing linear take-make-waste economies towards economic systems that are regenerative by design and minimize waste (Ghisellini et al., 2016; Kirchherr et al., 2023), have been recognized in previous work for their potential to contribute to such a second DT (Kern et al., 2020). Through the lens of the DT framework, a change in directionality towards the CE would fundamentally challenge prevailing operating principles of systems within industrial modernity, and its successful implementation could be considered one of the most radical changes to the way global commerce is organized since the Industrial Revolution (Preston, 2012). Thus, it suggests a transformative shift towards sustainability that transcends individual systems but entails the diffusion and implementation of its potential meta-rules, such as reduce,

[1] The use of the term "emerging" in this context aims to signal that the principles of the CE are still in development and acknowledges that the eventual configuration of related meta-rules and their transformative impact remain uncertain.

reuse, recycle, and recover across a wide range of systems (Andersen et al., 2024; Kern et al., 2020).

However, given the very limited number of research that analyses contemporary transition processes through the lens of the DT framework (for an exception see Kern et al., 2020; Penna et al., 2023), our current understanding of how meta-rules are diffused—interpreted, prioritized, and implemented—across multiple systems remains limited. I argue that this can be attributed to three specific areas of theoretical underdevelopment in the DT framework that limit its analytical power: First, it inadequately theorizes the role of negotiations and conflicts over emerging meta-rules, failing to capture the contestations over viable interpretations or adaptive tensions that may arise when meta-rules, such as those of the CE, are interpreted differently across multiple systems (Kemp et al., 2022; Kern et al., 2020). Second, it tends to overemphasize meta-rule alignment and evolutionary convergence, thus neglecting the potential for diverse, complex, and messy pathways of socio-technical change (Stirling, 2009, 2011, 2019) and leaving the possibility of divergent developmental pathways or conflicting meta-rule interpretations underexplored. Lastly, it remains vague when explaining how certain meta-rules become dominant (over other possible meta-rules) (Kemp et al., 2022; Kern et al., 2020). As a result, the framework lacks the conceptual means to thoroughly analyse how meta-rules may be (re)interpreted and thereby diffused in vastly different ways across different systems. Such an analysis is, however, crucial, as it could reveal a spectrum of directionalities influenced by system-specific factors rather than a singular, aligned directionality of transition, thus offering a more nuanced understanding of the underlying processes of a potential second DT (Kemp et al., 2022).

In response to the above, this chapter aims to improve our understanding of the dynamics of how meta-rules diffuse. It specifically focuses on the interpretive flexibility of meta-rules—their openness to multiple meanings and applications by different actors—and the implications of negotiation, conflict, and contestation, understood as power-laden actor struggles over the meaning and implementation of these meta-rules in light of divergent directionalities. The following two research questions are addressed:

(1) How are the meta-rules of CE diffused across varying systems?
(2) What are the implications of divergent meta-rule interpretations for the speed and scope of the CE and its potential to contribute to a second DT?

To answer these questions, an analytical framework is proposed that puts forward an institutional perspective on meta-rule diffusion, building on the concept of field multiplicity (Greenwood et al., 2010; Zietsma et al., 2017). This framework is applied to an empirical case study on the circularity developments within the Swedish heavy-duty vehicle (HDV) sector to explore how various actors along a nascent circular value chain of battery-electric vehicles interpret, prioritize, and potentially implement the emerging meta-rules of CE differently. In this sector, the transition towards zero-emission vehicles, material shortages, and emerging regulatory demands such as the End-of-Life Vehicles Directive have prompted original equipment manufacturers (OEMs) to investigate how CE principles can be integrated into their business operations. These developments have resulted in the early-stage formation of a circular value chain involving the energy, transportation, and waste systems. The early stage of development offers a concrete empirical basis to apply the analytical framework and examine how the emerging meta-rules of CE are being (re)interpreted and potentially enacted differently across different systems to reveal the implications that divergent directionalities may have for the potential of the CE to contribute to the second DT.

The rest of this chapter is structured as follows: the next section introduces the emerging meta-rules of the CE and draws from institutional theory to present an analytical framework for meta-rule diffusion. The section after this outlines the methodology of this paper and gives an overview of the chosen case. This is followed by an analysis of the recent circularity developments in the Swedish heavy-duty vehicle sector, a discussion of key insights from the empirical case and concluding remarks.

# Theoretical Background and Analytical Framework

## The Emerging Meta-Rules of the Circular Economy

In essence, the economic model of the CE aims to minimize virgin resource input, waste, emission, and energy leakage by slowing, closing, and narrowing material and energy loops. This can be accomplished by transforming the entire continuum of product life cycles, from design and manufacturing to consumption and usage, focusing on maintenance, repair, reuse, remanufacturing, refurbishing, and recycling to conserve resources and reduce waste (Bocken et al., 2016; Geissdoerfer et al., 2017). Therefore, the implementation of the CE is not limited to a single system but entails radical changes to production and consumption processes across a broad range of interconnected systems (Kern et al., 2020). Based on Kirchherr et al. (2017) and the prevailing EU CE policy framework at the time, Kern et al. (2020) identified four meta-rules with the potential to enable such radical changes: 'Reduce', 'Reuse', 'Recycle', and 'Recover'.

However, these emerging meta-rules, frequently also referred to as the R-framework, have since then been extended to include ten,[2] as evident in the "Circular Economy Action Plan" launched by the EU Commission in 2020 and its related "Circular Economy Monitoring Framework" (albeit at different levels of emphasize). These ten potential meta-rules are hierarchical and can be classified into three categories based on the size of the waste loop each represents, thus indicating the extent of resource recovery, with smaller loops retaining more value and requiring less energy: (1) Short loops focus on product design and manufacturing (R0, R1, and R2), (2) medium loops focus on life extension strategies (R3, R4, R5, R6, R7), (3) long loops focus on capturing and retaining value from the end-of-life phase of a product (R8, R9) (Larae & Tian,

[2] This increase of emerging meta-rules reflects the unsettled nature of the frenzy phase, where different visions and practices are tested, setting the stage for possible alignment or dominance of certain meta-rules in later phases.

2023). The complete list of emerging CE meta-rules with accompanying descriptions can be found in Table 8.1.

From a DT perspective, the CE in the European context emerged as a policy idea and went through an irruption phase, characterized by the EU's effort to consolidate and complement existing rules from different policy debates under one coherent narrative of the CE (Domenech & Bahn-Walkowiak, 2019; Kern et al., 2020). Currently, the CE can be described as stuck in the frenzy phase in the sense that an emerging meta-rule set has started to form. However, there is still contestation about the interpretation of the meta-rules (e.g., different EU member states CE initiatives vary drastically in their radicalness) (Kern et al., 2020; Madsen et al., 2022). The analytical focus of this chapter is the underlying dynamics of this current development phase with a specific interest in understanding how meta-rules diffuse across different systems and the implications of potential contestation or conflicts over viable meta-rule interpretations (such as those of the R-framework) for the scope

**Table 8.1** Emerging circular economy meta-rules following the 10R-framework

| Meta-rule | Description |
|---|---|
| R0-Refuse | The use of unnecessary products and materials is avoided entirely |
| R1-Reduce | The number of resources, products, energy consumed, and waste produced is minimized |
| R2-Reuse | Products are being used multiple times by different consumers before discarding them |
| R3-Repair | Damaged products are fixed to extend their life |
| R4-Refurbish | Products are updated and restored to a near-original condition for extended use |
| R5-Remanufacture | Products at the end of their first life are restored to their original condition |
| R6-Repurpose | Products at the end of their first life are altered to be used for a different purpose than originally intended |
| R7-Recycle | Used materials are processed to obtain new materials of the same/lower quality to create new products |
| R8-Recover | Materials or energy are reclaimed from waste that cannot be reused or recycled |
| R9-Rethink | Challenging current processes and products to innovate and improve sustainability |

Descriptions based on Muller et al. (2022) and the Ellen MacArthur Foundation "The circular economy glossary", 2021 and speed of the CE implementation and its potential to contribute to a second DT. This focus is justified because exploring the negotiation processes over meta-rule implementations between various systems is central to understanding how they align and develop a common directionality or misalign, thus diverge in their development trajectories in this critical phase. The rest of this section will proceed by putting forward a theoretical account of the underlying processes of how meta-rules diffuse by drawing on insights from institutional theory.

## Meta-Rule Diffusion from an Institutional Perspective

In recent years, transition scholars have increasingly integrated insights from institutional theory to expand our conceptual understanding of transition dynamics and allow for more in-depth empirical assessment of endogenous change processes of systems (Ertelt & Kask, 2024; Fuenfschilling & Truffer, 2014; Hacker & Binz, 2021). Under this institutional perspective on transitions, systems can be characterized as containing organizational fields (hereafter fields) (Fuenfschilling & Truffer, 2014; Geels, 2020; Geels & Schot, 2007)—sets of interacting actors that carry out interrelated activities guided by a common meaning system (Scott, 2013). While a certain level of continuous incremental change can be observed in such fields due to changes in actor position (Fligstein & McAdam, 2015) and ongoing struggles (Hoffman, 1999), more drastic field restructuring to enable transition would require the transformation of the underlying institutional field logics that make up the common meaning system of a field (Fuenfschilling, 2019; Fuenfschilling & Truffer, 2014).

In this chapter, institutional field logics (hereafter logics) are defined as the combination of technologies, shared beliefs, practices, rules, and norms that guide the behaviour and interactions of actors in a given field (Thornton et al., 2012) and thus lock the evolution of a system into a dynamically stable development trajectory (Hacker & Binz, 2021). These logics may be subject to change when actors have to respond to institutional pressures such as changing legislations (Beckert, 2010; Scott, 2008). However, actors within fields are not mere recipients of such pressures but active agents who interpret, enact, and potentially alter the logics within their fields (Beckert, 2010; Davis et al., 2005). Therefore, it is important to acknowledge the embedded agency of actors (Grin et al., 2010) who navigate, negotiate, and sometimes transform these logics (Apajalahti et al., 2018). For the theorisation of meta-rule diffusion, this implies that actors can align with, resist, or reshape emerging meta-rules, thereby influencing the speed and scope of diffusion within their fields.

The emerging meta-rule set of the CE originating from public policy debates (Kern et al., 2020) thus can be conceptualized as a form of regulative institutional pressure (Scott, 2008), forcing actors to re-evaluate the prevailing logics that govern their respective fields. This emerging set of meta-rules exhibits some level of partial alignment in the sense that in the public discourse, the meta-rules are starting to be recognized as a potential direction for system evolution (Kanger et al., 2022). I propose that it is important to acknowledge that the diffusion of an emerging meta-rule set within single systems is influenced by interactions of meta-rules with the existing logics that govern the fields of said systems. Each field responds differently to the emerging meta-rules because, within the fields, actors reinterpret the meta-rules through the lens of their field's prevailing logics.

Drawing from organizational institutionalism, I further suggest that an important factor that influences to what extent the full meta-rule set is diffused in a field is the degree of compatibility of the emerging meta-rules with existing logics (Besharov & Smith, 2014). Meta-rules that are highly compatible, thus, are diffused quickly by blending with prevailing logics. In contrast, less compatible ones will remain contested or become subsidiary meta-rules (i.e., less prioritized) (Besharov & Smith, 2014). Given the diverse logic at play across different systems, this may lead to variations in which meta-rules of the set are diffused the fastest. For example, within the automotive sector, OEMs have been running remanufacturing programmes for vehicle parts for several decades (Casper & Sundin, 2018); thus, the CE meta-rule remanufacture is identical to an existing logic. As a result, the meta-rule is diffused rapidly within the sector, and remanufacturing can be considered the most frequently implemented circular practice in the sector (Kalverkamp & Raabe, 2018;

Saidani et al., 2020). Additionally, the degree of interpretive flexibility, i.e., how specific a meta-rule is or whether it allows for much room for interpretation by affected fields, further increases the chances of divergent interpretations of an emerging meta-rule set across different systems (Greenwood et al., 2011). This interpretive flexibility is especially relevant in the context of the CE, which critiques frequently describe the concept as a "floating signifier", suggesting that the meaning of the CE is defined differently by various actors and, thus, implemented in different ways (Corvellec et al., 2020; Rödl et al., 2022).

As an emerging set of meta-rules begins to cross system boundaries during the frenzy phase, this leads to increasing interactions between multiple systems (Kanger & Schot, 2019; Schot & Kanger, 2018). These interactions prompt actors from different systems, who are each grounded in their distinct logic, to engage in negotiation over viable interpretations of these meta-rules and coordination efforts for meta-rule alignment. However, I propose that instead of instant alignment, the varying meta-rule interpretations across different fields of these systems lead to a state of field multiplicity. The concept of field multiplicity draws on the idea of institutional pluralism (Greenwood et al., 2010; Zietsma et al., 2017), and in this chapter, it is utilized to conceptualize a state where instead of meta-rule alignment across systems, multiple meta-rule interpretations co-exist as well as potentially conflict. During this state of field multiplicity, "fields become centres of debates" (Hoffman, 1999 p. 351), and viable meta-rule interpretations are negotiated at field-level across multiple systems. This gives rise to adaptive tensions understood here as pressures experienced by actors as they navigate diverging and potentially conflicting meta-rule interpretations and attempt to resolve cross-system misalignments (Bruneel et al., 2016; Zhao & Lounsbury, 2016), and as a result of these the direction of evolution becomes contested. It is important to note that such conflicts over meta-rule interpretations and a shared directionality are not necessarily conflicts between regime and niche actors, but the unfolding directionality is much rather a result of interactions among actors of different systems with varying degrees of incumbency (Jørgensen, 2012; Yang et al., 2022; Yap & Truffer, 2019).

Outcomes from such adaptive tensions between systems may lead to a certain interpretation of a specific meta-rule becoming dominant while other meta-rules of the set remain subsidiarily (Padgett & Powell, 2012). Alternatively, such negotiations may result in compromises that can (at least temporarily) reconcile conflicting meta-rule interpretations (Micelotta et al., 2017). The contestation and negotiation of a meta-rule set can also foster innovation (Dalpiaz et al., 2016; Fehrer & Wieland, 2021) or lead to fragmentation and complexity (Greenwood et al., 2011) as actors strive for coordination in the face of divergent system-specific meta-rule set interpretations. Ultimately, these outcomes contribute to the evolutionary trajectory of the second DT; however, their impact on the speed and scope of the CE remains an empirical question. Figure 8.1 illustrates the above and outlines a four-stage process of meta-rule diffusion.

It is important to note that this framework is not deterministic in the sense a state of field multiplicity may not only lead to one outcome, but rather, these interactions between systems should be viewed as sites of continuous negotiation and re-negotiation as conditions and actors, therefore also outcomes may change over time. In the rest of this paper, this framework is applied to the analysis of the nascent circular value chain of the Swedish HDV sector to study the meta-rule interpreted and negotiated process across the transport, energy, and waste systems as well as draw theoretical implications about divergent meta-rule interpretations for the speed and the scope of the of the CE and its potential to contribute to a second DT.

# Methods

## Case Description: The Swedish HDV Sector

To illustrate the usefulness of the proposed analytical framework, the nascent circular value chain of the Swedish HDV sector was chosen as a relevant case. The HDV sector includes trucks, buses, and commercial vehicles essential for logistics, construction, and public transportation. Home to some of Europe's largest OEMs, the sector significantly

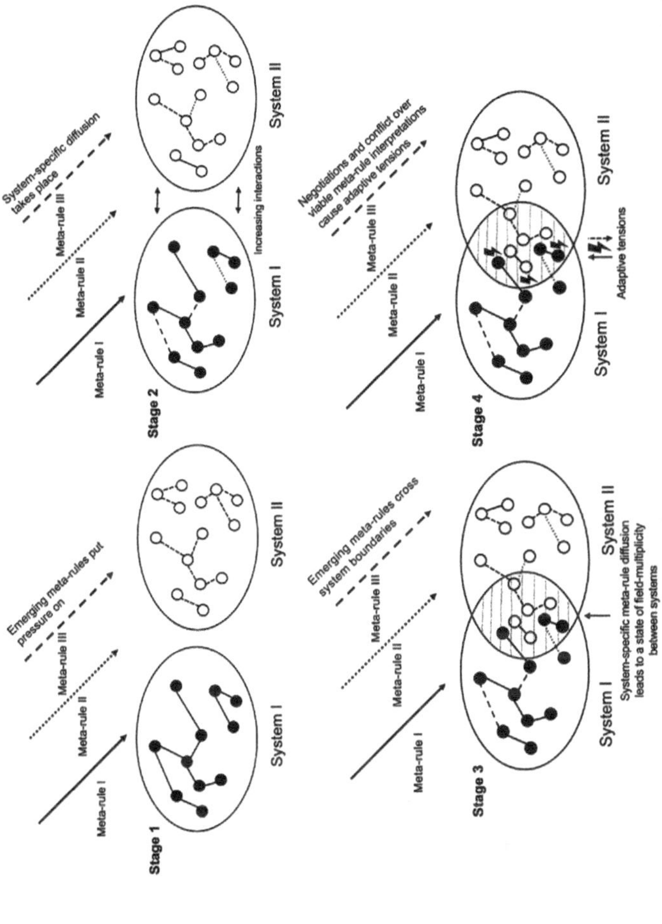


**Fig. 8.1** Analytical framework. Dots represent actors, and connectors represent interactions guided by prevailing field logics

contributes to Sweden's GDP, with revenues exceeding several billion euros annually. However, the sector accounts for over 25% of greenhouse gas emissions from road transport in the EU (European Commission, 2023) and 6.5% of all emissions in Sweden (Fossilfritt Sverige, 2024). In response, the Nordic country aims for a 50% share of zero-emission HDVs in new registrations before 2030 and a climate-neutral fleet by 2045 (Fossilfritt Sverige, 2024). Propelled by this regulatory pressure, the HDV sector is currently transitioning towards electrification. While various technological pathways exist for vehicle electrification, rapid technological advancement of batteries in terms of energy efficiency and power density, coupled with consistently decreasing costs of batteries (Link et al., 2024) and investments into the development of a charging infrastructure network, (Danese et al., 2021) have culminated in an increased focus on developing battery-electric trucks and buses among OEMs.

This increased focus on battery-electric vehicles (BEVs) has resulted in significant changes across the HDV value chain, impacting upstream, midstream, and downstream segments. Upstream, the need for critical raw materials such as lithium and other rare earth elements essential for battery cell production arises. Midstream electrification requires OEMs to develop manufacturing processes for electric drivetrains and batteries, alongside the development of a charging infrastructure network and the integration of energy supply to power BEVs. Downstream sales and service networks must be adapted to support BEVs, and new strategies for managing end-of-life vehicle components, particularly battery reuse and recycling, have been developed. Consequently, the electric HDV value chain spans and requires input from three systems: transport, energy, and waste. Table 8.2 outlines specific activities for each value chain segment in more detail.

The transition towards electrification has also expedited the integration of CE practices within the HDV sector. On the one hand, the EU Battery Regulation and the End-of-Life Vehicles Directive mandate the recovery and recycling of vehicle components and extend producer responsibility. On the other hand, recent economic challenges, such as global semiconductor shortages and potential geopolitical tensions over

**Table 8.2** Activities in the value chain of electric HDVs across different systems

| | Transport system | Energy system | Waste system |
|---|---|---|---|
| Upstream activities | – Vehicle design<br>– Mining and processing of raw materials for vehicle components and batteries | | |
| Midstream activities | – Battery cell manufacturing and battery pack assembly<br>– Electric drivetrain production and vehicle assembly | – Development and construction of charging infrastructure<br>– Ensuring grid capacity and provision of energy supply | – Management of production-related waste |
| Downstream activities | – Distribution and sales of HDVs<br>– After-sales services and customer support services | – Maintenance and operation of charging stations | – Disassembly and recycling of end-of-life vehicles and components |

raw material supplies, notably from China, both critical for manufacturing electric HDVs, have highlighted the need for resilient and circular supply chains.

## Data Collection and Analysis

To understand how the emerging meta-rules of the CE interact with prevailing logics in the HDV value chain, 32 interviews with key stakeholders were conducted between August 2023 and March 2024. Stakeholders, chosen via purposive sampling (Robinson, 2014), included representatives from OEMs, utility and grid companies, charging providers, recycling and waste management companies, and government actors. Interviews, conducted in-person and online, lasted about an hour and followed a semi-structured format.[3] The interview guide explored

[3] A full list of all interview participants can be found in the Appendix A.

existing logics related to CE principles, interpretations, and implementation practices of the 10Rs of CE, conflicts and challenges, and the impact of these tensions on circular value chain development. Specific questions also addressed whether and how CE principles were recognized or acted upon before and after the introduction of the EU's Circular Economy Action Plan in 2015. This plan marked an important institutionalization of CE discourse and was therefore used as the temporal reference point for analysis, nevertheless interviewees also referred to other regulatory developments such as the EU Taxonomy, the 2020 CE Action Plan, and the End-of-Life Vehicles Directive, which were considered during coding and interpretation as part of the broader diffusion process of CE meta-rules. Data triangulation was achieved using field notes from 12 industry events and company visits, annual reports, press releases, industry reports, media articles, and academic publications.

The above-described qualitative data was imported into MAXQD, a software for qualitative content analysis, and analysed using a thematic coding approach (Gibbs, 2007). An iterative abductive coding scheme development was utilized to ensure that the coding scheme relates to the key concepts from institutional theory outlined in Section "Theoretical Background and Analytical Framework". This meant that the author continuously iterated between data, emerging patterns, and existing theory as part of the proposed analytical framework, thus allowing for refinement of the codes to better capture the dynamics observed in the data.[4] The data was further analysed following a two-mode data structure approach, thus coding for actor type (i.e., belonging to either transport, energy, or waste system) as the associating variable and as the 10Rs of the CE as the mapped variable. Lastly, to gain a better understanding of which of these CE meta-rules may have already existed before the formalization of the CE as a policy concept, the coded data was divided into two time periods: one pre-the EU Commission's introduction of its circular economy action plan in 2015 and one post-2015.

[4] The final coding scheme can be found in Appendix B.

## Results and Analysis

### CE Meta-Rules Interpretation and Prioritization Across Different Systems

Based on the analysed data, it becomes apparent that prevailing logics across different systems along the Swedish value chain coincide with emerging meta-rules of the CE. Even before the launch of the EU circular economy action, prevailing logics that correspond to specific Rs can be identified for each system: In the Transport system, OEMs and suppliers have implemented 'Reduce' by decreasing HDV emissions and improving fuel efficiency, 'Reuse' by employing older vehicle components in new production, 'Repair' through maintenance services, 'Remanufacture' by rebuilding vehicle parts to original specifications, 'Refurbish' by restoring parts to near-new condition, and 'Recycle' by dismantling end-of-life HDVs for reusable parts and materials. The Energy system aligns with 'Reduce' through fuel efficiency and energy-efficient technologies, 'Reuse' by maximizing energy life cycle, and 'Recover' through waste-to-energy technologies. In the Waste system, 'Reduce', 'Reuse', 'Recycle', and 'Recover' coincide with minimizing landfill waste by sorting, processing, reusing materials, or recovering energy.

Beyond CE meta-rules that coincide with prevailing logics, Table 8.3 outlines stylised descriptions of how the implementation of the emerging meta-rule set is interpreted across the Transport, Energy, and Waste systems and illustrates system-specific adaptations. Moreover, the absence of clear descriptions for certain meta-rules in specific systems (as indicated through N/A in Table 8.3) points towards the potential of a selective diffusion of the emerging meta-rule set.

This selective diffusion becomes more apparent when examining the prioritization of different emerging meta-rules across the three systems. Figure 8.2 shows the distribution of statements related to each CE Rs as a percentage of each system. In the Transport system, extending vehicle component life prioritizes 'Remanufacture' and 'Refurbish'. Energy system stakeholders prioritize 'Reduce' and 'Recover', reflecting operational efficiency and energy recuperation. The Waste system focuses

**Table 8.3** Stylised descriptions of how the implementation of the emerging CE meta-rules is interpreted across different systems

| Emerging CE meta-rules | Transport system | Energy system | Waste system |
|---|---|---|---|
| R0-Refuse | Avoid the use of unnecessary or nonrecyclable vehicle components | Avoid the use of non-renewable electricity sources | N/A |
| R1-Reduce | Minimize resource consumption and maximize vehicle components' life cycle | Decrease overall electricity demand through smart grid and demand response charging | Implement practices that minimize nonrecyclable waste production |
| R2-Reuse | Use parts and components of old HDVs in the manufacturing of new vehicles | Reuse energy through vehicle-to-grid technologies and storage solutions | Find alternative uses for HDV materials without processing them into waste |
| R3-Repair | Offer services to end-users to repair damaged parts of HDVs to extend their life cycle | Offer on-site repairs to logistic depot charging stations | N/A |
| R4-Refurbish | Update and restore HDVs to a near-original condition for extended use | Upgrade charging stations to handle newer technology/ higher capacities | N/A |
| R5-Remanufacture | Restore vehicle parts to their original condition for a new life cycle | N/A | N/A |

(continued)

**Table 8.3** (continued)

| Emerging CE meta-rules | Transport system | Energy system | Waste system |
|---|---|---|---|
| R6-Repurpose | Alter retired vehicle parts, specifically batteries, for new applications, like stationary power sources | Alter existing fossil fuel infrastructure to charging infrastructure | Repurpose HDV components into non-automotive products/other industrial applications |
| R7-Recycle | Process vehicle components back to raw materials to be used in creating new vehicle parts | N/A | Invest in advanced technologies to break down vehicle parts into raw materials |
| R8-Recover | N/A | Convert nonrecyclable vehicle waste into electricity or heat | Harvest materials or energy from nonrecyclable vehicle waste |
| R9-Redesign | Design HDVs with modularity for easier upgrading, repurposing, and recycling | Incorporate future-proofing and energy-efficient design in infrastructure | Redesign waste processing to accommodate vehicle materials more efficiently |

on 'Recycle' and 'Reduce', turning used parts into raw materials and minimizing landfill waste. Overlaps in Fig. 8.2's radar plot suggest system-specific interpretations of the emerging CE meta-rules, with shared priorities on certain Rs like 'Reuse' in Transport and Energy systems. However, distinctions exist, as 'Remanufacture' and 'Repurpose' are not prioritized by the Energy system, nor is 'Repair' by the Waste system.

Comparing this prioritization to existing logics in the different systems, it appears that prioritized CE meta-rules align well with those logics that have been present before 2015. Figure 8.3 illustrates the distribution of statements related to each of the CE meta-rules (on the left)

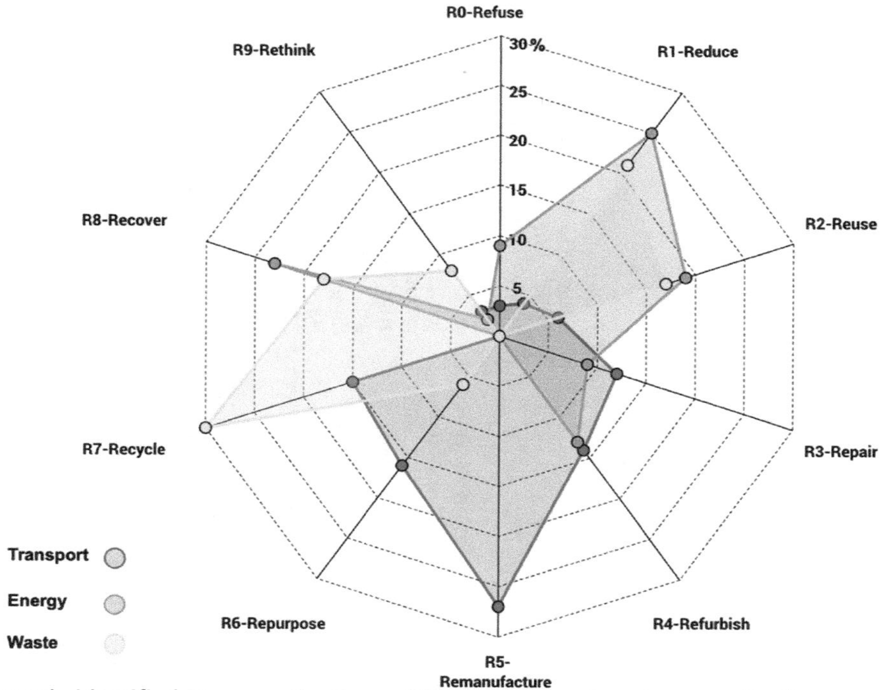


**Fig. 8.2** Most commonly identified Rs across the three different systems (numbers are % of statements analysed)

and field logics already prevailing in each system before the EU Commission introduced the circular economy action plan in 2015 (on the right). Although some diversification is observed compared to pre-2015 logics, all emerging CE meta-rules are not yet found within the Energy and Waste system.

## Three Divergent Directions Along the Emerging Circular Value Chain

The above outlined system-specific interpretations and clashing prioritizations of the CE meta-rules have given rise to three adaptive tensions between different systems along the nascent circular HDV value chain: First, downstream, once the vehicle battery loses at least some of its original capacity and the vehicle nears the end of its first life, OEMs and battery recyclers have a diverging understanding of which emerging meta-rules should be prioritized. OEMs advocate for a larger loop by refurbishing, remanufacturing, and repurposing batteries to prolong battery usage, delaying the entry of secondary raw materials back into the production cycle. Waste sector professionals prefer a smaller loop, advocating for immediate recycling to ensure rapid reclamation of secondary materials for new batteries. Vehicle battery recycling is nascent, so waste system actors stress the need for end-of-life vehicles to enter the recycling market to establish economies of scale.

Moreover, OEMs' repurposing of HDV batteries into secondary markets, like stationary energy storage, has created a second adaptive tension with energy system stakeholders, specifically utility companies, who prefer new, high-efficiency storage solutions aligning with smart grid technologies. Repurposed batteries challenge utilities' prioritization of 'Reduce', especially if repurposed vehicle batteries are employed as decentralized energy storage; utilities see this as challenging their prioritized meta-rule 'Reduce' as it significantly lowers the predictability of grid-supplied electricity demand and complicates grid management while also representing a direct competition to their central energy storage solutions. Lastly, the analysis revealed a third adaptive tension over the optimal use of resources between the energy and waste system. Waste

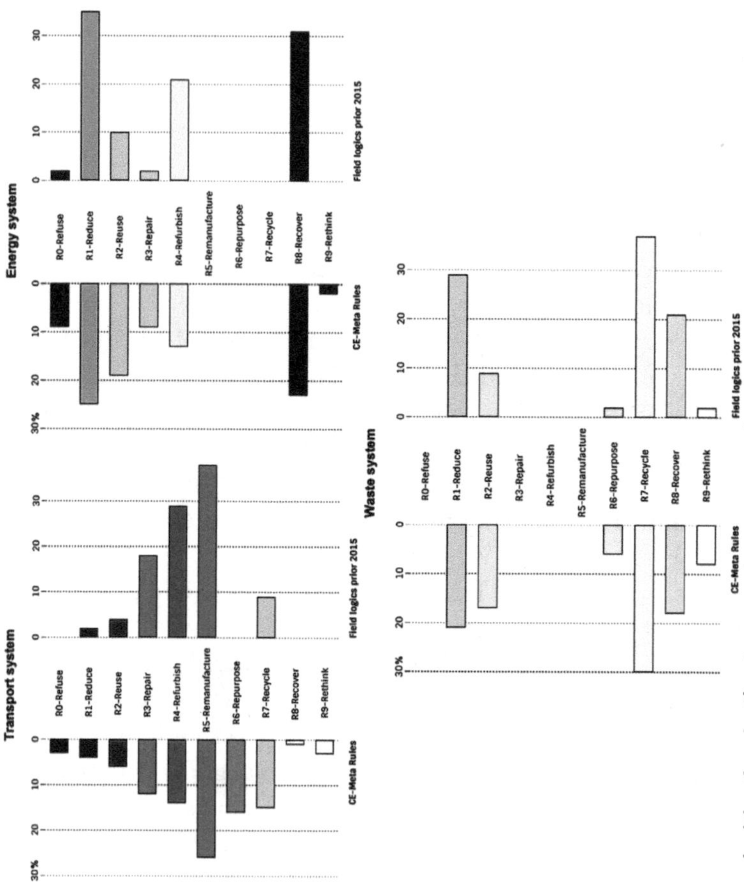


**Fig. 8.3** Meta-rule prioritization in the transport, energy, and waste system vs. pre-2015 prevailing logics

systems stakeholders prioritize 'Recycle' to recover secondary raw materials, while from the energy systems standpoint, specifically the energy production companies, the focus is on maximizing energy recovery by converting waste materials into energy through incineration. The different prioritizations of meta-rules 'Recycle' vs. 'Recover' thus lead to competition for the same waste streams and contestations over the optimal use of resources, i.e., raising questions about when it is more economically feasible and beneficial from an emissions perspective to recover instead of recycling.

Overall, the upstream–midstream–downstream analytical lens revealed that differences across systems are not limited to expected sectoral emphases (e.g., waste on recycling), but include competing strategic priorities and contested visions of circularity. Some tensions arise because actors prioritize fundamentally different pathways for the CE (e.g., centralized vs. decentralized systems or longer vs. shorter loops). In contrast, others stem from different interpretations of what a meta-rule actually means or requires in practice.

## Key Insights and Discussion

This section teases out key insights from the empirical results and discusses their implications for our understanding of meta-rule diffusion. First, the findings highlight that CE is not an entirely new concept; many of its emerging meta-rules are thus compatible with logics and related practices in the HDV sector, which have existed for several decades. This finding aligns with CE reviews that conclude that the CE as a policy framework builds on and integrates various practices from industrial ecology, systems theory, and ecological economics that have been applied to enhance resource efficiency in various sectors for many years (Kern et al., 2020; Reike et al., 2018; Winans et al., 2017). As a result, the diffusion—interpretation, prioritization, and implementation—of the emerging CE meta-rules across the Swedish HDV value chain is influenced by varying system-specific contexts. Existing logics that coincide with emerging CE meta-rules are symbolically rebranded as circular and given priority in their implementation. Less compatible meta-rules become subsidiaries while clashing meta-rules are not diffused (see Fig. 8.3) and are pushed towards other actors along the value chain. This finding contrasts with the more homogenized and coordinated view of meta-rule diffusion proposed in the DT framework, (Kanger & Schot, 2019; Schot & Kanger, 2018) as it emphasizes how different systems interpret, prioritize and implement meta-rules in ways that allow them to leverage their unique positions within the value chain and focus on those meta-rules that align closely with their operational capabilities.

Second, the findings indicate that interpretations of how emerging CE meta-rules should be implemented vary across different systems of the HDV value chain, highlighting the high interpretative flexibility of these meta-rules (see Table 8.3). This interpretative flexibility allows different systems to implement CE meta-rules in contextually relevant ways. This could lead to cumulative transformative changes across multiple systems, thus signalling a potential shared directionality that could contribute to a second DT. However, this flexibility also risks leading to incremental changes: Implementing CE meta-rules that align with existing logics in systems rather than fundamentally challenging them could reproduce elements of industrial modernity by reinforcing existing production and consumption patterns instead of disrupting them. For example, in the studied case, OEMs might prioritize refurbishing and remanufacturing vehicle parts to extend their life cycle without fundamentally altering the overall structure of production and consumption. This focus on extending the life of existing components, while beneficial in the short term, does not address the deeper systemic issues, such as reducing the number of vehicles produced or shifting towards consumption patterns that require less transportation. Such incremental adoption fails to address the root causes of unsustainability embedded in industrial modernity, such as the continuous pursuit of economic growth and efficiency at the expense of ecological health (Kanger et al., 2022). Consequently, without a fundamental shift in how economic activities are structured and valued, the full scope of systemic changes envisioned by the CE framework may not materialize, thus limiting its potential to contribute to a second DT.

Third, the prioritization of CE meta-rules that align most with existing logics may risk further entrenching the status quo in these systems and help to mask the inertia against adopting more radical yet necessary meta-rules. This selective prioritization based on compatibility with current logics perpetuates a fragmented approach to circularity that allows actors within these systems to focus on easily integrated changes rather than challenging the fundamental unsustainable aspects of industrial modernity. By aligning with existing practices, this fragmented diffusion of specific emerging meta-rules of the CE can create an illusion of progress and highlights the possibility that rather than interrelated changes across multiple systems, each system may make incremental changes in isolation. However, without a settled prioritization—where different systems are aligned and coordinated in implementing the emerging CE meta-rules based on a common understanding and shared goals (Zietsma et al., 2017)—these changes again may not aggregate into something that can contribute to a second DT.

Lastly, the results reveal the presence of adaptive tensions between different systems along the HDV value chain and the negotiations and conflicts inherent in the diffusion of the emerging CE meta-rules. These include tensions between OEMs and waste actors over delayed reuse vs. immediate recycling, between OEMs and energy utilities over decentralized repurposing vs. grid efficiency, and between waste and energy systems over recycling vs. energy recovery. The potential and impact of conflict are less emphasized in the DT framework, which often depicts the diffusion process as relatively smooth and coordinated through policy and international intermediaries (Kanger & Schot, 2019; Schot & Kanger, 2018). However, within the HDV value chain, these tensions, if unresolved, may slow down CE implementation, as competing priorities across systems could lead to stalled alignment of emerging meta-rules. This, in turn, may weaken the momentum of the second DT and limit its potential to address the double challenge.

## Concluding Remarks, Key Learnings, and Directions for Future Research

This chapter has offered an initial exploration of the evolution of an emerging sustainability paradigm that could challenge industrial modernity by focusing on the diffusion of the emerging CE meta-rules within the Swedish HDV sector. The study demonstrates that, while the emerging CE meta-rules have the potential to contribute to a second DT, their diffusion is shaped by interpretative flexibility and selective prioritization across the investigated systems. Consequently, the key contribution of this research is the improved theorisation of meta-rule diffusion in contemporary transitions: In contrast to the coordinated, homogeneous diffusion suggested by the original DT framework, this theorisation of how meta-rules highlights how system-specific interpretations and priorities can lead to conflicts and adaptive tensions. Central to this dynamic are the three identified tensions—between OEMs and waste actors over reuse versus immediate recycling, between OEMs and energy utilities over decentralized repurposing versus grid control, and between waste and energy systems over recycling versus energy recovery. These tensions illuminate how diffusion is shaped not only by institutional alignment but also by strategic positioning and competing visions for circularity. By making these tensions visible, the study highlights how contestation and negotiation shape the uneven and conflictual diffusion of CE meta-rules. Building on the theoretical ideas that underpin the concept of these adaptive tensions, the proposed analytical framework is able to show that the ongoing diffusion of CE meta-rules is not a smooth, uniform process but is marked by continuous negotiation and conflict.

The extended theorisation of meta-rule diffusion put forward in this chapter thus provides insights into the conflicts inherent to radical change processes across multiple systems. However, this work should be viewed as an initial building block that could enable more future research into the dynamics of contemporary meta-rule diffusion. Additional research across different empirical and geographical contexts, as well as over longer periods of time, will be necessary to deepen our understanding of these dynamics and to discern whether the emerging

CE meta-rules can indeed contribute to a transformation of industrial modernity.

Ending this chapter with the theme of this book in mind, two key learnings emerge for industrial transformations: first, genuine transformations towards circularity require moving beyond symbolic adoption and addressing structural barriers that limit more radical change, such as cross-system misalignments or short-term optimization logics. Second, coordinated cross-system engagement is essential, as tensions between actors (e.g., over reuse vs. recycling or decentralized vs. centralized energy storage) reveal that isolated strategies risk stalling sectoral-wide progress.

## Appendix A: Interview Information

| Number | Interviewee position | Company type | Interview type | Duration |
|---|---|---|---|---|
| #1 | Battery Business Development | OEM | Online | 1:03h |
| #2 | Battery Production | OEM | Online | 0:57h |
| #3 | Circular Business Developer | OEM | In-person | 1:08h |
| #4 | Battery Energy Storage System Sales | OEM | Online | 1:14h |
| #5 | Battery second life project management | OEM | Online | 1:05h |
| #6 | Product Development BETs | OEM | Online | 0:57h |
| #7 | Product Development BESS | OEM | Online | 1:02h |
| #8 | Battery Recycling Specialist | OEM | Online | 0:51h |
| #10 | Circular Parts Manager | OEM | Online | 1:13h |
| #11 | Sales and Services | OEM | In-person | 1:03h |
| #12 | Environment and Innovation Director | OEM | Online | 0:58h |

(continued)

(continued)

| Number | Interviewee position | Company type | Interview type | Duration |
|---|---|---|---|---|
| #13 | Circularity and Remanufacturing director | OEM | In-person | 1:08h |
| #14 | Product Development | Automotive supplier | Online | 1:02h |
| #15 | Circular Parts and Innovation manager | Automotive supplier | Online | 0:59h |
| #16 | Head of R&D | Automotive supplier | Online | 0:51h |
| #17 | Strategic Business Developer | Energy provider | In-person | 1:06h |
| #18 | Sustainability Director | Energy provider | Online | 1:00h |
| #19 | Head of Fuel-mix Optimization | Energy provider | Online | 0:47h |
| #20 | Head of E-mobility | Energy provider | Online | 1:12h |
| #21 | E-mobility Business Developer | Electric utility | Online | 1:02h |
| #22 | E-mobility Business Manager | Electric utility | In-person | 0:56h |
| #23 | Business Innovation Manager | Electric utility | Online | 1:08h |
| #24 | Sales Manager for BESS | Battery storage system developer | Online | 1:01h |
| #25 | Head of R&D | Battery storage system developer | Online | 0:54h |
| #26 | CEO | Battery storage system developer | Online | 1:05h |
| #27 | Secondary raw material supply chain manager | Mining association | In-person | 0:49h |
| #28 | Battery Recycling Specialist | Recycling | Online | 1:07h |
| #29 | Industry Manager Automotive | Recycling | Online | 1:02h |

(continued)

(continued)

| Number | Interviewee position | Company type | Interview type | Duration |
|---|---|---|---|---|
| #30 | Automotive Recycling Manager | Recycling | Online | 0:58h |
| #31 | Chief Operations Manager | Recycling | Online | 0:55h |
| #32 | Business Manager Batteries | Recycling | Online | 1:10h |

## Appendix B: Concepts, Definitions, and Codes Utilized for the Data Analysis

| Key concept | Codes/indicators |
|---|---|
| **CE meta-rule compatibility.** *The degree to which the 10Rs of the CE coincide with, support, or contradict prevailing field logics within a system* | **Symbolic diffusion.** Existing field logic is rebranded as one of the Rs without substantive changes to practise<br>**Selective prioritization.** The selective implementation of certain Rs of the CE that are compatible with existing field logics<br>**Conflict.** Challenges or resistance to the integration of one or more of the Rs due to clashes with existing field logics |
| **Interpretive flexibility of CE meta-rules**. *The extent to which the 10Rs of the CE can be adapted or interpreted differently across various systems* | **Variability of implementation.** The range of different practices that stakeholders describe when implementing the same R across different systems<br>**System-specific adaptations.** Unique modifications or prioritization of the 10Rs that are tailored to the needs and constraints of different systems |

(continued)

(continued)

| Key concept | Codes/indicators |
|---|---|
| **Field multiplicity.** *A state where multiple interpretations and/or applications of CE meta-rules co-exist across two or more interacting systems* | **Co-existing interpretations.** Instances where multiple approaches to implementing the 10Rs are identified across systems |
| **Adaptive tensions.** *The contestations and conflicts that arise when two or more systems negotiate over viable meta-rule interpretations* | **Conflicting interpretations.** Instances where disagreements on the meaning or implementation of the CE meta-rules become explicit<br>**Negotiation strategies.** Specific approaches or tactics stakeholders employ to negotiate and reconcile conflicting interpretations of CE meta-rules<br>**Contestation over interpretations.** Ongoing debates regarding how CE meta-rules should be understood and implemented |

# References

Andersen, A. D., & Geels, F. W. (2023). Multi-system dynamics and the speed of net-zero transitions: Identifying causal processes related to technologies, actors, and institutions. *Energy Reseach & Social Science, 102*, 103178. https://doi.org/10.1016/j.erss.2023.103178

Andersen, A. D., Mäkitie, T., Steen, M., & Wanzenböck, I. (2024). *Integrating industrial transformation and sustainability transitions research through a multi-sectoral perspective* (TIK Working Paper 20240206). University of Oslo. https://doi.org/10.13140/RG.2.2.34164.73602

Apajalahti, E.-L., Temmes, A., & Lempiälä, T. (2018). Incumbent organisations shaping emerging technological fields: Cases of solar photovoltaic and electric vehicle charging. *Technology Analysis & Strategic Management, 30*, 44–57. https://doi.org/10.1080/09537325.2017.1285397

Beckert, J. (2010). How do fields change? The Interrelations of institutions, networks, and cognition in the dynamics of markets. *Organisation Studies, 31*, 605–627. https://doi.org/10.1177/0170840610372184

Besharov, M. L., & Smith, W. K. (2014). Multiple institutional logics in organizations: Explaining their varied nature and implications. *Academy of Management Review, 39*, 364–381.

Bocken, N. M. P., de Pauw, I., Bakker, C., & van der Grinten, B. (2016). Product design and business model strategies for a circular economy. *Journal of Industrial and Production Engineering, 33*, 308–320. https://doi.org/10.1080/21681015.2016.1172124

Bruneel, J., Moray, N., Stevens, R., & Fassin, Y. (2016). Balancing competing logics in for-profit social enterprises: A need for hybrid governance. *Journal of Social Entrepreneurship, 7*, 263–288. https://doi.org/10.1080/19420676.2016.1166147

Casper, R., & Sundin, E. (2018). Addressing today's challenges in automotive remanufacturing. *Journal of Remanufacturing, 8*, 93–102. https://doi.org/10.1007/s13243-018-0047-9

Corvellec, H., Böhm, S., Stowell, A., & Valenzuela, F. (2020). Introduction to the special issue on the contested realities of the circular economy. *Culture and Organization, 26*, 97–102. https://doi.org/10.1080/14759551.2020.1717733

Dalpiaz, E., Rindova, V., & Ravasi, D. (2016). Combining logics to transform organizational agency: Blending industry and art at Alessi. *Administrative Science Quarterly, 61*, 347–392. https://doi.org/10.1177/0001839216636103

Danese, A., Garau, M., Sumper, A., & Torsæter, B. N. (2021). Electrical infrastructure design methodology of dynamic and static charging for heavy and light duty electric vehicles. *Energies, 14*, 3362. https://doi.org/10.3390/en14123362

Davis, G. F., McAdam, D., Scott, W. R., & Zald, M. N. (Eds.). (2005). *Social movements and organization theory*. Cambridge University Press. https://doi.org/10.1017/CBO9780511791000

Domenech, T., & Bahn-Walkowiak, B. (2019). Transition towards a resource efficient circular economy in Europe: Policy lessons from the EU and the Member States. *Ecological Economics, 155*, 7–19. https://doi.org/10.1016/j.ecolecon.2017.11.001

Ertelt, S.-M., & Kask, J. (2024). Home field advantage: Examining incumbency reorientation dynamics in low-carbon transitions. *Environmental Innovation and Societal Transitions, 50*, 100802. https://doi.org/10.1016/j.eist.2023.100802

European Commission. (2023). *$CO_2$emission standards for Heavy-Duty Vehicles* [WWW Document]. European Commission https://ec.europa.eu/commission/presscorner/detail/en/qanda_23_763. Accessed 30 December 2023.

Fehrer, J. A., & Wieland, H. (2021). A systemic logic for circular business models. *Journal of Business Research, 125*, 609–620. https://doi.org/10.1016/j.jbusres.2020.02.010

Fligstein, N., & McAdam, D. (2015). *A theory of fields*. Oxford University Press.

Fossilfritt Sverige. (2024). *Automotive industry—heavy transport* [WWW Document]. Foss. Sver. https://fossilfrittsverige.se/en/roadmap/the-automotive-industry-heavy-transport/. Accessed 21 March 2024.

Freeman, C., & Louca, F. (2001). *As time goes by: the information revolution and the industrial revolutions in historical perspective*. Oxford University Press Inc.

Fuenfschilling, L. (2019). An institutional perspective on sustainability transitions. In *Handbook of sustainable innovation* (pp. 219–236). Edward Elgar Publishing.

Fuenfschilling, L., & Truffer, B. (2014). The structuration of socio-technical regimes—Conceptual foundations from institutional theory. *Research Policy, 43*, 772–791. https://doi.org/10.1016/j.respol.2013.10.010

Geels, F. W. (2002). Technological transitions as evolutionary reconfiguration processes: A multi-level perspective and a case-study. *Research Policy, 31*, 1257–1274. https://doi.org/10.1016/S0048-7333(02)00062-8

Geels, F. W. (2004). From sectoral systems of innovation to socio-technical systems: Insights about dynamics and change from sociology and institutional theory. *Research Policy, 33*, 897–920. https://doi.org/10.1016/j.respol.2004.01.015

Geels, F. W. (2020). Micro-foundations of the multi-level perspective on socio-technical transitions: Developing a multi-dimensional model of agency through crossovers between social constructivism, evolutionary economics and neo-institutional theory. *Technological Forecasting & Social Change, 152*, 119894. https://doi.org/10.1016/j.techfore.2019.119894

Geels, F. W., & Schot, J. (2007). Typology of sociotechnical transition pathways. *Research Policy, 36*, 399–417. https://doi.org/10.1016/j.respol.2007.01.003

Geissdoerfer, M., Savaget, P., Bocken, N. M. P., & Hultink, E. J. (2017). The circular economy—A new sustainability paradigm? *Journal of Cleaner Production, 143*, 757–768. https://doi.org/10.1016/j.jclepro.2016.12.048

Ghisellini, P., Cialani, C., & Ulgiati, S. (2016). A review on circular economy: The expected transition to a balanced interplay of environmental and economic systems. *Journal of Cleaner Production, 114*, 11–32. https://doi.org/10.1016/j.jclepro.2015.09.007

Gibbs, G. (2007). *Analyzing qualitative data*. SAGE Publications, Ltd, 1 Oliver's Yard, 55 City Road, London England EC1Y 1SP United Kingdom. https://doi.org/10.4135/9781849208574

Greenwood, R., Díaz, A., Li, S., & Cespedes-Lorente, J. (2010). The multiplicity of institutional logics and the heterogeneity of organizational responses. *Organization Science, 21*, 521–539. https://doi.org/10.1287/orsc.1090.0453

Greenwood, R., Raynard, M., Kodeih, F., Micelotta, E. R., & Lounsbury, M. (2011). Institutional complexity and organizational responses. *Academy of Management Annals, 5*, 317–371. https://doi.org/10.5465/19416520.2011.590299

Grin, J., Rotmans, J., & Schot, J. (2010). *Transitions to sustainable development: New directions in the study of long term transformative change* (1st ed.). Routledge. https://doi.org/10.4324/9780203856598

Hacker, M. E., & Binz, C. (2021). Navigating institutional complexity in socio-technical transitions. *Environmental Innovation and Societal Transitions, 40*, 367–381. https://doi.org/10.1016/j.eist.2021.09.003

Hoffman, A. J. (1999). Institutional evolution and change: Environmentalism and the U.S. chemical industry. *Academy of Management Journal, 42*, 351–371. https://doi.org/10.2307/257008

Jørgensen, U. (2012). Mapping and navigating transitions—The multi-level perspective compared with arenas of development. *Research Policy, 41*, 996–1010. https://doi.org/10.1016/j.respol.2012.03.001

Kalverkamp, M., & Raabe, T. (2018). Automotive remanufacturing in the circular economy in Europe: Marketing system challenges. *Journal of Macromarketing, 38*, 112–130. https://doi.org/10.1177/0276146717739066

Kanger, L. (2022). The spatial dynamics of deep transitions. *Environmental Innovation and Societal Transitions, 44*, 145–162. https://doi.org/10.1016/j.eist.2022.06.005

Kanger, L., & Schot, J. (2019). Deep transitions: Theorizing the long-term patterns of socio-technical change. *Environmental Innovation and Societal Transitions, 32*, 7–21. https://doi.org/10.1016/j.eist.2018.07.006

Kanger, L., Schot, J., Sovacool, B. K., van der Vleuten, E., Ghosh, B., Keller, M., Kivimaa, P., Pahker, A.-K., & Steinmueller, W. E. (2021). Research frontiers for multi-system dynamics and deep transitions. *Environmental*

*Innovation and Societal Transitions, 41*, 52–56. https://doi.org/10.1016/j.eist.2021.10.025
Kanger, L., Tinits, P., Pahker, A.-K., Orru, K., Tiwari, A. K., Sillak, S., Šeļa, A., & Vaik, K. (2022). Deep Transitions: Towards a comprehensive framework for mapping major continuities and ruptures in industrial modernity. *Global Environmental Change, 72*, 102447. https://doi.org/10.1016/j.gloenvcha.2021.102447
Kemp, R., Pel, B., Scholl, C., & Boons, F. (2022). Diversifying deep transitions: Accounting for socio-economic directionality. *Environmental Innovation and Societal Transitions, 44*, 110–124. https://doi.org/10.1016/j.eist.2022.06.002
Kern, F., Sharp, H., & Hachmann, S. (2020). Governing the second deep transition towards a circular economy: How rules emerge, align and diffuse. *Environmental Innovation and Societal Transitions, 37*, 171–186. https://doi.org/10.1016/j.eist.2020.08.008
Kirchherr, J., Reike, D., & Hekkert, M. (2017). Conceptualizing the circular economy: An analysis of 114 definitions. *Resources, Conservation & Recycling, 127*, 221–232. https://doi.org/10.1016/j.resconrec.2017.09.005
Kirchherr, J., Yang, N.-H.N., Schulze-Spüntrup, F., Heerink, M. J., & Hartley, K. (2023). Conceptualizing the circular economy (revisited): An analysis of 221 definitions. *Resources, Conservation & Recycling, 194*, 107001. https://doi.org/10.1016/j.resconrec.2023.107001
Larae, M., & Tian, D. (2023). *R-strategies for a circular economy*. URL https://www.circularise.com/blogs/r-strategies-for-a-circular-economy. Accessed 23 February 2024.
Link, S., Stephan, A., Speth, D., & Plötz, P. (2024). Rapidly declining costs of truck batteries and fuel cells enable large-scale road freight electrification. *Nature Energy, 9*(8), 1032–1039. https://doi.org/10.1038/s41560-024-01531-9
Madsen, S., Miörner, J., & Hansen, T. (2022). Axes of contestation in sustainability transitions. *Environmental Innovation and Societal Transitions, 45*, 246–269. https://doi.org/10.1016/j.eist.2022.11.001
Micelotta, E., Lounsbury, M., & Greenwood, R. (2017). Pathways of institutional change: An integrative review and research Agenda. *Journal of Management, 43*, 1885–1910. https://doi.org/10.1177/0149206317699522
Muller, L. N. P. E. S., Delai, I., & Alcantara, R. L. C. (2022). Circular value chain practices for developing resource value retention options. *Journal of Cleaner Production, 359*, 131925. https://doi.org/10.1016/j.jclepro.2022.131925

Padgett, J. F., & Powell, W. W. (2012). *The emergence of organizations and markets*. Princeton University Press.

Pahker, A.-K., Kanger, L., & Tinits, P. (2024). Where is the deep sustainability turn most likely to emerge? An Industrial Modernity Index. *Technological Forecasting & Social Change, 201*, 123227. https://doi.org/10.1016/j.techfore.2024.123227

Penna, C. C. R., Schot, J., & Steinmueller, W. E. (2023). Transformative investment: New rules for investing in sustainability transitions. *Environmental Innovation and Societal Transitions, 49*, 100782. https://doi.org/10.1016/j.eist.2023.100782

Perez, C. (2002). *Technological revolutions and financial capital*. Edward Elgar.

Preston, F. (2012). *A global redesign? Shaping the circular economy*. Chatham House.

Reike, D., Vermeulen, W. J. V., & Witjes, S. (2018). The circular economy: New or refurbished as CE 3.0?—Exploring controversies in the conceptualization of the circular economy through a focus on history and resource value retention options. *Resources, Conservation & Recycling, 135*, 246–264. https://doi.org/10.1016/j.resconrec.2017.08.027

Robinson, O. C. (2014). Sampling in interview-based qualitative research: A theoretical and practical guide. *Qualitative Research in Psychology, 11*, 25–41. https://doi.org/10.1080/14780887.2013.801543

Rödl, M. B., Åhlvik, T., Bergeå, H., Hallgren, L., & Böhm, S. (2022). Performing the circular economy: How an ambiguous discourse is managed and maintained through meetings. *Journal of Cleaner Production, 360*, 132144. https://doi.org/10.1016/j.jclepro.2022.132144

Saidani, M., Yannou, B., Leroy, Y., & Cluzel, F. (2020). Dismantling, remanufacturing and recovering heavy vehicles in a circular economy—Technico-economic and organisational lessons learnt from an industrial pilot study. *Resources, Conservation & Recycling, 156*, 104684. https://doi.org/10.1016/j.resconrec.2020.104684

Schot, J., & Kanger, L. (2018). Deep transitions: Emergence, acceleration, stabilization and directionality. *Research Policy, 47*, 1045–1059. https://doi.org/10.1016/j.respol.2018.03.009

Scott, W. R. (2013). *Institutions and organizations: Ideas and interests* (4th ed). SAGE.

Scott, W. R. (2008). *Institutions and organizations: Ideas and interests*. SAGE.

Stirling, A. (2019). How deep is incumbency? A 'configuring fields' approach to redistributing and reorienting power in socio-material change. *Energy*

*Research & Social Science, 58*, 101239. https://doi.org/10.1016/j.erss.2019.101239

Stirling, A. (2011). Pluralising progress: From integrative transitions to transformative diversity. *Environmental Innovation and Societal Transitions, 1*, 82–88. https://doi.org/10.1016/j.eist.2011.03.005

Stirling, A. (2009). *Direction, distribution and diversity! Pluralising progress in innovation, sustainability and development*. STEPS Centre, University of Sussex.

The circular economy glossary [WWW Document]. (n.d.). https://www.ellenmacarthurfoundation.org/topics/circular-economy-introduction/glossary. Accessed 31 March 2024.

Thornton, P. H., Ocasio, W., & Lounsbury, M. (2012). *The institutional logics perspective: A new approach to culture*. Oxford University Press. https://doi.org/10.1093/acprof:oso/9780199601936.001.0001

Winans, K., Kendall, A., & Deng, H. (2017). The history and current applications of the circular economy concept. *Renewable & Sustainable Energy Review, 68*, 825–833. https://doi.org/10.1016/j.rser.2016.09.123

Yang, K., Schot, J., & Truffer, B. (2022). Shaping the directionality of sustainability transitions: The diverging development patterns of solar photovoltaics in two Chinese provinces. *Regional Studies, 56*, 751–769. https://doi.org/10.1080/00343404.2021.1903412

Yap, X.-S., & Truffer, B. (2019). Shaping selection environments for industrial catch-up and sustainability transitions: A systemic perspective on endogenizing windows of opportunity. *Research Policy, 48*, 1030–1047. https://doi.org/10.1016/j.respol.2018.10.002

Zhao, E. Y., & Lounsbury, M. (2016). An institutional logics approach to social entrepreneurship: Market logic, religious diversity, and resource acquisition by microfinance organizations. *Journal of Business Venturing, 31*, 643–662. https://doi.org/10.1016/j.jbusvent.2016.09.001

Zietsma, C., Groenewegen, P., Logue, D. M., & Hinings, C. R. (2017). Field or fields? Building the scaffolding for cumulation of research on institutional fields. *Academy of Management Annals, 11*, 391–450. https://doi.org/10.5465/annals.2014.0052

# 9

# Captains of Industry? Value Chain Relations and Sustainability Reorientation of Firms

Taran Thune and Tuukka Mäkitie

## Introduction

In this chapter, we contribute to the literature by exploring how firms are influenced by the sustainability-oriented behaviour of other firms (Bohnsack et al., 2020; Bulah et al., 2023). In many mature industries, industry relationships are typically organized in the form of complex value chains where lead firms interact with suppliers, users, and competitors. Value chains string together an array of heterogeneous firms from many sectors supplying diverse types of components, materials, and services to deliver a particular product or service (Andersen et al., 2024). Mature value chains are often dominated by lead firms that play important roles

T. Thune (✉) · T. Mäkitie
Centre for Technology, Innovation and Culture, University of Oslo, Oslo, Norway
e-mail: tuukka.makitie@sintef.no

T. Mäkitie
Department of Technology Management, SINTEF Digital, Trondheim, Norway



M. Steen et al. (eds.), *Sustainability Transitions and Industrial Transformation*,
https://doi.org/10.1007/978-981-95-4893-4_9

in directing other firms, including suppliers, through multiple forms of competition, collaboration, contractual agreements, and informal networks. Value chains are characterized by relationships of dependence, where suppliers occupy less powerful positions and are dependent on lead firms.

The sustainability transitions literature has recently started to engage with multi-sector and value chain perspectives (Andersen et al., 2020; Rosenbloom, 2020), but limited attention has been put on firms in different parts of value chains (Chizaryfard & Karakaya, 2022). Compared with literature on firms and sustainability transitions, the firms in a value chain perspective puts emphasis on the relationships between firms in the different parts of a value chain, for example, between technology providers (such as manufacturers of heavy-duty trucks) and technology users (transport companies). Different positionalities and the types of links between firms condition firm responses to sustainability. The established relationships between firms in a value chain are thus likely to shape whether and how these firms reorient to more sustainable markets (Steen & Weaver, 2017). In this chapter, we add to the growing literature on multi-system dynamics (Andersen & Geels, 2023) and firms and industries in sustainability transitions (Magnusson & Werner, 2022) by paying explicit attention to how firm relationships in value chains influence sustainability reorientation.

For this purpose, we draw upon theoretical insights in organization science on bandwagons (Abrahamson & Rosenkopf, 1993) and isomorphism (DiMaggio & Powell, 1983). We are interested in understanding mimetic behaviour and if value chain positions and relationships influence how and when firms develop sustainability-oriented behaviour (Schrage & Rasche, 2022). More specifically, we propose that leader–follower relationships in value chains may shape the sustainability-oriented behaviour of firms (Geels, 2021). By firm behaviour we mean "strategies, decisions and actions taken to achieve goals" and "sustainability-oriented behaviours" is when firm behaviour is intended to achieve goals of environmental sustainability. We are particularly interested in strategic sustainability reorientation (Stalmokaitė & Hassler, 2020), by which we mean strategic actions firms take to move from unsustainable markets to more sustainable ones.

To address the research question '*how is sustainability reorientation shaped by inter-firm relations in value chains?*' we have performed an empirical study of sustainability reorientation in firms that are part of the Norwegian oil and gas (O&G) value chain. This is an interesting context, as firms in this complex value chain are actively engaging in strategic reorientation and relabelling from oil to the more generic label "energy companies". Several companies have made considerable investments in renewable energy, particularly in solar, hydropower, and wind energy (Pick, 2019). At the same time, sustainability reorientation in this context is highly contested, and several studies highlight the discrepancy between high ambitions and the actual depth and scope of change (Nilsen, 2017; Bach 2019; Ferns et al., 2019; Kenner & Heed, 2021; Schupfer & Soppe, 2024).

The Norwegian O&G value chain is believed to be entering a phase of decline and transformation because of the mounting climate change concerns and few new commercial-scale O&G findings on the Norwegian Continental Shelf (Thune et al., 2018). In previous research, we have found that several firms had already employed proactive environmental strategies, including diversification to green markets such as offshore wind power (Andersen & Gulbrandsen, 2020; Hansen & Steen, 2015; Mäkitie et al., 2018; 2019; Thune & Mäkitie, 2018). However, diversification efforts have been intermittent, meaning that several firms have moved back and forth between O&G and offshore wind power. Also, almost half of the companies were still focused only on the O&G market. Hence, we observe firms pursuing various kinds of reorientation strategies amidst the increasing institutional and economic pressures towards sustainability, ranging from reorientation to green markets, to minor diversification and engagement with environmental topics, all the way to continued full commitment to the O&G market.

In this chapter, we present a case study of sustainability reorientation in two firms in the O&G value chain: an operator firm and a large supply firm. Drawing on data on how the firms have dealt with sustainability issues during a period of 15 years, the chapter addresses whether the positions and relationships between firms in value chains shape sustainability reorientations of incumbent firms. To develop a conceptual framework for the analysis, we draw on two separate theoretical traditions. First, we position in the chapter in current literature on firms and industries in transitions. Second, we draw on perspectives from organizational theory to explain why inter-organizational relationships, and particularly between lead and follower firms in value chains, influence whether and how firms adopt new business practices—in our case sustainability reorientation.

## Theoretical Perspectives

### Sustainability Transitions and Firms

Sustainability transitions research has typically focused on analysing change processes at system level, such as in energy, transport, or food systems (Markard et al., 2012). Transition processes are however seen to be "shot through with agency" (Geels, 2011: 29). The literature has been exploring actor perspectives, strategies, and behaviour in sustainability transitions (Farla et al., 2012; Markard & Truffer, 2008), including those of, incumbent firms (Karltorp & Sandén, 2012; Magnusson & Werner, 2022; van Mossel et al., 2018).

The literature on incumbents has tended to focus on particularly proactive firms (i.e., the 'leaders') (Berggren et al., 2015; Mäkitie, 2020; Wesseling et al., 2015) or on the other extreme—the particularly inert ones (the 'laggards') who may even seek to purposefully slow down transitions (Bansal & Roth, 2000; Smink et al., 2015). Most established firms can be placed between the two poles (Turnheim & Sovacool, 2020) as followers (Van Mossel et al., 2018; Hockerts & Wüstenhagen, 2010) or loungers (O'Regan et al., 2008). Due to risks involved in being a leader in a transition, many firms may adopt a more cautious approach (Bohnsack et al., 2020), and prefer to take follower roles.

To understand the different positions firms take in sustainability reorientation, we need to understand the business context (Schrage & Rasche, 2022). Despite growing attention to incumbent in the literature, it is yet immature in analysing the role of value chain relationships in shaping firm behaviour in sustainability transitions (Bohnsack et al., 2020; Kungl, 2024). Recent literature has identified the potential key role that interactions between sectors along technology value chains may have in driving or hindering sustainability transitions (Andersen & Markard, 2020; Mäkitie et al., 2022; Malhotra et al., 2019; Stephan et al., 2017). This literature has highlighted the vital role of such multi-system dynamics for acceleration of sustainability transitions (Markard et al., 2020). However, this technology and sector-level analysis has yet elaborated little on the underlying processes at the firm level (Chizaryfard & Karakaya, 2022). There is thus a need for literature that helps to explain how inter-sectoral dynamics at the value chain level may be shaped by inter-organizational relationships at the firm level. In the next section, we turn attention to this issue.

## Bandwagons and Adoption of Innovations

Isomorphism (DiMaggio & Powell, 1983) is a phenomenon of interest for both innovation and organization scholars trying to account for why firms often adopt similar innovations. Both research traditions assume that the drivers for adoption of novel ideas are different for early adopters, who develop innovations to solve problems or to explore new opportunities, compared to followers who often adopt innovations "for other reasons" Abrahamson and Rosenkopf (1993) refer to this as the "bandwagon effect". Abrahamson & Rosenkopf argue that rational or economic perspectives cannot account for follower-behaviour. When increasing numbers of organizations adopt an innovation without a proper assessment of the value for the firm, but adopt it due to the fact that a large number of other organizations are doing it—a "bandwagon cycle" has started.

What factors influence firms to adopt an innovation when there are not clear economic rationalities behind it? First, a fundamental perspective in this literature is that an organization identifies with a collective and are influenced by the actions of other organizations in their environment. These might be industries, regional clusters, or other forms of collective identities, such as clubs, associations, and the like. DiMaggio and Powell (1983) particularly discuss organizational fields, or different organizations that "constitute a recognized area of institutional life: key suppliers, resource and product consumers, regulatory agencies, and other organizations that produce similar services or products". This broader definition means that an impetus for modelling behaviour might not stem from firms that are similar (such as in a sector), but also from other kinds of organizations in the environment (e.g., a value chain). Being part of a collective provides information about the intentions, actions, and outcomes of others, regardless of whether one sees the information and comparison through the competitive or institutional perspectives. However, there is evidence that suggests that fields that are cohesive (i.e., more densely connected) are more likely to diffuse similar kind of knowledge and hence spur more mimetic behaviour. This entails that in organizational fields that are tightly connected and have developed joint institutions, such as interest organizations, standards, professional networks, etc., mimetic behaviour across firms are more likely to occur.

A second driver for adoption is uncertainty. When faced with a new situation or a new problem, organizations are likely to search for examples and model themselves after other organizations in their surroundings that have responded to the situation. Organizations do not search at random, but are more likely to adopt an innovation if it already has been adopted by an organization that is seen as successful or significant (Haunschild & Miner, 1997). However, organizations are also more likely to adopt an innovation if the model organization is perceived as similar or being part of ones "reference group" (Greve, 1995).

A third driver for adoption is when uncertainty stems from dependence on other organizations for significant resources. Organizations that are dependent on other organizations for resources, and particularly if the power to control resources is centralized, the more likely it is that organizations will adopt an innovation promoted by the leading firm. Resource dependency and normative pressures can also coincide, for instance through requirements to adopt certain practices or standards to become eligible for a grant or a contract (Boxenbaum & Jonsson, 2017). In situations of dependence, decoupling strategies are also more likely, where firms engage in symbolic adoption strategies, with limited actual change (ibid.).

The adoption pressures are not independent, they might pull in different directions, but also coincide. If pressures to adopt coincide, we can assume that followers are most likely to adopt innovations if, i) they are part of a cohesive organizational field, ii) there is a high degree of uncertainty, iii) there are significant organizations that occupy powerful positions and can dominate the "rules of the game". With these perspectives in mind, we will explore sustainability reorientation by firms as a form of innovation that is adopted and will look at how this is influenced by relationships in value chains.

## Case Study and Methods

O&G is a relevant context to study how value chain relationships shape sustainability behaviour. The O&G value chain is characterized by a complex web of horizontal and vertical inter-firm relationships (Acha & Cusmano, 2005) and highly concentrated power on the apex of the value chain (i.e., the operators, or oil firms). The majority of economic activities and jobs are not in the operator firms that own and operate assets, but are found inside the diverse service and supply industry (Blomgren et al., 2015). These firms are dependent on the lead operator firms that put out contracts and tenders for technologies and services. Through financial and contractual ties, operators shape strategies and behaviours of its suppliers. The first-tier supply and service firms do the planning, engineering, and building of equipment, and operate and provide services on O&G production units. The first-tier companies work with a range of sub-suppliers that are responsible for specialized services and equipment.

In this context, we have performed a study of sustainability reorientation of firms. In this chapter, we focus particularly on two large firms: an operator firm (Equinor) and a first-tier supply firm (BigService). We have collected data on both firms' approaches to dealing with sustainability issues over time. Specifically, we have collected data on strategies, investments, projects, and organizational development that each firm has taken to deal specifically with sustainability concerns. We distinguish between events connected to "*sustainability orientation*" (plans, strategies, communications that signal attention and commitment to sustainability issues) and "*sustainability actions*" (investments, projects, and organizational developments). We look specifically at what point in time events occur, the context in which they occur, and changes over time. Data is used to construct an event database for each company over a period of 15 years in Excel. The data sources used to create the event database are public documents: annual plans, investor reports, sustainability plans, press releases, and other forms of media information searchable through media database Retriever, and other forms of information available on websites, current and past. Since the two companies are publicly listed, basic information about strategies, financial commitments, investments, and other actions (projects, technology development, R&D, etc.) are easily traceable through such documents. In addition, interviews with industry experts and representatives from the two companies have been performed to support the document analysis, and to help make sense of the events and their implications over time.[1]

## Empirical Study

In this section we present our analysis of the sustainability orientation and actions in Equinor and BigService. First, we trace important developments over time in each firm followed by a comparative analysis and discussion of how relationships in value chains may influence sustainability reorientation of companies and the value chain as a whole.

[1] Documentary data included in the analysis: EQUINOR: Annual reports from 2007 to 2023, sustainability reports (later integrated in annual reports) from 2007 to 2023, media releases (in archive on the webpage and searchable through media service Retriever). BigService: Annual reports from 2011 to 2023 (before that other company, later merged to form one integrated company), sustainability reports 2011–2023, media releases in news archive and Retriever. Many interviews have been carried out in four related projects: SIVAC, INTRANSIT, NTRANS, and OFFLEX. In BigService (10 engineers and 3 managers), Equinor (2 managers and 1 engineer), as well as a range of industry experts, policy analysts, and researchers (20 + ). In this chapter, interviews are used as supplementary material to guide the analysis. Information about data material can be obtained by contacting the corresponding author.

## Equinor

Equinor (Statoil until 2018) is a publicly listed O&G company. The Norwegian state is the majority owner (2/3). The company originated from running operations on the NCS but has moved increasingly to various geographic areas across the world. Despite its increasingly international profile, it is the central operator firm on the NCS, and thus a key customer for the O&G supplier industry in Norway (and abroad). Equinor has major influence on firm behaviour across the O&G value chain. In addition to being a lead firm in the O&G value chain, Equinor has for decades engaged with sustainability topics, with large investments in renewable energy, particularly offshore wind and in carbon reduction technologies. In renewables, small projects, and scattered R&D work without a seeming strategic orientation, had been in operations since the late 1990s, and with generous public funding, Equinor and partners have also matured a range of carbon capture technologies.

In 2007, after a merger with Norsk Hydro, Equinor announced to seek further development of technologies in carbon capture and storage (CCS) and pursuing commercially sound projects in renewable energy. In 2009 Equinor decided to limit its focus in low-carbon technologies to technologically related fields where they could use their existing knowledge and business positions, emphasizing especially offshore wind power and CCS, and divesting from, e.g., onshore wind. This strategy over time led to development of several offshore wind power projects in the UK and the US. In all its offshore wind projects, Equinor frequently used existing suppliers from their O&G operations, citing the good existing relationships as one of the reasons for such selection, which also led Norwegian supply companies to develop capabilities for the wind market at that time.

The sustainability strategy took a new turn in 2015 as the company decided to establish a business area called New Energy Solutions, focusing on creating profitable business opportunities in renewable energy and other low-carbon solutions. The same year, the company decided also to invest in the first floating wind power farm in the world, the 30 MW Hywind Scotland. The company also began increasingly paying attention to the carbon emissions in its O&G operations, and in 2015 joined the Oil and Gas Climate Initiative of oil companies established for accelerating such ambitions. These activities were institutionalized through a Climate Roadmap for 2030 issued in 2017. This plan presented an aim to reduce the emissions in O&G production by 20% by 2030, and that by the same time 15–20% of the company's investments would support "profitable renewables and low-carbon solutions".

A change in gear in engaging with sustainability took place in 2018. The sustainability orientation was made explicit when Statoil announced rebranding its name to Equinor, stating that "the name Equinor reflects the company's strategy and development towards becoming a broad energy company". In 2020, Equinor updated its Climate Roadmap by announcing to reduce the net carbon intensity of its produced energy by 50% by 2050.

In this Climate Roadmap, the company says to embed climate in its decision-making and seeking to become an "offshore wind major", but also increasingly diversify to other renewable energy technologies, and engage, e.g., in low-carbon maritime shipping fuels, such as hydrogen-produced ammonia. The stronger strategic attention to energy also increasingly translated to concrete sustainability actions, pursuing new onshore and offshore wind projects, e.g., in Poland, the USA, the UK, South America, and East Asia, but also by developing solar power projects, especially in South America. It also made investments in a range of renewable energy companies and have diversified its portfolio of renewables projects to different energy technologies and markets. In 2019, Equinor decided to invest in the Hywind Tampen project, an 88 MW floating wind power farm providing electricity for two oil fields in the North Sea. Hywind Tampen started operating in 2024 and is the largest floating wind park in the world.

In 2021, the company also made an investment decision in its carbon transport and storage joint venture with Shell and Total, named Northern Lights. The first phase of Northern Lights' infrastructure is largely publicly funded by the Norwegian state as part of the state's Longship project, seeking to realize the first full-scale CCS value chain. The joint venture offers an "open source" infrastructure for carbon transport and storage parts of the CCS value chain, where captured carbon from various point sources will be transport with a ship to an onshore terminal in Western Norway, and then pumped to a permanent storage offshore on the NCS. As one of the first CCS value chains outside of O&G production, Northern Lights is one of the first providers of large-scale carbon transport and permanent storage as a service. In 2022, Equinor was awarded with two more operatorships of carbon storage in NCS, Smeaheia and Polaris fields, where the former is foreseen to be connected with carbon pipeline from continental Europe, serving large point-source emitters.

In 2021 Equinor developed a new strategy and aimed to be "a leader in the energy transition in the path to net-zero" and in 2022 an "energy transition plan" was mandated, where concrete goals and targets for transitioning the company was set down in a "roadmap to a net-zero future". According to the plan, Equinor has three pillars—reducing carbon emissions in O&G production (50% by 2030), increased investments in renewables, particularly offshore wind, and industrialize low-carbon solutions. The company aimed for 30% of capital expenditure investments in renewables and low-carbon solutions by 2025, and according to the latest annual report (2023), the current level is 20% of gross capital expenditure investments in 2022, up from 4% in 2020.

To synthesize, Equinor began to integrate sustainability into its strategy in 2007–2017, but also had considerable exploratory activities dating back to the 1990s. Climate change was especially seen to increase the market opportunities in renewable energy and carbon capture, and strategy of the company was thus accordingly shifted to seeking to exploit such opportunities alongside continued O&G activities. Moreover, our interviews indicated that the engagement with sustainability topics was also motivated by gaining more positive image as an employer, and by the requirements of social acceptance and maintaining the "licence to operate". However, the renewable projects in the early phase were still few and dwarfed by the scale of activities in O&G. After 2018, the renamed company Equinor engaged in climate change to a larger degree, having more offshore wind projects in various parts of the world, entering other new renewables markets, thus moving beyond its initial focus on only technologically related markets like offshore wind and CCS. Moreover, the company has engaged in more multifaceted action, engaging in discussions regarding Equinor's views on climate change issues through various events and media campaigns and more public paying attention to the company's own carbon emissions and that of its suppliers. Such changes show how the company continuously sought to adapt its strategy in engaging with the growing climate change pressure, illustrating the uncertainty that climate change poses for O&G companies like Equinor. The last years' "transition plan" illustrates the complexity of the current strategy, where the company pursues multiple technologies and markets outside O&G, as well as protecting its core business in O&G.

## BigService

BigService is a large supply company providing various products and services for O&G companies, and for whom Equinor is an important customer. It is a "first-tier" supply firm, meaning that it often seeks to acquire large and encompassing contracts with O&G companies, and uses its own competences in completing and coordinating tasks, but also a network of sub-supplier companies with specialized skill sets to complete sub-sets of tasks. The company therefore acts as a "system-integrator", tying together the needs and expectations of O&G companies with the capabilities in the value chain, available either in-house or in the sub-supplier companies. The company was influenced by other firms—both higher up and lower down in the value chain. The company also has multiple subsidiary companies, providing numerous services and products to different market segments. In this analysis, we focus on BigService's first-tier supplier role and not the activities of the subsidiary companies.

BigService has engaged climate mitigation solutions in a substantial way mainly during the last seven years. The company had scattered low-carbon technology development (especially around carbon capture) in the O&G market since the 2000s. In early 2010s, the company considered entering renewable energy, particularly in the offshore wind market, to exploit its technological proves in floating and subsea technologies. The company had a portfolio of innovation projects in renewables and low-carbon solutions, but did not prioritize industrializing them, as the market was not seen as stable or lucrative enough, or they lacked alliance partners and customers to exploit their technologies in new markets. Weak incentives, lack of political support and unstable framework conditions was (and continues to be) seen as negative for future market prospects by the firm.

However later, unlike Equinor, BigService moved quickly in developing their orientation and actions around sustainability. In 2017, the company officially recognized the ongoing transition and the need for decarbonization in the energy sector and has since begun to explore the markets around renewable energy technologies, particularly offshore wind, low-carbon solutions, and more recently, in hydrogen projects. In offshore wind, activities around 2017–2018 were modest, but this technology has since then become an important focus area for the company, actively pursuing projects and contracts globally, also with success. Meanwhile, the low-carbon technology development in the O&G market has picked up speed, culminating in 2018–2020 in several awarded contracts.

The company's exploration phase with sustainability has since then quickly escalated into partial strategic reorientation. Such a change has been made explicit by the new strategic ambitions to become a "leading company in the energy transition" and "a supplier accelerating the transition to sustainable energy production". In 2021, the company set specific targets for its transition and stated that it sought to transform its revenue streams to about 30% revenues from renewables and low-carbon technologies in near future (by 2025). Around the same time (2021/22), the new lines of business developed into separate companies, for offshore wind, hydrogen, carbon capture, and a "green" investment company.

In its attempt to engage in renewable energy markets, BigService has found itself sometimes competing against Equinor, while at the same time acting as a supplier for them. BigService described such situations as delicate and sought to avoid confrontations that could affect relationships in other markets' value chains, such as in O&G. In offshore wind, BigService has been supplier to several large wind development projects by Equinor, such as Hywind Tampen, but also partnered with, e.g., Siemens, to develop solutions for offshore wind projects in the US and UK. BigService was also a supplier in publicly funded CCS projects in different capacities, and developed proprietary technologies for carbon capture for onshore industrial plants in Norway and the UK. As a supply company, collaborating with other firms to exploit new opportunities is key, and it recently entered strategic alliances and joint ventures in the wind and carbon capture sectors (2022 and 2024) and have projects with many partners and markets across the globe.

Hence, despite mainly disregarding climate change topics until a few years ago, BigService is now seeking to achieve a transformation comparable to that of Equinor. This reorientation has been driven by the recognition of a fundamental long-term reorientation of energy markets, and the emerging commercial opportunities in renewable energy and low-carbon technologies. According to the latest annual reports, the company is well on its way to achieve the target of 30% of revenues from renewables and transitional energy solutions by 2025.

1.1 Summary of Timelines and Events.

Figures 9.1 and 9.2 provide an overview of the main events in the timelines of the two firms seen together. Equinor was an early mover and has had engagement in renewables and low-carbon solutions for over three decades. There have been strategic shifts over time. Particularly since 2017 we can observe a strategic reorientation of the company and expansion of activities and investments, building on technological strengths developed over time, and expanding to other renewables markets. We also observe that BigService was a late(r) mover, but quickly picked up pace and built up a portfolio of activities in the period 2017–20. It also followed in the technological trajectories developed by Equinor in offshore wind and carbon capture. Both companies developed concrete ambitions for transitions around 2021–22 in the form of roadmaps and transition plans, with similar goals and targets.

In the next section, we will discuss these developments considering the framework of leader–follower relations.

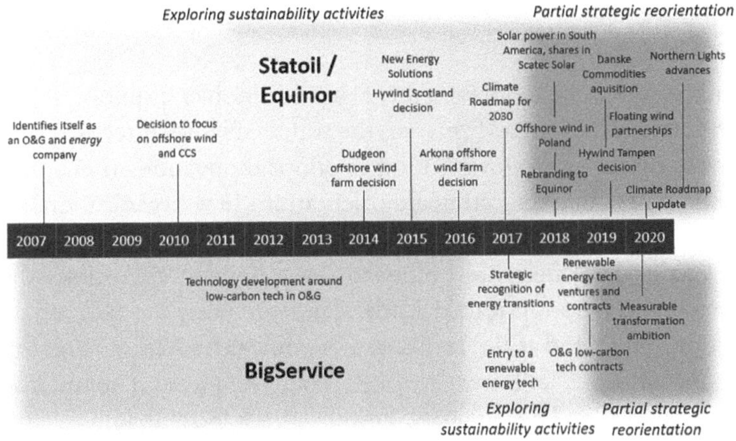


**Fig. 9.1** Timeline of main developments in the two firms (2007–2020)

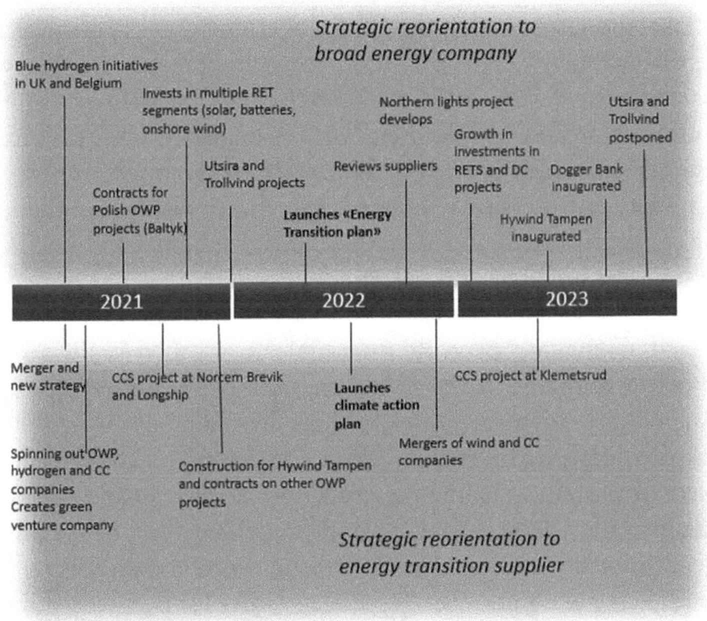


**Fig. 9.2** Timeline of main developments in the two firms (2021–2023)

## Analysis

In this chapter, we address how a lead firm like Equinor influences the behaviour of other firms in the value chain—directly and indirectly. As discussed, followers tend to adopt innovations (new products, entry into new markets, strategic orientation, new organizational developments, etc.) that have already been adopted by other firms in their value chain. The literature points to the following conditions that are likely to increase adoption: if leaders and followers are part of a cohesive community and if the leader is a "significant other" for the follower and can influence followers through coercive means (Abrahamson & Rosenkopf, 1983; DiMaggio & Powell, 1983; Boxenbaum & Jonsson, 2017). In addition, a high degree of uncertainty is associated with adoption. In our case study, these conditions are met, and we discuss them in turn.

To take the latter condition first, the companies experience a high degree of uncertainty. The long-term economic prospects of Norwegian companies in the O&G value chain are highly uncertain as the NCS is a maturing oil region with limited assets to be produced. This economic uncertainty coincided with increased uncertainty in the "institutional environment" of the firms. Increasingly, the public and policymakers see oil as controversial, and there has been public debates on the need to set an "expiry date" on oil production. Embracing sustainability as a strategic issue, diversification into renewable energy and activities connected to making oil carbon-neutral are ways to build a case for continued investment in the O&G market in the future. Investments in sustainability-oriented innovations are by many actors in the O&G seen as providing them with a "licence to operate". Thus, leaders and followers of the value chain alike experience economic and institutional uncertainty, and they are looking and modelling their behaviour to fit new institutional conditions and long-term economic prospects.

For BigService, Equinor is a "significant other" and a successful firm to imitate. As the empirical story of the two companies describe, BigService developed substantial interest and concrete activities in sustainability-oriented innovations (including new green technologies and markets) only after Equinor had developed a strategic initiative in renewable energy, in addition to its main operations in O&G. We interpret this as a "mimetic change", where BigService experienced uncertainty and modelled their behaviour after a significant other firm that had dealt with the same uncertainty in a specific way. Equinor's strategic commitment to new energy markets in 2017/18 reduced the uncertainty and functioned as a signal for other firms in the value chain, such as BigService, to take low-carbon markets more seriously. This might also have functioned as a competitive catalyst for change, as suppliers become "afraid of losing out" on new business opportunities (Liebermann & Montgomery, 1988).

The results suggest that firms in the value chain can be influenced by the behaviour of others, and particularly of firms that occupy a powerful position in the value chain. Due to both the structure of the value chain and the position that Equinor has in it, Equinor influences other companies in multiple ways. The most obvious example is that Norwegian O&G suppliers have been able to enter, e.g., offshore wind power market as suppliers for Equinor's offshore wind power projects. Supply companies, including BigService, remarked that Equinor's offshore wind activities represented a new and alternative market to utilize the suppliers' competence. "Piggybacking" on Equinor into new terrain offered suppliers what they saw as an easy route to experiment with new solutions and business models in new markets (Hansen & Steen, 2015; Steen & Weaver, 2017). Equinor's offshore wind projects offered suppliers an opportunity to gain experience and get a portfolio of "reference projects". Thus, the influence of Equinor on its suppliers has been more like a carrot than a stick and providing new business opportunities has been the most effective way to influence supplier behaviour.

The analysis indicated that BigService's outlook to renewable and low-carbon markets was influenced by its customers' activities in such themes, evidence was also found to point to the importance of generally improving market opportunities in the renewables market, and the perceived paradigmatic energy transition going on in the sector. The quick adoption and development of substantial activities in green technologies and new markets of BigService underpin this interpretation and indicate that BigService's actions were at least partly motivated by the "fear of losing out".

Equinor as a powerful player in its value chain also influences suppliers in general and BigService in particular through its ability to direct and control its suppliers. Equinor is the most important customer for supply companies such as BigService (as are other oil companies). Therefore, standards, qualification criteria, and contracts with operators are effective means to influence suppliers (Alexander, 2020). We have not investigated the use of coercive mechanisms such as tenders and contracts (that are not publicly available) but note that these are likely to play a role in supply companies' behaviour, including their engagement in sustainability. We do however see that BigService started to set specific sustainability targets and reporting of its activities, which may partly be due to new demands from clients and stakeholders as manifested in contracts and tenders. It is interesting to observe that both companies developed and implemented sustainability roadmaps with concrete targets for growth in renewable and low-carbon segments as well as emission reduction at the same time and with almost identical targets.

## Discussion

The study provides insights into incumbents' roles in transitions. Prior research on firms and industries in sustainability transitions has addressed incumbents as leaders or laggards, where most firms may indeed be best characterized as "followers" (Turnheim & Sovacool, 2020; van Mossel et al., 2018). Understanding what motivates followers to adopt sustainability-oriented innovations is important, as most firms are followers and not first movers (Bohnsack et al., 2020). This chapter has contributed to this end by highlighting a value chain perspective on incumbents' efforts to reorient (Schrage & Rasche 2022; Chizaryfard & Karakaya, 2022).

Moreover, while the recent literature on value chain interactions in sustainability transitions has focused on the level of technology value chains and subsystems (Andersen & Markard, 2020; Mäkitie et al., 2022), we contribute to the literature on multi-system interactions with a strategic perspective that highlights the role of market-based and institutional linkages between firms in value chains. More specifically, we draw on organization science to propose three actor level processes that may drive or hinder sustainability reorientations within value chains through inter-organizational relationships: mimetic behaviour under uncertainty, field cohesion, and power/dependence relations.

The exploration of these processes may help to, we argue, understand the role that value chain relationships may play in shaping firm behaviour in sustainability transitions. In particular, a better understanding of micro-dynamics may help to explain the shape and direction that novel technology value chains may take (Andersson et al., 2021). Novel technology value chains, for example those around renewable energy, zero-emission transport, and agri-food technologies, may get populated by diversifying firms that follow lead firms into these novel product-markets. This, then, not only affects which actors are likely to take leading positions in the emerging new value chain, but also the direction of technology development (e.g., leading to path dependence in technological knowledge), and the kind of institutions and business models that are established. The features of novel technology value chains may thus become based on the prior inter-firm relationships in other (older) value chains. We thus argue that perspectives from organization science, such as on bandwagons and isomorphism, can provide explanatory power to the analysis of multi-system interaction processes, providing more nuance to inter-organizational factors that influence, for example, the mechanisms of multi-system interactions (Andersen & Geels, 2023) and actor behaviour (Löhr & Chlebna, 2023).

## Conclusion

In this chapter, value chain relationships and how they may influence companies' behaviour in sustainability transitions has been addressed. We find that uncertainty about the market outlook, competitive pressures, and dependence are key factors, but also that the cohesiveness of the value chain and the privileged position that lead firms play as "captains of industries" are important and often overlooked in the literature. The behaviours of firms in a value chain are of course also influenced by policy, regulatory changes, and social factors that we have not addressed, but information about broader macro-changes is also channelled through behaviours of lead firms in the value chain that can make sustainability reorientation possible, viable, and expected. Our work has only scratched the surface in exploring such potential effects and further research should seek to assess such relationships between firms in various parts of a value chain, rather than in just few selected companies. Moreover, while our analysis has focused on describing and explaining the observed firm behaviour, further research should also seek to explore the specific effects that such relationships may have for sustainability transitions.

**Acknowledgements** The empirical study was funded by the Research Council of Norway to research centre INTRANSIT. We thank the handling editor and two reviewers for useful comments to the chapter.

## References

Abrahamson, E., & Rosenkopf, L. (1993). Institutional and Competitive Bandwagons: Using Mathematical Modeling as a Tool to Explore Innovation Diffusion. *The Academy of Management Review, 18*(3), 487–517. https://doi.org/10.2307/258906\

Acha, V., & Cusmano, L. (2005). Governance and co-ordination of distributed innovation processes: Patterns of R&D co-operation in the upstream petroleum industry. *Economics of Innovation and New Technology, 14*(1–2), 1–21. https://doi.org/10.1080/1043859042000228651

Alexander, R. (2020). Emerging roles of lead buyer governance for sustainability across global production networks. *Journal of Business Ethics, 162*, 269–290.

Andersen, A. D., & Geels, F. (2023). Multi-system dynamics and the speed of net-zero transitions: Identifying causal processes related to technologies, actors, and institutions. *Energy Research & Social Science*, 102.

Andersen, A. D., & Gulbrandsen, M. (2020). The innovation and industry dynamics of technology phase-out in sustainability transitions: Insights from diversifying petroleum technology suppliers in Norway. *Energy Research & Social Science, 64*, 101447. https://doi.org/10.1016/j.erss.2020.101447

Andersen, A. D., Mäkitie, T., Steen, M., & Wanzenböck, I. (2024). *Integrating industrial transformation and sustainability transitions research through a multi-sectoral perspective*. Available at SSRN. https://dx.doi.org/10.2139/ssrn.4766925

Andersen, A. D., & Markard, J. (2020). Multi-technology interaction in socio-technical transitions: How recent dynamics in HVDC technology can inform transition theories. *Technological Forecasting and Social Change, 151*, 119802. https://doi.org/10.1016/j.techfore.2019.119802

Andersson, J., Hellsmark, H., & Sandén, B. (2021). The outcomes of directionality: Towards a morphology of sociotechnical systems. *Environmental Innovation and Societal Transitions, 40*, 108–131.

Bach, M. (2019). The oil and gas sector: from climate laggard to climate leader? *Environmental Politics, 28*(1), 87–103.

Bansal, P., & Roth, K. (2000). Why companies go green: A model of ecological responsiveness. *The Academy of Management Journal, 43*(4), 717–736. https://doi.org/10.2307/1556363

Berggren, C., Magnusson, T., & Sushandoyo, D. (2015). Transition pathways revisited: Established firms as multi-level actors in the heavy vehicle industry. *Research Policy, 44*(5), 1017–1028. https://doi.org/10.1016/j.respol.2014.11.009

Blomgren, A., Quale, C., Austnes-Underhaug, R., Harstad, A. M., Fjose, S., Wifstad, K., Mellbye, C., Amble, I. B., Nyvold, C. E., Steffensen, T., Viggen, J. R., Iglebæk, F., Arnesen, T., & Hagen, S. E. (2015). *Industribyggerne 2015* (Rapport IRIS 2015/031, Issue. http://gammelweb.iris.no/internet/student.nsf/199f312efd2a0cacc125680e00635b85/5adee19e7e2f45d3c12580ca00417e81/$FILE/Industribyggerne%202015%20Rapport%20IRIS%202015%20031%20-%20160315.pdf

Bohnsack, R., Kolk, A., Pinkse, J., & Bidmon, C. M. (2020). Driving the electric bandwagon: The dynamics of incumbents' sustainable innovation. *Business Strategy and the Environment, 29*(2), 727–743. https://doi.org/10.1002/bse.2430

Boxenbaum, E., & Jonsson, S. (2017). Isomorphism, Diffusion and Decoupling: Concept Evolution and Theoretical Challenges. In R. Greenwood, C. Oliver, T. B. Lawrence, & R. E. Meyer (Eds.), *The SAGE Handbook of Organizational Institutionalism* (pp. 77–97). SAGE Publications Ltd. https://doi.org/10.4135/9781526415066

Bulah, B. M., Tziva, M., Bidmon, C., & Hekkert, M. P. (2023). Incumbent entry modes and entry timing in sustainable niches: The plant-based protein transition in the United States, Netherlands, and United Kingdom. *Environmental Innovation and Societal Transitions, 48*, 100735. https://doi.org/10.1016/j.eist.2023.100735

Chizaryfard, A., & Karakaya, E. (2022). The value chain dilemma of navigating sustainability transitions: A case study of an upstream incumbent company. *Environmental Innovation and Societal Transitions, 45*, 114–131. https://doi.org/10.1016/j.eist.2022.10.002

DiMaggio, P. J., & Powell, W. W. (1983). The iron cage revisited: institutional isomorphism and collective rationality in organizational fields. *American Sociological Review, 48*(2), 147–160. https://doi.org/10.2307/2095101

Farla, J., Markard, J., Raven, R., & Coenen, L. (2012). Sustainability transitions in the making: A closer look at actors, strategies and resources. *Technological Forecasting & Social Change, 79*(6), 991–998. https://doi.org/10.1016/j.techfore.2012.02.001

Ferns, G., Amaeshi, K., & Lambert, A. (2019). Drilling their own graves: How the European oil and gas supermajors avoid sustainability tensions through mythmaking. *Journal of Business Ethics, 158*, 201–231.

Geels, F. W. (2011). The multi-level perspective on sustainability transitions: Responses to seven criticisms. *Environmental Innovation and Societal Transitions, 1*(1), 24–40. https://doi.org/10.1016/j.eist.2011.02.002

Geels, F. W. (2021). From leadership to followership: A suggestion for interdisciplinary theorising of mainstream actor reorientation in sustainability transitions. *Environmental Innovation and Societal Transitions, 41*, 45–48.

Greve, H. R. (1995). Jumping ship: The diffusion of strategy abandonment. *Administrative Science Quarterly, 40*(3), 444–473. https://doi.org/10.2307/2393793

Hansen, G. H., & Steen, M. (2015). Offshore oil and gas firms' involvement in offshore wind: Technological frames and undercurrents. *Environmental Innovation and Societal Transitions, 17*, 1–14. https://doi.org/10.1016/j.eist.2015.05.001

Haunschild, P. R., & Miner, A. S. (1997). Modes of interorganizational imitation: The effects of outcome Salience and Uncertainty. *Administrative Science Quarterly, 42*(3), 472–500. https://doi.org/10.2307/2393735

Karltorp, K., & Sandén, B. A. (2012). Explaining regime destabilisation in the pulp and paper industry. *Environmental Innovation and Societal Transitions, 2*, 66–81. https://doi.org/10.1016/j.eist.2011.12.001

Kenner, D., & Heede, R. (2021). White knights, or horsemen of the apocalypse? Prospects for Big Oil to align emissions with a 1.5 C pathway. *Energy Research & Social Science, 79*, 102049.

Kungl, G. (2024). Challenges of the current discourse on incumbent firms in sustainability transitions. *Energy Research & Social Science, 108*, Article 103367. https://doi.org/10.1016/j.erss.2023.103367

Lieberman, M. B., & Montgomery, D. B. (1988). First-mover advantages. *Strategic Management Journal, 9*(S1), 41–58.

Löhr, M., & Chlebna, C. (2023). Multi-system interactions in hydrogen-based sector coupling projects: System entanglers as key actors. *Energy Research & Social Science, 105*, 103282.

Magnusson, T., & Werner, V. (2022). Conceptualisations of incumbent firms in sustainability transitions: Insights from organisation theory and a systematic literature review. *Business Strategy and the Environment, 32*(2), 903–919. https://doi.org/10.1002/bse.3081

Malhotra, A., Schmidt, T. S., & Huenteler, J. (2019). The role of inter-sectoral learning in knowledge development and diffusion: Case studies on three clean energy technologies. *Technological Forecasting and Social Change, 146*, 464–487. https://doi.org/10.1016/j.techfore.2019.04.018

Markard, J., Geels, F. W., & Raven, R. (2020). Challenges in the acceleration of sustainability transitions. *Environmental Research Letters, 15*(8), 081001.

Markard, J., Raven, R., & Truffer, B. (2012). Sustainability transitions: An emerging field of research and its prospects. *Research Policy, 41*(6), 955–967. https://doi.org/10.1016/j.respol.2012.02.013

Markard, J., & Truffer, B. (2008). Actor-oriented analysis of innovation systems: Exploring micro–meso level linkages in the case of stationary fuel cells. *Technology Analysis & Strategic Management, 20*(4), 443–464. https://doi.org/10.1080/09537320802141429

Mäkitie, T. (2020). Corporate entrepreneurship and sustainability transitions: Resource redeployment of oil and gas industry firms in floating wind power. *Technology Analysis & Strategic Management, 32*(4), 474–488. https://doi.org/10.1080/09537325.2019.1668553

Mäkitie, T., Andersen, A. D., Hanson, J., Normann, H. E., & Thune, T. M. (2018). Established sectors expediting clean technology industries? The Norwegian oil and gas sector's influence on offshore wind power. *Journal of Cleaner Production, 177*, 813–823. https://doi.org/10.1016/j.jclepro.2017.12.209

Mäkitie, T., Hanson, J., Steen, M., Hansen, T., & Andersen, A. D. (2022). Complementarity formation mechanisms in technology value chains. *Research Policy, 51*(7), Article 104559. https://doi.org/10.1016/j.respol.2022.104559

Mäkitie, T., Normann, H. E., Thune, T. M., & Sraml Gonzalez, J. (2019). The green flings: Norwegian oil and gas industry's engagement in offshore wind power. *Energy Policy, 127*, 269–279. https://doi.org/10.1016/j.enpol.2018.12.015

Nilsen, T. (2017). Innovation from the inside out: Contrasting fossil and renewable energy pathways at Statoil. *Energy Research & Social Science, 28*, 50–57.

O'Regan, N., Sims Martin, A., & Gallear, D. (2008). Leaders, loungers, laggards: The strategic-planning-environment- performance relationship re-visited in manufacturing SMEs. *Journal of Manufacturing Technology Management, 19*(1), 6–21. https://doi.org/10.1108/17410380810843426

Pick, M. J. (2019). The renewable energy strategies of oil majors–From oil to energy? *Energy Strategy Reviews, 26*, 100370.

Rosenbloom, D. (2020). Engaging with multi-system interactions in sustainability transitions: A comment on the transitions research agenda. *Environmental Innovation and Societal Transitions, 34*.

Schupfer, H., & Soppe, B. (2024). Greening from within: The role of organisational purpose shift in building internal legitimacy for fossil fuel incumbents' green innovation. *Industry and Innovation*, 1–24.

Schrage, S., & Rasche, A. (2022). Inter-organizational paradox management: How national business systems affect responses to paradox along a global value chain. *Organization Studies, 43*(4), 547–571.

Smink, M. M., Hekkert, M. P., & Negro, S. O. (2015). Keeping sustainable innovation on a leash? Exploring incumbents' institutional strategies. *Business Strategy and the Environment, 24*(2), 86–101. https://doi.org/10.1002/bse.1808

Stalmokaitė, I., & Hassler, B. (2020). Dynamic capabilities and strategic reorientation towards decarbonisation in Baltic Sea shipping. *Environmental Innovation and Societal Transitions, 37*, 187–202.

Steen, M., & Weaver, T. (2017). Incumbents' diversification and cross-sectorial energy industry dynamics. *Research Policy, 46*(6), 1071–1086. https://doi.org/10.1016/j.respol.2017.04.001

Stephan, A., Schmidt, T. S., Bening, C. R., & Hoffmann, V. H. (2017). The sectoral configuration of technological innovation systems: Patterns of knowledge development and diffusion in the lithium-ion battery technology in Japan. *Research Policy, 46*(4), 709–723. https://doi.org/10.1016/j.respol.2017.01.009

Thune, T., & Mäkitie, T. (2018). Chapter 11. Versatile competences and product market diversification among oil and gas supply firms. In T. Thune, O. A. Engen, & O. Wicken (Eds.), *Petroleum industry transformations: Lessons from Norway and Beyond*. Routledge.

Thune, T., Wicken, O., & Engen, O. A. (2018). *Petroleum industry transformations: Lessons from Norway and Beyond*. Routledge.

Turnheim, B., & Sovacool, B. K. (2020). Forever stuck in old ways? Pluralising incumbencies in sustainability transitions. *Environmental Innovation and Societal Transitions, 35*, 180–184. https://doi.org/10.1016/j.eist.2019.10.012

van Mossel, A., van Rijnsoever, F. J., & Hekkert, M. P. (2018). Navigators through the storm: A review of organization theories and the behavior of incumbent firms during transitions. *Environmental Innovation and Societal Transitions, 26*, 44–63. https://doi.org/10.1016/j.eist.2017.07.001

Wesseling, J. H., Niesten, E. M. M. I., Faber, J., & Hekkert, M. P. (2015). Business Strategies of Incumbents in the Market for Electric Vehicles: Opportunities and Incentives for Sustainable Innovation. *Business Strategy and the Environment, 24*(6), 518–531. https://doi.org/10.1002/bse.1834

# 10

# Navigating Paradoxes in Sustainability Transitions: How Incumbent Firms (Can) Foster Circular Economy

Eli Fyhn Ullern and Sigrid Damman

## Introduction

Climate change, resource scarcity, and loss of biodiversity pose significant threats to sustainable development. Achieving net-zero by 2050 is crucial for our future, as it represents a critical threshold to limit the most severe environmental challenges (UN, 2024). Against this backdrop, the circular economy (CE) is gaining prominence in policy, academia, and industry as a pathway to reduce pressure on natural resources and extend the value of materials and products (Geissdoerfer et al., 2017). However, the global circularity rate declined to 7.2% in 2024 (Fraser et al., 2024).

E. F. Ullern (✉)
Department of Industrial Economics and Technology Management, Norwegian University of Science and Technology (NTNU), Trondheim, Norway
e-mail: eli.fyhn.ullern@sintef.no

E. F. Ullern · S. Damman
Department of Technology Management, SINTEF Digital, Trondheim, Norway


M. Steen et al. (eds.), *Sustainability Transitions and Industrial Transformation*,
https://doi.org/10.1007/978-981-95-4893-4_10

In the EU, more than 8 billion tonnes of materials are used annually, yet only 11.5% come from secondary materials (European Environment Agency, 2023). This highlights the need for more circular practices.

Industries that rely on plastics derived from fossil resources face pressing concerns, as linear consumption contributes significantly to $CO_2$-emissions and pollution (Johansen et al., 2022). Although the latest discussions on the global plastic treaty have not yielded a final agreement, they signal future implications for both plastic producers and consumers. The EU has defined specific targets for mandatory recycled content, innovation, and investment to enable more sustainable plastics (European Commission et al., 2024), and an increasing number of countries are taking similar steps, e.g., the Circular Economy Action Plan for Norway identifies plastics as one of the seven prioritized value chains (Norwegian Government, 2021). Adopting CE strategies, such as incorporating recycled post-consumer plastics[1] (PCP) waste in circular business models and value chains (Dijkstra et al., 2020), offers industries an opportunity to increase value creation while fostering innovation and sustainability transition.

Within socio-technical transition studies, sustainability transitions are conceptualized as long-term, multi-dimensional, and fundamental transformations towards more sustainable production and consumption patterns (Markard et al., 2012). These shifts necessitate alterations in various elements, including markets, technologies, infrastructures, and user practices (Köhler et al., 2019). From a socio-technical perspective, drivers and barriers are shaped by the interaction between processes at different levels of structuration (Geels, 2002). Studies on CE and innovation often identify drivers and barriers to transitioning, particularly linked to changes in regulations, markets, technologies, or new ways of organizing to adopt CE strategies (De Jesus & Mendonça, 2018). Exploring how actors navigate these complex system changes is generating increasing interest in the field of sustainability transitions, as it provides insights into the dynamics that hinder or foster transitions.

[1] Post-consumer plastics refer to materials that have fulfilled their intended purpose or can no longer be used ISO (2008). Plastics–Guidelines for the recovery and recycling of plastics waste. *ISO 15270: 2008 (E).*

The transition to CE requires shifting away from the dominant take-make-dispose approach, towards business models and value chains that enhance value by slowing (i.e., extending a product's useful life), closing (creating more circular flows through recycling), and narrowing resource loops (i.e., using fewer resources per product) (Bakker et al., 2014; Bocken et al., 2016). Incumbent firms, a category commonly used for large companies with a long history and significant political and economic power (Kungl, 2025), are often major producers and users of materials, playing a crucial role in this transition. In the European furniture industry, firms are increasingly driven to use recycled PCP to reduce their environmental footprint (Fact.MR, 2019). This shift is further supported by regulatory frameworks like the European Green Deal, the New Circular Economy Action Plan[2], and the Ecodesign for Sustainable Products Regulation (ESPR).[3] Additionally, Extended Producer Responsibility (EPR) schemes will place increased responsibility on manufacturers using plastic as an input factor. Given that approximately 80% of a product's environmental impact is determined during its design phase (European Commission, 2016), the choice of materials and design for longevity has a significant impact on the overall environmental footprint of manufacturing firms. For example, by adopting circular product design to promote more sustainable resource use in business models (Bocken et al., 2016). Yet, few incumbent firms have succeeded in replacing existing business models with scaled circular business models (Frishammar et al., 2025).

While the use of recycled PCP to increase circularity is gaining recognition, the ability to scale this strategy depends on access to a stable and reliable supply. This highlights the critical importance of developing new forms of collaboration with actors across existing value chains and industries to more effectively balance supply and demand (Fehrer & Wieland,

---

[2] EU Commission (2020). Retrieved 26.06.2024 from: https://eur-lex.europa.eu/legal-content/EN/TXT/?qid=1583933814386&uri=COM:2020:98:FIN

[3] EU Commission (2024). Retrieved 26.11.2024 from: https://commission.europa.eu/energy-climate-change-environment/standards-tools-and-labels/products-labelling-rules-and-requirements/ecodesign-sustainable-products-regulation_en

[3] 3 EU Commission (2024). Retrieved 26.11.2024 from: https://commission.europa.eu/energy-climate-changeenvironment/standards-tools-and-labels/products-labelling-rules-and-requirements/ecodesign-sustainable-products-regulation_en

2021; Hahladakis & Iacovidou, 2019). However, utilizing resources already in circulation poses a dual challenge for the manufacturing sector: extending product life cycles while also incorporating post-consumer resources from other industries into new products (European Commission, 2016). Indeed, as incumbent firms explore new circular business models, they may encounter a range of tensions and paradoxes (Daddi et al., 2019; Dagilienė & Varaniūtė, 2023; De Angelis, 2021).

Paradox theory is suited to enhance our understanding of these dynamics, as it explores the co-existence of "contradictory yet interrelated elements that exist simultaneously and persist over time" (Smith & Lewis, 2011, p. 382). A paradox lens may help both practitioners and researchers illuminate how organizations experience and respond to tensions (Jarzabkowski et al., 2013; Jarzabkowski et al., 2019), such as balancing short-term and long-term goals (Slawinski & Bansal, 2015) or reconciling the use of recycled materials with the competitiveness of products (Daddi et al., 2019). Furthermore, paradox theory seems suitable to explore how incumbents experience drivers and barriers across scales in the transition to CE. While paradox theory has been applied as a lens to CE innovation, particularly when studying business models (Dagilienė & Varaniūtė, 2023; De Angelis, 2021; Pascucci et al., 2024), we believe that the full potential of this perspective remains untapped in the field of sustainability transitions. Applying paradox theory to understand incumbent actors is highly relevant, as it offers a lens to understanding the complex and often contradictory forces these actors encounter at the interface of sectors and socio-technical systems.

Incumbents have been widely studied in the field of transitions, and there are many theoretical conceptions of such actors. Against this background, we follow Kungl (2025), who defines incumbents narrowly as "the firms that, at a certain point in a sustainability transition process, hold a central position for the reproduction of an existing socio-technical system" (Kungl, 2025, p. 10). Incumbency has been defined in different ways (Kungl, 2024), e.g., as "profit-seeking actors that are 'established' and 'positioned' in markets" (Steen & Weaver, 2017), and "actors that are deeply entrenched in the socio-technical regime" (Galvan et al., 2020, p. 79). Despite the variety of perspectives, there has been a tendency to view incumbents as coherent actors (Magnusson & Werner, 2023).

Recent studies, however, point to the wide range of strategies incumbents apply, differentiating between defensive and proactive strategies (Saleh et al., 2025). Notably, Magnusson and Werner (2023) recently suggested paradoxes as a promising avenue for future research to capture the heterogeneity of how incumbent firms navigate sustainability transitions. By applying a paradox lens, we can gain deeper insights into transition dynamics and the strategies incumbent firms use to navigate these complex changes (Turnheim & Sovacool, 2020).

To address the actions that influence transition dynamics, Schaltegger et al. (2023) call for more action-oriented research on how incumbents can contribute to transition beyond organizational boundaries. We respond to this call through a longitudinal, qualitative case study of Flokk, a European furniture manufacturing company adopting CE principles. Our chapter aims to bridge paradox research with the sustainability transitions literature. We contribute to deepening the understanding of transition dynamics by illustrating how incumbents navigate CE transition through strategy development and exploration of transformation opportunities in interaction with other actors. We hope to help researchers and industry practitioners understand:

1. How a paradox lens may inform research on incumbents in transition towards CE
2. How strategies at the firm level become a key driver of interactions across sectors and may drive system-level change towards CE

The chapter is structured as follows: In the next section, we provide a brief background on paradox theory and the role of incumbents in transition processes. Following section outlines the research design. The case study employs an applied research approach, engaging with the firm through a series of workshops and observations over time. We then present the analysis and discuss how incumbent firms navigate paradoxical tensions in transition to CE and highlight how a paradox perspective enriches the understanding of incumbent firms in industrial transformation. Finally, we conclude with implications and suggest avenues for future research.

## Bridging Paradox Research and Sustainability Transitions

### The Paradox Lens

The paradox perspective has gained traction in organization and management studies over the past 25 years (Schad et al., 2016), offering a valuable lens for addressing grand challenges that are interwoven and complex by nature (Smith et al., 2017). A paradox is defined as "contradictory yet interrelated elements that exist simultaneously and persist over time" (Smith & Lewis, 2011, p. 382). These elements can be inherent in organizations and systems, and socially constructed based on actors'perceptions (Hahn & Knight, 2021; Schad & Bansal, 2018; Smith & Lewis, 2011).

While traditional approaches treat tensions as trade-offs or dilemmas, the paradox lens emphasizes how firms navigate interdependent demands through a both/and approach (Jarzabkowski et al., 2013; Smith & Lewis, 2011). This is undoubtedly true when it comes to sustainability, which Van der Byl and Slawinski (2015, p. 55) argue is "rife with tensions" due to conflicting economic, social, and environmental demands that unfold across different scales and timeframes (Hahn et al., 2018). Besides, these tensions often lie beyond the control of a single actor (Smith et al., 2017), making them particularly complex to navigate. Indeed, tensions between core business and societal needs may lead to defensive strategies (Pinkse et al., 2019), such as protecting established business models and operations. However, tensions can also help companies identify potential ways to address complex issues and inform their decision-making (Van der Byl & Slawinski, 2015), e.g., through sensemaking (van Bommel, 2018). For instance, research on product design companies shows that paradoxes can simultaneously fuel and frustrate innovation, underscoring the need to navigate paradoxes (Andriopoulos & Lewis, 2010).

Although the paradox literature has focused on individual and organizational levels of analysis, adopting a systems perspective is essential for enhancing our understanding of how tensions arise and how they can best be managed (Carmine & De Marchi, 2023; Hahn et al., 2018). Incumbent firms face seemingly contradictory demands in CE transitions, e.g., in terms of calls for using recycled materials and requirements for qualities and product properties that match existing designs and processes while performing equally or better than existing solutions based on virgin resources (Franco, 2017). Recently, literature on CE has started to engage with the paradox literature to explore paradoxical tensions (Daddi et al., 2019; Dagilienė & Varaniūtė, 2023; De Angelis, 2021; Tosi et al., 2024). In particular, these studies have explored how the development of circular business models poses tensions between competing demands and objectives. For instance, illuminating how incumbent firms face the paradox of disrupting their existing profitable business model while not changing may pose a risk in a changing market (Kuhlmann et al., 2023). However, there is limited research on the tensions that arise in practice for incumbent firms in transition to CE (Dagilienė & Varaniūtė, 2023; Pascucci et al., 2024).

## Incumbent Strategies in Transition Processes

Incumbent actors are at the core of transition processes, shaping change across micro-, meso-, and macro levels through interactions that drive change (Geels, 2002). Within transition studies, incumbents have predominantly been depicted as actors who prevent or slow down transitions and are thus associated with regime stability (Smink et al., 2015). However, previous research illustrates that incumbents also contribute actively to sustainability transitions. Case studies from various capital-intensive industries show that incumbents can strategically hedge or diversify towards niche innovations (Bergek et al., 2013; Berggren et al., 2015; Penna & Geels, 2015; Steen & Weaver, 2017), e.g., through new business models (Bidmon & Knab, 2018). Overall, recent contributions shed light on a variety of behaviours and roles, depending on markets, value chain segments, and organizational capabilities. Thus, Turnheim and Sovacool (2020) argue for a more pluralized understanding of incumbencies and the conditions when these may contribute to transformative pathways, calling for more emphasis on 1) a multiplicity of incumbent actor types, 2) a variety of actor strategies within (and across) organizational populations, 3) the transient nature of strategic positioning (over time), and 4) the varied resources incumbents may deploy to support transformative change. For example, a recent study highlights external factors, such as policy support and internal competition, alongside internal factors, including economic feasibility and strategic mindsets, that shape incumbents‘strategies (Gregory & Geels, 2024).

To better understand incumbent behaviour, Magnusson and Werner (2023) propose two spectra: innovation/defending and collaborating/competing. Although studies highlight the role of incumbents in driving industrial transformation, such as in the case of electric vehicles (Bohnsack et al., 2020), they also risk cherry-picking certain incumbents‘active roles and neglecting conservative behaviour (Magnusson & Werner, 2023). Besides, Turnheim and Geels (2019) point especially to the role of incumbent actors from neighbouring regimes and their early involvement in niche innovation. A key question, therefore, is how paradoxes may foster such interactions across established value chains and markets.

In the following section, we apply these theoretical insights to an empirical case study of Flokk. By exploring the paradoxical tensions Flokk encounters and the CE strategies they employ in response, we aim to illustrate how a paradox lens can be fruitful to understand what fosters incumbent strategies in CE transitions.

## Research Setting

### The Furniture Manufacturer Flokk

We conducted a qualitative, in-depth case study (Langley, 1999; Yin, 2009) of Flokk, a European furniture manufacturing company. A case study is suitable for illustrating the complexities that incumbent firms in the manufacturing industry encounter in transitioning to CE. Furniture is one of the prioritized industries in the EU Circular Economy Action Plan (European Commission, 2024) and provides a unique setting for studying this emergent phenomenon due to firms' reliance on high-quality materials combined with design and aesthetics.[4] For example, recycled materials must meet strict quality requirements, be safe, and be aesthetically appealing. While the interviews and workshops highlighted barriers the furniture manufacturing industry is facing, such as the challenge of ensuring a steady supply of high-quality recycled materials and managing reverse logistics, similar challenges are also relevant to other industries undergoing CE transition. These issues, particularly those related to materials, are common across sectors like automotive, electronics, and construction, where integrating recycled materials presents both challenges and opportunities for innovation.

Flokk is headquartered in Norway and operates production sites across Europe and the USA. The company has a portfolio of nine office furniture brands and is currently one of the leading actors in this market. We chose Flokk as a case study due to its long history of working with sustainability and using recycled materials since 1995, along with ambitious targets for using renewable energy, implementing circular design criteria, and reducing the use of virgin materials. As an incumbent manufacturing firm, 90% of Flokk's environmental footprint is attributed to its products, with 95% of the footprint stemming from materials consumption. Plastics, such as PP (Polypropylene), PA (Polyamide), and PES (Polyester), account for the majority of structural parts in furniture production. As furniture products do not emit pollutants during the use phase, the design and choice of materials play an important role. Flokk is guided by five design principles to enable circularity: low weight, fewer components, the right choice of materials, long lifespan, and design for disassembly. This chapter focuses on Flokk's work in exploring how new business models and value chains enable circular products based on 100% post-consumer recycled materials.

Our engagement with the company was part of a research and innovation project called "Circular Seating" (2018–2020), where the aim was

[4] EU Commission (2020). Retrieved 26.06.2024 from: https://eur-lex.europa.eu/legal-content/EN/TXT/?qid=1583933814386&uri=COM:2020:98:FIN

[3] EU Commission (2024). Retrieved 26.11.2024 from: https://commission.europa.eu/energy-climate-change-environment/standards-tools-and-labels/products-labelling-rules-and-requirements/ecodesign-sustainable-products-regulation_en

to overcome external barriers related to material quality and ensure a consistent delivery of recycled plastics to meet the company's goal of 100% recycled and recyclable materials. The project was a collaboration between practitioners from environmental management and business development, and a multidisciplinary research team including material specialists, engineers, and social scientists. While this chapter primarily builds on Circular Seating, we also followed Flokk's activity related to CE and plastics through a subsequent project ("POCOplast", 2020–2024), in which seven practice partners and two research institutions investigated new pathways for the sustainable use of PCP from aquaculture. This provided an opportunity to follow Flokk's strategic work on CE over time.

## Data Collection and Analysis

Data were collected through multiple methods, including stakeholder workshops, desk study of policy briefs and white papers, interviews, and participant observation in monthly online meetings. This triangulation was important to gain insight into how Flokk's strategies were formed in interaction between priorities and perspectives of different organizational departments, as well as pressures and tensions arising from their external environment, e.g., linked to changing policies and customer requirements.

The primary source of data for this chapter was a series of four iterative workshops in 2018–2019 (Fig. 10.1), with a dedicated project group of practitioners and researchers.

In the first workshop, we discussed the overall project aims related to Flokk's circularity targets. Between the first and the second workshops, we focused on assessing drivers and barriers to CE transition from an organizational perspective. We reviewed policy briefs and white papers to deepen our understanding of the external drivers and barriers within Flokk's broader landscape. This informed a PESTLE analysis, which provided a comprehensive overview of the external factors impacting Flokk's operations. To complement the PESTLE analysis, we conducted

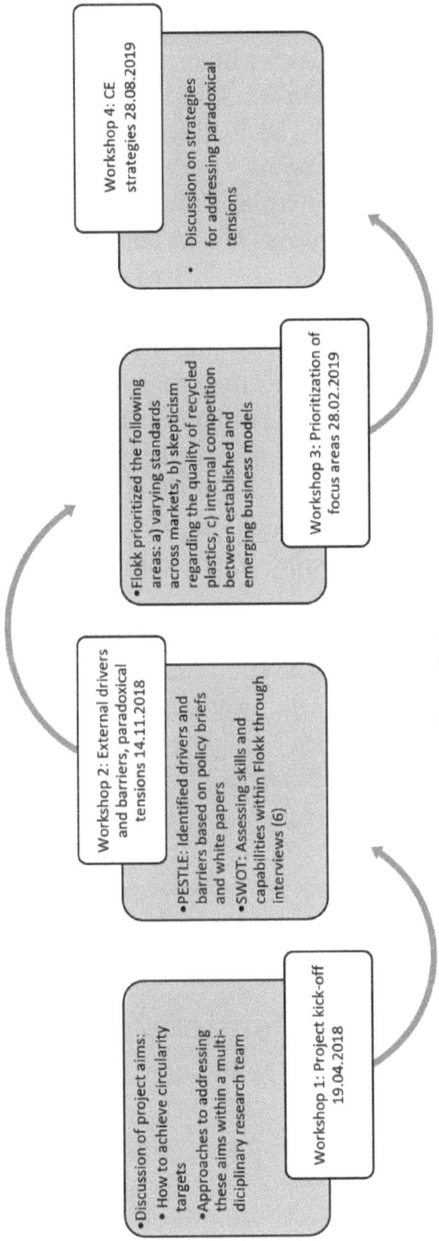


**Fig. 10.1** Overview of primary data collection and analysis

six interviews with key informants within the firm, including sustainability advisors, procurement officers, the Vice President for Environment, and the Senior Vice President of Products and Brands. These interviews aimed to identify Flokk's skills and capabilities in responding to these drivers and barriers. The findings from these interviews were integrated into a SWOT analysis to identify internal strengths and weaknesses. This process resulted in an analysis of external drivers and barriers and identified gaps between Flokk's current situation and an ideal scenario where circularity targets are achieved.

The second workshop built on these insights and explored how Flokk could address the identified drivers and barriers. We introduced paradoxical tensions as a lens to discuss the interrelationship between drivers and barriers, where researchers facilitated discussions on how Flokk could manage these paradoxical tensions. In the third workshop, Flokk prioritized focus areas based on the identified drivers and barriers. The final workshop centred on the implications for strategies, focusing on the paradoxical tensions and the various roles incumbents can take in transition processes.

In June 2021, we conducted a follow-up interview with Flokk, focusing on their search for new value chains for homogeneous sources of various recycled plastics. Additionally, 20 interviews with value chain actors were conducted to deepen our understanding of changes in Flokk's external environment, particularly regarding the sustainable use of plastics in industry in the Norwegian context. Finally, in June 2023, a workshop was held at Flokk to discuss the sustainable use of PCP from aquaculture and Flokk's long-term goals for scaling the use of recycled materials.

## Exploring Paradoxical Tensions and Incumbent Strategies

In "Circular Seating", Flokk identified three focus areas for further exploration in reaching their circularity targets: i) internal competition between established and emerging business models, ii) scepticism towards the quality of recycled plastics in the market, and iii) the divergence in standards across markets. Consistent with prior research on CE, the analysis of external drivers and barriers ranged across technical, regulatory, and organizational dimensions (De Jesus & Mendonça, 2018). In the following, we briefly highlight paradoxical tensions that became salient during the project. Furthermore, we present the identified strategies to navigate these through i) business model exploration, ii) positioning in a public policy process, iii) strategic acquisitions, iv) external collaborations, and v) R&D project partnerships.

## Paradoxical Tensions in CE Transition Processes

### Balancing Established and Emerging Business Models

To explore how CE could be adopted at the firm level, Flokk decided to explore two alternative business models. One paradoxical tension Flokk experienced was balancing these seemingly different business models. On the one hand, the company aimed to adapt its established business model by increasing the use of recycled PCP in its products. To achieve a 100% recycled materials product portfolio, particular emphasis was placed on ensuring a steady supply of recycled materials with different material properties but consistent quality, from sources such as packaging waste, the automotive industry, the agricultural sector, and the aquaculture industry. For example, while access to recycled PP had been well-established through a long-term collaboration with a Swedish supplier, recycled plastics with other material properties were harder to obtain. On the other hand, Flokk explored the concept of product-as-a-service by extending the product lifetime by selling office chairs as a service. These two business models presented distinct challenges. For example, one barrier to increasing the share of recycled PCP was customers' skøepticism about products made from recycled materials. To address this, Flokk relied on established dealers to educate customers on the benefits of recycled PCP. Simultaneously, Flokk paradoxically experienced that pilot projects for product-as-a-service accelerated product turnover and challenged Flokk's existing revenue model and organizational set-up.

Balancing the exploration of two alternative business models typically reveals tensions between exploration and exploitation (Andriopoulos & Lewis, 2009). For example, the pilot on "seating-as-a-service" highlighted tensions between long-term sustainability and short-term financial performance (Slawinski & Bansal, 2015). Consequently, balancing between existing and new business models for CE might lead to so-called cannibalism between different initiatives (Guide Jr & Vanwassenhove, 2001). Regardless of whether they sell their products as a service or substitute virgin materials with recycled PCP, the total environmental impact also depends on how the products are used after they are sold, including the possibility of repairing and creating furniture that customers will keep for as long as possible.

## Navigating Limited Value Chains for Recycled Post-Consumer Materials

A second paradoxical tension arose from the limited availability of established value chains for recycled PCP, particularly for colored materials. By substituting virgin materials with recycled ones, Flokk aimed to address the sustainability challenge of plastic waste while reducing its dependence on virgin materials over time. Since most of the environmental footprint of furniture production is determined by the choice of materials, circular material streams are crucial for enhancing its sustainability. Moreover, interviews emphasized that 2/3 of their environmental footprint stemmed from upstream value chain activities, making value chain aspects crucial.

A key challenge for Flokk was securing a consistent supply of recycled material that met its quality and safety requirements, with the opportunity of scaling across different material qualities. For example, incorporating recycled materials impacted product design, as materials had to be proven safe for long-term use. Flokk therefore established a long-term collaboration with one of its suppliers in Sweden. This collaboration fostered the development of expertise and capabilities to design new products based on specific material specifications for recycled PP. This approach contrasts with prevailing industry practice, where design typically precedes material considerations, and manufacturers work to specified designs rather than beginning with the properties and limitations of available (recycled) materials.

However, the limited availability of high-quality recycled PCP, particularly in specific colors, restricted Flokk's design options and aesthetic considerations for new products. Therefore, Flokk continuously searched for potential sources of recycled PCP. This effort revealed the paradoxical tension where Flokk's reliance on recycled PCP in different material qualities was inhibited by limited value chains for recycled materials. Flokk aimed to increase the use of recycled material sources with local origin, and therefore engaged to stimulate the Norwegian market. Despite the known challenges of high-quality PCP waste in other industries, such as aquaculture, the limited overview of the volumes and qualities limited Flokk's ability to secure materials for new products. On the other hand, using recycled PCP as a substitute for virgin materials was considered essential for Flokk's long-term viability. Interviews also highlighted that using recycled PCP is crucial, as virgin material will become more costly and less available in the future. Thus, achieving net-zero commitments while relying on an inconsistent supply of secondary materials presented a notable tension between environmental commitments and market limitations.

### Responding to Diverging Standards across Markets

The third paradoxical tension was related to the variety of standards Flokk had to comply with in different markets. As Flokk expanded its sustainability efforts, they were confronted with various environmental certifications and standards. Simultaneously, Environmental Product Declarations (EPDs) were perceived as drivers towards circular business models and value chains. Interestingly, Flokk highlighted that standards and certifications often fail to reward innovation exceeding requirements. Furthermore, standards and environmental assessment methods vary across markets, potentially hindering the development of circular value chains, for example, for PCP used in aquaculture and subsequently recycled for applications in furniture production or the construction industry. This paradoxical tension became salient as standards intended to promote circular business models and sustainability can unintentionally impede innovation. For instance, in procurement processes, the diverse approaches to measuring and verifying environmental data presented challenges that complicated their use and reliability. A key example is the term "recyclable", which has different interpretations. While some standards define "recyclable" in technical terms, Flokk argued that they should also include the necessary infrastructure, meaning a system that supports further recycling. As Flokk operates across several geographical locations, they encountered varying degrees of demand for circularity. For example, in the Netherlands, CE solutions were a strict requirement in public tenders.

## Strategies to Navigate Paradoxical Tensions

While transitioning to a CE is associated with paradoxical tensions, we observed that Flokk actively navigated these through organizational learning and innovation (Smith & Lewis, 2011). In this section, we illustrate how Flokk positioned itself through various strategies, showing the variety of roles incumbent firms may take in transition (Turnheim & Sovacool, 2020). Unlike previous literature that emphasizes incumbents prioritizing their interests and inhibiting new technology and transition (Smink et al., 2015), we find that Flokk responded to pressures and addressed CE transition through five strategies: i) business model exploration ii) positioning in a public policy process, iii) strategic acquisitions, iv) external collaborations, and v) novel R&D partnerships across sectors.

Flokk actively worked to increase circularity in its business models by incorporating recycled PCP as material for new products. This was closely linked to sourcing processes within the firm. For example, Flokk participated in a public policy process led by the Norwegian Agency for Public Management and eGovernment (Difi), aiming to establish criteria for sustainable procurement promoting CE. Although Flokk does not directly participate in public tenders, its products are included in proposals from dealers. By engaging in the policy process, Flokk was able to influence the criteria, thereby stimulating demand for recycled products. As a result, a guide promoting CE in public furniture procurement was developed. The involvement of industry representatives, researchers, and employer associations helped Flokk address issues regarding the standards used in procurement processes and environmental trade-offs in the decision criteria. Moreover, Flokk has been encouraging competition from other actors by promoting the use of EPDs. This long-standing work on EPDs has led them to actively share their experiences across markets. Moreover, the strategic acquisition of a furniture firm in the Netherlands provided new knowledge and skills to explore different business models and helped Flokk strengthen its market presence in a country with a more developed CE. This expansion provided new insights into circularity and inspired Flokk to explore similar opportunities in Norway.

Furthermore, we observed that Flokk actively collaborated with external partners to identify potential new suppliers and evaluate the quality of recycled material sources. A notable example was the discovery of discarded snowplow markers along Norwegian roads as a potential source of recycled plastics. These markers have an average lifespan of only three winter seasons, contribute to 190 tonnes of plastic waste annually, and were previously sent to incineration or landfill. Recognizing this waste fraction as a potential feedstock, Flokk initiated a pilot project that repurposed it into 200 new office chairs (see Fig. 10.2). This evolved into a partnership with the Norwegian Public Roads Administration, their main contractor, and the supplier of snowplough markers. The partnership also includes actors in snow plowing operations to ensure that discarded markers are collected.

While this developed into a stable source of recycled PCP, Flokk continued to explore other potential sources of larger volumes. As concerns regarding ocean plastics were rising on the public agenda, the company sought collaboration to enable a sustainable value chain for PCP from aquaculture (e.g., via POCOplast). Although this has not yet led Flokk to source materials from the aquaculture industry, circular value chains for recycled PCP from aquaculture are emerging (Damman et al., 2022). Flokk also engaged in a circular accounting cluster to address the financial challenges of a product-as-a-service business model.

Fig. 10.2 Picture of discarded snowplow markers and the chair HÅG Capisco Puls (Picture: Flokk)

By collaborating with other actors, Flokk overcame the barriers to this business model by navigating the contradictory demands of profitability and circularity within their organization. For example, they learned how gathering data over time could provide the necessary information to mitigate the risk of scaling this business model. These efforts illustrate how interaction with external partners presents an opportunity to learn how to manage contradictory demands (Smith & Lewis, 2011).

## Navigating Paradoxical Tensions in CE Transition Processes

The case of Flokk provides valuable insights into how paradoxical tensions can act as both barriers and drivers to CE. By exploring these tensions, we highlight the heterogeneity of responses of an incumbent firm during an ongoing CE transition. In particular, we contribute to a deeper understanding of the various strategies that incumbents employ, along with the broader implications for transition dynamics (Turnheim & Sovacool, 2020). While these strategies can foster increased interactions across sectors, we argue that Flokk's innovation activities cannot necessarily be classified as belonging to one of the typologies within the field of sustainability transitions. Our findings suggest that incumbents within furniture manufacturing face a core dependency on material resources, such as post-consumer waste, from other actors. We discuss how the identified CE strategies may foster novel interactions across sectors and present three key insights and implications for research and practice on incumbents in CE transition.

First, we highlight how a paradox lens helps us understand the challenges that the transition to CE poses for incumbent firms. The case of Flokk illustrates how adopting CE principles often entails reconciling contradictory demands (see, e.g., Daddi et al., 2019), such as meeting ambitious sustainability goals for incorporating recycled plastics while navigating constraints like the availability and quality of recycled materials. Applying a paradox lens (Smith & Lewis, 2011), we find that although these tensions can impede CE transition, they can also drive innovation. For example, access to plastic material is critical for Flokk, as it is one of the key input factors for producing office chairs. While material constraints present clear challenges, failing to adopt a circular approach can pose a risk to the long-term survival of incumbent companies like Flokk due to regulatory pressures and material resource scarcity. This aligns with Andriopoulos and Lewis (2010), who noted that paradoxical tensions can both promote and frustrate innovation in product design companies. This is particularly relevant to manufacturing firms, where regulations increasingly require companies to take responsibility for their products throughout their entire life cycle, e.g., through policy schemes such as ESPR. Together with changing customer demands, this makes their value chain increasingly complex. The case of Flokk illustrates how they navigate these complexities through a both/and approach (Smith & Lewis, 2011) in business model innovation, resulting in various strategies. In this way, the paradox perspective enriches our understanding of the complexities incumbents face during transition to CE.

Second, the identified CE strategies recognize the multiplicity of strategies incumbents play during transitions (Turnheim & Sovacool, 2020). For example, Flokk has explored the creation of new sustainable production and consumption patterns (e.g., via product-as-a-service), and the destruction of unsustainable ones (e.g., shifting from linear to circular design principles (Schaltegger et al., 2023)). Our findings suggest that although the external landscape put pressure on Flokk, change was also driven from within. Flokk's active involvement in a public policy process highlights how Flokk connects internal strategies and capabilities with external pressures and opportunities. While such initiatives may be seen as strategic positioning for incumbents, they can also contribute to institutional infrastructure and create windows of opportunity for other actors. Flokk's role as an intermediary in CE transition is evident, e.g., in their participation in R&D projects aiming to develop new value chains for recycled plastics. Such activities align with previous research on intermediaries' contributions to niche development (Sovacool et al., 2020), which emphasizes the importance of cross-regime interactions in niche innovation (Turnheim & Geels, 2019).

The diversity of identified CE strategies challenges established interpretations and typologies in transition studies. In light of previously proposed spectra for understanding incumbents'behaviour (Magnusson & Werner, 2023), we find that Flokk's responses to paradoxical tensions in relation to circular business models lean more towards 'innovating' than 'defending', and more towards 'collaborating' than 'competing'. They also appear more proactive than defensive (Saleh et al., 2025). For example, Flokk actively searched for sources of recycled PCP across traditional industry boundaries. On the one hand, incorporating post-consumer recycled materials can be viewed as an innovative approach, as it contributes to industrial transformation by addressing challenges like industrial waste. However, it can also be viewed as a defensive strategy that preserves the status quo of the existing business model. Furthermore, while product-as-service is often framed as a radical CE strategy (see, e.g., Arekrans et al., 2023), this case suggests that such business models do not necessarily lead to more sustainable resource use. Initially, adopting this model required innovation in terms of Flokk's value creation activities. However, insights from the pilot of product-as-a-service revealed that it increased product turnover, reinforcing existing business models rather than transforming them. In contrast, incorporating recycled PCP led to novel collaborations and broader CE adoption, e.g., through partnership with the public road administration.

Moreover, it is important to recognize how incumbents'activities are framed in transition processes across different research fields, at varying analytical levels, and across different time frames in empirical studies (Magnusson & Werner, 2023). For instance, in studies of CE and innovation, recycling has been called a "low-hanging fruit" that does not exhibit radical innovation (Arekrans et al., 2023). Yet, this case illustrates how this can drive novel couplings across sectors. Consequently, we argue that strategies for transition to CE cannot necessarily be strictly interpreted as defensive versus proactive (Saleh et al., 2025). Due to dependencies on other actors and resources, such as the supply of secondary materials, cross-sector interactions may primarily reflect the underlying interdependencies that characterize CE transition. For example, Flokk depends on high-quality, reliable sources of post-consumer waste from other sectors, illustrating that the organizational boundaries of incumbents are shaped by the dependencies and complementarities with external actors (Magnusson & Werner, 2023).

Third, this chapter illustrates how CE strategies at the firm level can foster interactions across sectors and potentially lead to system-level change. Flokk's search for recycled PCP exemplifies how materials can catalyze novel collaborations across various industries. For example, their search for new material sources led to a partnership across sectors to recycle discarded snowplow markers. Although their efforts to incorporate PCP from Norwegian aquaculture have not yet led to changes in Flokk's value chain, their demand for high-quality recycled PCP has driven change in adjacent industries, such as aquaculture. This illustrates how CE strategies become a driver for cross-sectoral interactions, particularly where increasing the use of secondary materials is a key challenge. While transition studies have focused primarily on sectors such as energy, agriculture, water, and construction, it can be argued that manufacturing firms face distinct opportunities and challenges. Manufacturing tends to be less locked into heavy, long-term infrastructure investments and place-bound resources and is more exposed to changing markets and customer preferences than sectors that provide critical functions for society. Therefore, manufacturing also encounters more uncertainty regarding materials and value chain development. This uncertainty may increase in transitions to CE, as the management of circular value chains is more complex and systemic than linear ones (Viva et al., 2020). For incumbent firms, this involves navigating paradoxes at the intersection between internal tensions and external dependencies. For managers, it underscores the need for a better overview of material flows to guide them through transition processes that involve both transforming current business practices and shaping a circular future.

Although the paradox lens sheds new light on the complexity faced by incumbent firms in the transition to CE, this study also has limitations. For example, there is a certain risk of neglecting areas where incumbents possess a more defensive approach (Magnusson & Werner, 2023). While we have highlighted how transition processes reveal paradoxical tensions for Flokk, we have not explored how tensions may be experienced across different organizational units. The identification of paradoxical tensions is based on how these are emphasized at a specific point in time. Following how these tensions change over time could further yield insights into the temporal dynamics of transitions (Kungl, 2024). Furthermore, although emphasizing the interactions between Flokk and other sectors, this study is based on a single case study of an incumbent. By doing so, we risk underestimating the evolving nature of paradoxical tensions in interaction with other actors and their impact on organizational and sectoral dynamics. Indeed, understanding these tensions as part of inter-organizational systems could have strengthened this contribution, as we find that navigating these interrelated paradoxical tensions largely depends on interaction with other actors (Heucher et al., 2024; Jarzabkowski et al., 2019). Extending research on paradoxes in sustainability transitions across sectors is thus a promising avenue for future research.

## Concluding Remarks

By exploring Flokk's CE strategies, this chapter has demonstrated how paradoxical tensions can act as barriers and drivers in transition processes. These tensions, stemming from competing demands such as meeting ambitious sustainability targets while navigating material constraints, offer a glimpse into the complexity of the CE transition for incumbents.

Drawing on paradox theory, the chapter offers three main insights. First, we contribute to the ongoing debate on incumbents in sustainability transition by illustrating how paradoxical tensions drive innovation and influence transition dynamics. Second, our case further emphasizes the need for nuanced approaches to studying incumbents, moving beyond generalized typologies to address their heterogeneous roles and contributions to sustainability transitions. Lastly, we have shown how CE strategies serve as a key driver of novel interactions across value chains and industries. We illustrate how the identified CE strategies highlight the role of materials in shaping CE transitions, leading to interactions that transcend the pre-existing boundaries of organizations.

Understanding transition dynamics is also of practical relevance to the transformation of business models and value chains. This includes the need for an overview of potential circular material streams and the systemic interdependencies that shape them. Such understanding is increasingly important as access to recycled materials becomes more competitive. This also points to the importance of facilitating new couplings between sectors to enable the development of secondary material markets. As incumbents continue to struggle with increasing circularity, policy instruments are important for promoting these markets, such as targets for incorporating recycled hard plastics and measures like circular public procurement criteria to foster transition.

## References

Andriopoulos, C., & Lewis, M. W. (2009). Exploitation-exploration tensions and organizational ambidexterity: Managing paradoxes of innovation. *Organization Science, 20*(4), 696–717.

Andriopoulos, C., & Lewis, M. W. (2010). Managing innovation paradoxes: Ambidexterity lessons from leading product design companies. *Long Range Planning, 43*(1), 104–122.

Arekrans, J., Ritzén, S., & Laurenti, R. (2023). The role of radical innovation in circular strategy deployment. *Business Strategy and the Environment, 32*(3), 1085–1105.

Bakker, C. A., Den Hollander, M., Van Hinte, E., & Zijlstra, Y. (2014). *Products that last: Product design for circular business models*. TU Delft Library.

Bergek, A., Berggren, C., Magnusson, T., & Hobday, M. (2013). Technological discontinuities and the challenge for incumbent firms: Destruction, disruption or creative accumulation? *Research Policy, 42*(6–7), 1210–1224.

Berggren, C., Magnusson, T., & Sushandoyo, D. (2015). Transition pathways revisited: Established firms as multi-level actors in the heavy vehicle industry. *Research Policy, 44*(5), 1017–1028.

Bidmon, C. M., & Knab, S. F. (2018). The three roles of business models in societal transitions: New linkages between business model and transition research. *Journal of Cleaner Production, 178*, 903–916.

Bocken, N. M., De Pauw, I., Bakker, C., & Van Der Grinten, B. (2016). Product design and business model strategies for a circular economy. *Journal of Industrial and Production Engineering, 33*(5), 308–320.

Bohnsack, R., Kolk, A., Pinkse, J., & Bidmon, C. M. (2020). Driving the electric bandwagon: The dynamics of incumbents' sustainable innovation. *Business Strategy and the Environment, 29*(2), 727–743.

Carmine, S., & De Marchi, V. (2023). Reviewing paradox theory in corporate sustainability toward a systems perspective. *Journal of Business Ethics, 184*(1), 139–158.

Daddi, T., Ceglia, D., Bianchi, G., & de Barcellos, M. D. (2019). Paradoxical tensions and corporate sustainability: A focus on circular economy business cases. *Corporate Social Responsibility and Environmental Management, 26*(4), 770–780.

Dagilienė, L., & Varaniūtė, V. (2023). Transitioning to a circular economy: Paradoxical tensions of the circular business model. *Organization & Environment, 36*(4), 559–589.

Damman, S., Lund, H. B., Kenzhegaliyeva, A., & Ullern, E. F. (2022). Fra oppdrettsplast til verdi Drivere og barrierer for sirkulær økonomi basert på hardplast fra norsk havbruk.

De Angelis, R. (2021). Circular economy and paradox theory: A business model perspective. *Journal of Cleaner Production, 285*, 124823. https://doi.org/10.1016/j.jclepro.2020.124823

De Jesus, A., & Mendonça, S. (2018). Lost in transition? Drivers and barriers in the eco-innovation road to the circular economy. *Ecological Economics, 145*, 75–89.

Dijkstra, H., van Beukering, P., & Brouwer, R. (2020). Business models and sustainable plastic management: A systematic review of the literature. *Journal of Cleaner Production, 258*, 120967.

European Commission. (2016). Closing the loop. New circular economy package. Retrieved from https://www.europarl.europa.eu/RegData/etudes/BRIE/2016/573899/EPRS_BRI(2016)573899_EN.pdf

European Commission, Centre, J. R., Pierri, E., Egle, L., Milios, L., & Saveyn, H. (2024). *EU-wide end-of-waste criteria for plastic waste—JRC technical proposals*. Publications Office of the European Union. https://doi.org/10.2760/9234350

European Environment Agency. (2023). *Accelerating the circular economy in Europe. State and outlook 2024.* (EEA Report 13/2023, Issue. P. o. o. t. E. Union.

Fact.MR. (2019). *Plastic Furniture Market*. https://www.factmr.com/report/3574/plastic-furniture-market

Fehrer, J. A., & Wieland, H. (2021). A systemic logic for circular business models. *Journal of Business Research, 125*, 609–620.

Franco, M. A. (2017). Circular economy at the micro level: A dynamic view of incumbents' struggles and challenges in the textile industry. *Journal of Cleaner Production, 168*, 833–845.

Fraser, M., Conde, Á., & Haigh, L. (2024). Circularity gap report 2024.

Frishammar, J., Parida, V., Panda, D., & Kaipainen, J. (2025). On the right path to circularity or running around in circles? A fresh perspective on circular business model barriers. *Business Strategy and the Environment*.

Galvan, M. G., Cuppen, E., & Taanman, M. (2020). Exploring incumbents' agency: Institutional work by grid operators in decentralized energy innovations. *Environmental Innovation and Societal Transitions, 37*, 79–92.

Geels, F. W. (2002). Technological transitions as evolutionary reconfiguration processes: A multi-level perspective and a case-study. *Research Policy, 31*(8–9), 1257–1274.

Geissdoerfer, M., Savaget, P., Bocken, N. M., & Hultink, E. J. (2017). The circular economy–a new sustainability paradigm? *Journal of Cleaner Production, 143*, 757–768.

Gregory, J., & Geels, F. W. (2024). Unfolding low-carbon reorientation in a declining industry: A contextual analysis of changing company strategies in UK oil refining (1990–2023). *Energy Research & Social Science, 107*, 103345.

Guide, V. D. R., Jr., & Vanwassenhove, L. N. (2001). Managing product returns for remanufacturing. *Production and Operations Management, 10*(2), 142–155.

Hahladakis, J. N., & Iacovidou, E. (2019). An overview of the challenges and trade-offs in closing the loop of post-consumer plastic waste (PCPW): Focus on recycling. *Journal of Hazardous Materials, 380*, 120887.

Hahn, T., Figge, F., Pinkse, J., & Preuss, L. (2018). A paradox perspective on corporate sustainability: Descriptive, instrumental, and normative aspects. *Journal of Business Ethics, 148*(2), 235–248.

Hahn, T., & Knight, E. (2021). The ontology of organizational paradox: A quantum approach. *Academy of Management Review, 46*(2), 362–384.

Heucher, K., Schrage, S., & Abosag, I. (2024). When global is not enough: Applying a paradox lens to sustainability transitions in Interorganizational systems. *Journal of International Management, 30*(5), 101186.

ISO. (2008). Plastics–Guidelines for the recovery and recycling of plastics waste. *ISO 15270: 2008 (E).*

Jarzabkowski, P., Bednarek, R., Chalkias, K., & Cacciatori, E. (2019). Exploring inter-organizational paradoxes: Methodological lessons from a study of a grand challenge. *Strategic Organization, 17*(1), 120–132.

Jarzabkowski, P., Lê, J. K., & Van de Ven, A. H. (2013). Responding to competing strategic demands: How organizing, belonging, and performing paradoxes coevolve. *Strategic Organization, 11*(3), 245–280.

Johansen, M. R., Christensen, T. B., Ramos, T. M., & Syberg, K. (2022). A review of the plastic value chain from a circular economy perspective. *Journal of Environmental Management, 302*, 113975.

Köhler, J., Geels, F. W., Kern, F., Markard, J., Onsongo, E., Wieczorek, A., Alkemade, F., Avelino, F., Bergek, A., & Boons, F. (2019). An agenda for sustainability transitions research: State of the art and future directions. *Environmental Innovation and Societal Transitions, 31*, 1–32.

Kuhlmann, M., Bening, C. R., & Hoffmann, V. H. (2023). How incumbents realize disruptive circular innovation-overcoming the innovator's dilemma for a circular economy. *Business Strategy and the Environment, 32*(3), 1106–1121.

Kungl, G. (2024). Challenges of the current discourse on incumbent firms in sustainability transitions. *Energy Research & Social Science, 108*, 103367.

Kungl, G. (2025). Incumbent firms in sustainability transitions–different conceptions, heterogeneous roles and ideal types. *Environmental Innovation and Societal Transitions, 57*, 101010.

Langley, A. (1999). Strategies for theorizing from process data. *Academy of Management Review, 24*(4), 691–710.

Magnusson, T., & Werner, V. (2023). Conceptualisations of incumbent firms in sustainability transitions: Insights from organisation theory and a systematic literature review. *Business Strategy and the Environment, 32*(2), 903–919.

Markard, J., Raven, R., & Truffer, B. (2012). Sustainability transitions: An emerging field of research and its prospects. *Research Policy, 41*(6), 955–967.

Pascucci, F., Pizzichini, L., Sabatini, A., Temperini, V., & Mueller, J. (2024). Knowledge-based dynamic capabilities for managing paradoxical tensions in circular business model innovation: An empirical exploration of an incumbent firm. *Journal of Knowledge Management, 28*(11), 255–282.

Penna, C. C., & Geels, F. W. (2015). Climate change and the slow reorientation of the American car industry (1979–2012): An application and extension of the Dialectic Issue LifeCycle (DILC) model. *Research Policy, 44*(5), 1029–1048.

Pinkse, J., Hahn, T., & Figge, F. (2019). Supersized tensions and slim responses? The discursive construction of strategic tensions around social issues. *Academy of Management Discoveries, 5*(3), 314–340.

Saleh, R., Auktor, G. V., & Brem, A. (2025). Incumbency and sustainability transitions: A systematic review and typology of strategies. *Energy Research & Social Science, 122*, 104000.

Schad, J., & Bansal, P. (2018). Seeing the forest and the trees: How a systems perspective informs paradox research. *Journal of Management Studies, 55*(8), 1490–1506.

Schad, J., Lewis, M. W., Raisch, S., & Smith, W. K. (2016). Paradox research in management science: Looking back to move forward. *Academy of Management Annals, 10*(1), 5–64.

Schaltegger, S., Loorbach, D., & Hörisch, J. (2023). *Managing entrepreneurial and corporate contributions to sustainability transitions* (Vol. 32, pp. 891–902). Wiley Online Library.

Slawinski, N., & Bansal, P. (2015). Short on time: Intertemporal tensions in business sustainability. *Organization Science, 26*(2), 531–549.

Smink, M. M., Hekkert, M. P., & Negro, S. O. (2015). Keeping sustainable innovation on a leash? Exploring incumbents' institutional strategies. *Business Strategy and the Environment, 24*(2), 86–101.

Smith, W., Erez, M., Jarvenpaa, S., Lewis, M. W., & Tracey, P. (2017). *Adding complexity to theories of paradox, tensions, and dualities of innovation and change: Introduction to organization studies special issue on paradox, tensions, and dualities of innovation and change*. SAGE Publications Sage UK.

Smith, W. K., & Lewis, M. W. (2011). Toward a theory of paradox: A dynamic equilibrium model of organizing. *Academy of Management Review, 36*(2), 381–403.

Sovacool, B. K., Turnheim, B., Martiskainen, M., Brown, D., & Kivimaa, P. (2020). Guides or gatekeepers? Incumbent-oriented transition intermediaries in a low-carbon era. *Energy Research & Social Science, 66*, 101490.

Steen, M., & Weaver, T. (2017). Incumbents' diversification and cross-sectorial energy industry dynamics. *Research Policy, 46*(6), 1071–1086.

The Norwegian Government. (2021). Norway's strategy for developing a green, circular economy. Oslo: The Norwegian Government. Retrieved from https://www.regjeringen.no/contentassets/a116f209e493471bb26c81cf645152a3/kld_strategi_sirkularokonomi_sammendrag_eng_0507.pdf

Tosi, D., Gusmerotti, N. M., Testa, F., & Frey, M. (2024). How companies navigate circular economy paradoxes: An organizational perspective. *Journal of Environmental Management, 353*, 120269.

Turnheim, B., & Geels, F. W. (2019). Incumbent actors, guided search paths, and landmark projects in infra-system transitions: Re-thinking strategic niche management with a case study of French tramway diffusion (1971–2016). *Research Policy, 48*(6), 1412–1428.

Turnheim, B., & Sovacool, B. K. (2020). Forever stuck in old ways? Pluralising incumbencies in sustainability transitions. *Environmental Innovation and Societal Transitions, 35*, 180–184.

UN. (2024). *For a livable climate: Net-zero commitments must be backed by credible action*. https://www.un.org/en/climatechange/net-zero-coalition

van Bommel, K. (2018). Managing tensions in sustainable business models: Exploring instrumental and integrative strategies. *Journal of Cleaner Production, 196*, 829–841.

Van der Byl, C. A., & Slawinski, N. (2015). Embracing tensions in corporate sustainability: A review of research from win-wins and trade-offs to paradoxes and beyond. *Organization & Environment, 28*(1), 54–79.

Viva, L., Ciulli, F., Kolk, A., & Rothenberg, G. (2020). Designing circular waste management strategies: The case of organic waste in Amsterdam. *Advanced Sustainable Systems, 4*(9), 2000023.

Yin, R. K. (2009). *Case study research: Design and methods (Vol. 5)*. sage.

# 11

# Digitalization and Platformization for Sustainability: The Case of the Oil and Gas Industry

Katja Maria Hydle and Hannah Schupfer

## Introduction

Organizations are under increasing pressure to deliver in an environmentally sustainable manner to contribute to the United Nations (UN) Sustainability Development Goals (SDG). Environmental sustainability involves biodiversity, energy efficiency, and reducing greenhouse gas emissions, which imply that companies in carbon-intensive industries, such as Oil and Gas (O&G), would radically need to change their technologies and behaviours to reduce emissions (Li et al., 2021). The energy sector accounts for over 70% of the global greenhouse gas emissions and also severely harms biodiversity in ecosystems, which exposes that

K. M. Hydle (✉) · H. Schupfer
Department of Informatics, University of Oslo, Oslo, Norway
e-mail: katjahy@uio.no

H. Schupfer
Department for Strategy, International Management, and Entrepreneurship, King's College London, London, UK


M. Steen et al. (eds.), *Sustainability Transitions and Industrial Transformation*,
https://doi.org/10.1007/978-981-95-4893-4_11

O&G companies need to divest from conventional, dirty technologies and engage in greener, more sustainable solutions (IPCC, 2014).

Engaging in radical innovation for sustainable solutions often involves digital technologies. 'Digital sustainability' by George et al., (2020, p. 1000) is defined as "the organizational activities that seek to advance the sustainable development goals through creative deployment of technologies that create, use, transmit, or source electronic data". The digital technologies, digital innovation, value creation, and activities for sustainability are in focus in this chapter, specifically digitalization and platformization with and for sustainability solutions.

A focus on digitalization for sustainability underscores the possibilities offered by digital sustainability (Christmann et al., 2024; Tabares et al., 2025). Researchers tend to refer to a twin transition of both sustainability and digitalization. "Successfully managing the green and digital 'twin' transitions is the cornerstone for delivering a sustainable, fair, and competitive future" (Muench et al., 2022, p. iv).

However, the use of digital technologies for sustainable solutions, might be used incrementally, meaning that digitalization may accentuate existing unsustainable technologies, which, in turn will lead to system optimization instead of transformation (John et al., 2022; Mäkitie et al., 2023). According to Mäkitie et al. (2023), only radical innovations possess the capacity to generate the significant structural transformations in production and consumption patterns that are essential for achieving sustainability targets.

In this chapter, we delve into the ways incremental change unfolds within companies through digitalization and platformization in the O&G industry. We explore digitalization at the organizational level along four theoretical foundations of automating, informating (Zuboff, 1985), representing, and mediating (Jonsson et al., 2018). Conversely, platformization centres on data, illustrated by our empirical material through the traits of standardizing, opening, and investing among companies in an ecosystem. Both digitalization and platformization enhance value creation and capture with implications for sustainability management. We end with a discussion of implications for transitions in one of the hard-to-abate sectors.

We illustrate the processes of digitalization and platformization in O&G companies, suppliers, and industrial platform companies, stemming from two different qualitative research projects from 2018 to 2024. One research project was with five leading manufacturing companies offering advanced products as software services to the O&G industry. Empirical data were gathered through semi-structured, in-depth interviews with a total of 66 key employees, as well as by attending 26 workshops and meetings (Aas et al., 2020, 2023; Hydle et al., 2021, 2022). The other research project was with a global, industrial AI Platform-as-a-Service (PaaS) company, providing services and platform solutions to the O&G industry. Between 2019 and 2024, the first author conducted 60 interviews with managers and employees, including those from O&G companies, Original Equipment Manufacturers, a cloud company, and various partner organizations. Of these, 26 interviews were conducted with the second author, specifically addressing sustainability, involving informants from four different continents—North America, Southeast Asia, the Middle East, and Latin America—with a major emphasis on Europe.

## Digitalization

Digitalization refers to the transformation of socio-technical structures from non-digital artefacts or relationships to digital ones. Digitalization involves "organizing new socio-technical structures with digital artifacts as well as the changes in artifacts themselves" (Yoo et al., 2010, p. 6). Hence, digitalization is about *using* digital technologies and the processes of *converting* physical formats and information to digital formats. Digitalization also involves the change of organizational structures to accommodate new ways of working with the use of digital technologies. Industry digitalization entails a transition from changing manual activities and processes into digital ones while altering work organization. A manager in an O&G company illustrates the digitalization process for oil well and drill management as follows:

> The tools that the reservoir engineers use from the software suppliers are, to a large extent, digital… We have now built a digital tool that will work across the reservoir and drilling, where you then place and design [oil] wells in one and the same tool so that you can have much more frequent interactions … The most important thing is that you get to do more interactions and try out more strategies... So that's an example of such a digitalization process.

## Automating, Informating, Representing, and Mediating

Digital technologies and their application in industry range from big data, the cloud, the Internet of Things (IoT), smart connected products, smart contracts, artificial intelligence (AI), analytics, automation, remote monitoring, and predictive maintenance (Parida et al., 2019). Four characteristics of digital technologies are useful for understanding digitalization (Hydle et al., 2021): "*Automating*" and "*informating*" explain digital work (Zuboff, 1985), while "*representing*" and "*mediating*" represent the interactive and collaborative aspects (Jonsson et al., 2018). *Automating* refers to how information technology is used to automate tasks, reproducing human skills, with precision and uniform production (Zuboff, 1985). For automating processes or activities, data are registered by machines or equipment. For industry, automating repetitive activities such as fault detection and monitoring processes is highly useful. *Informating* is the process of translating events, objects, and descriptions into information that can be viewed and understood by workers and used for decision-making (Zuboff, 1985). Informating occurs because of computer-based automating and through deliberate choice to use such information in industry. Digital *representing* refers to "how IT is used to monitor and produce digital content" (Jonsson et al., 2018, p. 218) and exposes the content. Digital *mediating* is "how IT can be used for digitally mediated cooperative work" (Jonsson et al., 2018, p. 218). In industry, examples such as condition-based maintenance expose *representing* through information obtained from the monitored machines and equipment and *mediating* how maintenance workers and/or data analysts collaborate when using digital technology.

An illustration of digitalization for O&G companies is a software and product called "StayAlert", developed and provided by a service supplier to the O&G industry for the process of drilling for oil and gas. Hydle et al. (2022, p. 305) disclose the following quote for StayAlert:

> There is a sea of different services for information input. Some are digitized, some are analogue, others have yellow Post-it patches on the wall. So, it is a wonderful collection of sources of error.... This product has taken aim at gathering these on a platform that allows all 'stakeholders' to get the information and collect information on the platform, to retrieve and collect exactly what they need... The system will provide necessary information that is customized wherever they are, at any time.

StayAlert as a software and product exposes *automating* of processes, *informating* of events to be seen by everyone involved in the drilling operation, *representing* through monitoring the drilling processes, and *mediating* through how all actors involved can collaborate by using StayAlert. For industry, digitalization is thus a means to change analogue information into digital, to relate machines and equipment to one another using digital technologies, to *automate* tasks, to *informate* workers, to *represent* the machines, equipment, and related processes and information in use, and to *mediate* cooperative work between industrial workers, data analysts, and decision-makers.

## Creating and Capturing Value in Companies

Digitalization in industry is not a means in itself, but is understood as a facilitator for value-creating and value-capturing opportunities, such as optimization of production, efficient operations, reduced downtime, and enhanced control of activities, processes, and operations. One middle manager in an O&G company explained:

> It is extremely inefficient in how to handle data, in how to make decisions, especially in an oil and gas company. We have a huge potential to

> get more value out of the decisions we make, and to some extent also reduce operational costs. We who work with digitalization, we work a lot with cost reduction to the extent that we are talking about the operating part of our company, to cut maintenance where possible, streamline maintenance and prevent unnecessary downtime, i.e., predict. It is often part of cutting costs, but it's really more about increasing earnings. There is a lot in digital that goes into creating good decision support to minimize the chances of making a mistake when making a decision, and then, in a way, get much more value out of the investments and decisions you make.

Following Parida et al. (2019), there are three digitalization functions involved in value creation and value capture in industry. First, operational data is collected through sensors, minimizing the need for human intervention. Second, connectivity allows for seamless sharing of data across digital devices via wireless communication networks. Third, analytics transform the available data into valuable insights and actionable directives while also enhancing knowledge through cloud storage and optimization. These functionalities open doors to various opportunities for value creation, leading to a progression from monitoring to control and optimization, ultimately enabling autonomous products through digitalization (Parida et al., 2019). A manager from a software-as-a-service (SaaS) company providing digitalization services to O&G companies, explains their product for energy reduction for offshore oil platforms: "We're optimizing heavy asset processes. So, by doing the operations more efficiently, you're already using less resources and emitting less".

## Embedding Energy and Reducing Emission in Operations

Digitalization represents potential for economic, sustainable, and social benefits, through increased process efficiency and safety (Parida et al., 2019). Parida et al. (2019) find that environmental benefits can be attained by implementing both gradual efficiency enhancements and transformative organizational approaches. The functionalities that support sustainability encompass a broad spectrum, including enhanced product design, predictive maintenance, and product tracking. Additionally, sustainable practices can be fostered through operational industrial efficiencies that minimize the environmental impact as compared to traditional industry methods. For instance, by monitoring operational data, it becomes possible to reduce scrap rates, minimize equipment deterioration, reduce downtime through annual maintenance, and instead use condition-based maintenance. One informant from a company enabling digitalization for O&G customers explained the sustainability benefits:

> They want to operate more efficiently; they want to reduce their maintenance costs, but there's always a sustainability benefit if you go to a needs-based maintenance scheme as opposed to just a scheduled maintenance scheme. You are going to reduce your flights, reduce the number of people on board, and reduce the number of equipment or machinery that you throw away unnecessarily.

Digitalization for O&G companies is mainly a transformation to enhance internal efficiency and reduce costs, with an internally oriented focus. The sustainability part is merely a benefit of digitalization. This is also found in other industries such as steel, where a study reveals that enabling technologies and AI rather reinforced path dependencies through incremental innovation, than to introduce novel technologies for the decarbonization transition of the steel industry (John et al., 2022). This is illustrated by one employee from a SaaS company describing the relation between optimizing oil production and accounting for GHG emissions:

> It's basically embedding the sustainability or emission information into the production operations part. It's one more variable that you have control over when you do your optimization. So, you can do your operation; maximize oil pull. That's one way that you could do it. Or you could also, if you add an emission variable to it, put a limit and say 'maximize my production, but make sure that my total emissions don't exceed x', and then it's gonna maximize production but stay to that limit.

Thus, digitalization in O&G may contribute to extending the life of the sector, where incremental innovation is for operating purposes, with sustainability benefits also being incremental, instead of radical innovation for attaining sustainability purposes (Mäkitie et al., 2023). A former employee from a global O&G company explained the transition by focusing on production while being as clean as possible:

> In my previous company [an oil and gas operator], the conversation was more around, 'how can we make sure that our operation is as clean as possible even though we are producing oil and gas? How can we try to reduce or produce as effective and clean as possible?' And investing the earnings back into green energy was kind of the focus there.

Hence, digitalization for O&G companies is an internal organizational exercise. It is a way for companies to gain control over their operations, reduce costs, and increase efficiency. This is achieved through automating, informating, representing, and mediating internally while focusing on creating and capturing value within the companies. Additionally, digitalization also brings sustainability benefits such as improved energy efficiency and reduced greenhouse gas emissions. While sustainability is not the primary goal of digitalization, it is an important benefit, since the main focus remains on enhancing industrial operations.

The internal company orientation focusing on digitalization changes to involve external actors and ecosystems through platformization. Monteiro (2022) contends that although much of the research on digitalization in O&G focuses solely on changes within organizations, it is essential to consider the entire ecosystems to comprehend how such transformations impact the entire industries.

## Platformization

Digital platforms have become increasingly present across a variety of industries, including O&G (Bode et al., 2019; McIntyre et al., 2021). Several researchers emphasize the role of platforms for tackling grand challenges (Bode et al., 2019; Hellemans et al., 2021) by bringing together multiple companies and actors to collaborate. Digital platforms serve as intermediary systems that facilitate transactions between two or more parties (Gawer & Cusumano, 2014). According to Gawer and Cusumano (2014, p. 417), industrial digital platforms are defined as "products, services, or technologies that act as a foundation upon which external innovators, organized as an innovative business ecosystem, can develop their own complementary products, technologies, or services".

The process of introducing and transforming information systems into digital platforms is called platformization. According to Törmer and Henningsson (2020), platformization refers to the socio-technical process of transforming a large-scale information system into a platform architecture. This architecture consists of a stable core functionality, flexible peripheral components, and standardized interfaces for component interactions. The authors illustrate this process by the internal transformation of a group's IT landscape into a company-wide digital platform structure. Rolland et al. (2018) explore platformization from a different perspective, where external actors heavily influence the process. These actors include the platform provider, the market for third-party complementary products, and other organizations utilizing the same platform. Rolland et al. (2018) analyse the implications of platformization on the internal IT landscape of a media company that adopted an industry news platform. The study delves into the intricate relationships between the platform, which is part of a larger ecosystem, and the organization's existing infrastructure and work processes.

## Standardizing, Opening, and Investing

Understood from an O&G company, platformization has been an evolving process with the three characteristics of standardizing (as opposed to customizing), opening up, and investing, as explained by a manager in an O&G company:

> I think that this whole thing has been a kind of evolving process for us. I think that number one is that we have an allergy to customization because we understand our size, so we don't want to create products that are only internally. And then the second is openness. And the third, we need to

> have skin in the game. You need to come and say, 'if I'm developing that with you, I'm also investing.' I also see that there is a value to do that. So, those are the main three things we're looking for to bring in people to our ecosystem when it comes to kind of developing new things.

For O&G companies, digital platforms imply internal digital transformation, such as large changes to internal processes across functions, collaboration between different units and professional groups, and enhanced cooperation with other firms by using data. Data emerges as a novel form of resource for decision-makers, where employees engage with the different datasets at their disposal through data engagement, governance, and access. However, it is not just about collecting, accumulating, and using data. Monteiro (2022) highlights that the utilization and interpretation of data are influenced by the methods, procedures, and technologies used to produce it. The data is made accessible through an industrial digital platform and can be used for sustainability purposes and drivers. A sustainability manager is explaining the data-driven use for emulsion management:

> People make decisions, and data is used for support or for algorithms and ultimate decisions. Data is not doing anything; it's always a tool. But data-driven sustainability means, 'where do we get impact, where do we get contribution so that it makes an impact on given sustainability drivers', whether it's emissions, whether it's waste management, or whether it's energy efficiencies. So, 'which data do we need, what assumptions do we have, and what decision can we make, where we actually have the control over the action?'. You need to be specific on what you can do, what the assumption is, and what do you control and how do you get support through data. And then linking that to whatever sustainability drivers you may have. That's what I like to put under 'data-driven'.

The platform ecology of interdependent firms and inter-organizational relations is often referred to as ecosystems (Hou & Shi, 2021; Jacobides et al., 2018). The data, and the inherent value creation across several organizations accessing and using the data, imply that individual companies become dependent on other companies taking part in the ecosystem. Data engagement thus happens outside of the formal boundaries of one single organization since value chains have transformed into ecosystems due to digitalization. Ecosystems, as a term, refer to firms' activities that depend on each other (Jacobides et al., 2018). However, there are different types of ecosystems: Innovation ecosystems, which emphasize a specific innovation or a value proposition along with all the actors involved (Adner, 2016); business ecosystems, which focus on a single organization and its business environment (Teece, 2007); and platform ecosystems, which organize the actors around a platform (Gawer & Cusumano, 2014; McIntyre et al., 2020). In this chapter, we focus on platform ecosystems.

## Co-creating and Co-capturing Value in Ecosystems

Studies on platform ecosystems focus on how value co-creation occurs between ecosystem actors (Autio & Thomas, 2014; Jacobides et al., 2018). The platform ecosystem does not necessarily reflect organizations that belong to an industry, but rather organizations that participate to provide their part in the value creation process. Hence, many organizations form part of a platform ecosystem, while data management across companies is not necessarily a straightforward process, as explained by a middle manager in an O&G company.

> We have requirements that we really want to be as open as possible. So, our boss [CEO] is very proactive in dealing with the fact that data sharing across the Norwegian continental shelf does not work at all; data is not owned per se by the operator, but the data is owned by a partnership. It is the oil field partnership that owns the information that comes in from the oil field platform. We only manage it on their behalf. It is not certain that everyone is necessarily equally eager and open to sharing that data. Then there is another problem with sensitive information. We are a little nervous about letting that data flow completely uncontrolled and freely because we can quickly get a slap on the wrist from the Financial Supervisory Authority.

Industrial platforms connect IoT-enabled facilities and machines, collect data from equipment, operations, and processes, and provide platform services such as integrated work management, predictive maintenance, autonomous services, and an overview of remote operations.

> In the beginning, we got new advanced algorithms. Like this: "Now this pump is going to break down. Do it like this", or you got a new one that said: "You can actually produce 10% more in this period of time". You get a lot of value extraction, but it doesn't change the way you work. It just gives you a single message. Now we're trying to switch to becoming more data-driven in everything we do. So instead of doing maintenance once every six months, you look at your data and look at the status of the pump.

## Monitoring, Reducing, and Reporting for Sustainability

The development of complementary platform services and products to add value for customers happens in an ecosystem (Hou & Shi, 2021; Jacobides et al., 2018). Complementary platform services for sustainability are solutions or applications on top of the digital platform, as project managers in the platform company explained:

> We also do a lot of solutions, like dashboards...which means they are algorithms and small code-type products that we've created with a small group of people. On the platform, we take all their equipment data; all is plugged in and feeding data into the digital platform, and then on top, we write code that enables dashboards that allow you to see your emissions data.
> Depending on their operations, they are able to reduce the amount of energy they use by being able to monitor it in real time. So, if they're monitoring their emissions and their energy needs in real time, as well as their fuel usage in real time, there are things that they can do to actually significantly reduce their energy usage.

Platform ecosystems play a crucial role in the creation of sophisticated platform services within industrial sectors (Jovanovic et al., 2022).

Advanced platform services like preventive maintenance and GHG emission reporting are built around industrial assets. Through dashboards or applications for sustainability, O&G companies can report their different emissions from a plant, an oil platform, different machines, or for entire companies, as explained by a middle manager in the digital platform company:

> ...this way we can automatically produce carbon reports at the plant-, machine, and customer level. So, you can choose; if you have the machine level, you can choose how you add them up, if you do it according to the product or the whole plants.
> We helped them first with reporting, and reporting is also visualising energy management data. And that was successful, so now we are going ahead to work with them further to reduce their energy usage.

Additionally, industrial platform ecosystems typically exhibit a narrow scope and size, coupled with high costs for industrial equipment, suggesting that platform coopetition is more beneficial than platform competition (Jovanovic et al., 2022). Coopetition among platforms facilitates collaboration and the integration of technologies into multiplatform ecosystems (MPEs), enabling shared value co-creation and co-capture (Mohamed et al., 2023). Mohamed et al. (2023) highlight that the leading platform within an MPE assumes the role of ensuring market dominance for the entire ecosystem. Regarding sustainability reporting, one of the largest global tech firms approached the digital platform company for O&G to co-create a platform of platforms and ecosystem of ecosystems to report on emissions for the industry:

> 'Cloud Company' and their Sustainability Multiplatform don't do this real-time thing at all. All they're doing is sort of reporting... they come to us, and they're looking for something that's a lot more granular where they can look at their sustainability data in real time and make decisions in the moment as they're operating.
> Cloud for sustainability is quite high-level, you report e.g., energy consumption for a whole company, so it's really high level, whereas we can actually pull the energy use of parts of a factory, so we're on the much lower level, so that's the optimisation.

The data from one platform is feeding into a larger platform to be used by the industrial actors to report to governments worldwide while having control over their worldwide operations and assets. In addition, O&G companies can insert future emission reduction goals and manage and track progress.

> I know that the Sustainability Multiplatform is meant to not just do the reporting. They're starting to generate what are your 2050 goals, and then it will do a gap analysis of where you are compared to your goal. So, if you've pledged to reduce your carbon intensity or greenhouse gas emissions, it will also report where the gap is, where you are and where you're supposed to be. And it will do this year on year, so, it'll track your progress towards those goals. I think that the visibility will help customers or operators to make the right decision, because they'll realize that they're not progressing.

The implications of such MPEs for an industry are that the digital platforms and the platform companies enable emission data to be reported and reduced by the industrial ecosystem of actors. The data from operations is used to report and reduce emissions. The O&G companies can report their sustainability metrics in an accurate manner and can even forecast and backcast their emissions and the necessary reduction in real time.

## Discussion and Concluding Remarks

This chapter exposes the digitalization and platformization transitions of O&G companies and ecosystems for handling sustainability challenges. Digitalization focuses on intraorganizational change, through automating, informating, representing, and mediating, with value creation and capture internally, where sustainability is embedded in production operations. Platformization focuses on industrial and ecosystem change through standardizing, opening, and investing, with value co-creation and value co-capture between actors in an ecosystem. This involves data centricity, which enables the monitoring, reducing,

**Table 11.1** Incremental innovations

| Transition | Characteristics | Centricity | Value | Sustainability |
|---|---|---|---|---|
| Digitalization | Automating, informating, representing, mediating | Organization and function | Value creation and value capture internally in companies | Embedding energy and reducing emission in production operations |
| Platformization | Standardizing, opening, investing | Data | Value co-creation and co-capture in platforms and ecosystems | Monitoring, reducing, and reporting energy usage and GHG emissions |

and reporting of energy usage and GHG emissions (Table 11.1). Thus, digitalization and platformization lead to incremental improvements for sustainability.

Although there are blurred lines between digitalization and platformization, platform ecosystems' data centricity provides different possibilities for sustainability management than the digitalization of individual companies. For individual O&G companies and the entire industry, neither digitalization nor the platform transition has reached its full potential. Challenges with data governance, ecosystem value co-creation and co-capture, as well as the large changes involved for employees, make the digitalization and platformization transition a slow-moving process. In line with George et al. (2020), digital technologies are employed for handling sustainability challenges, although the O&G industry's transition's impact on sustainability is in its sparse beginning.

The findings expose what Mäkitie et al. (2023) denote as incremental twin innovation, meaning that incremental digital and sustainable innovations are connected, with minimal transformation effects for sustainability. A radical twin innovation, with significant transformation effects for sustainability, opens new pathways for sustainability transitions by using digital technology for system reconfiguration (Mäkitie et al., 2023). When reflecting on what it would take to move from incremental to radical improvements in O&G, the questions are whether digitalization and platformization can truly drive radical innovation towards sustainability, and under what conditions this could occur?

Investigating how digitalization can lead to incremental or radical innovation, we highlight that organizations control their sustainability impact through digitalization. Depending on the company's choice, this can result in minimal efforts to reduce emissions or, on the contrary, transform to the maximum possibilities. This points to economic and market values that often drive organizations, rather than societal values, potentially sacrificing economic gains. Using digitalization and platformization to fundamentally change organizations and their organizing leads to a higher sustainable impact than using digital technologies for efficiency purposes. The companies control how far they will use these intermediate tools for sustainability transitions. A radical shift for individual organizations involves to really understand the emissions, where these come from and change accordingly in short and long term to assure continuous maximization of sustainability. Thus, radical innovation at the organizational level demands strategic drive and large organizational changes.

At an ecosystem and industrial level, the conditions necessary to drive radical innovation may already exist. Platforms enable more of the industry to change since there are many actors involved. However, it is still up to the orchestrator to facilitate the collaborative impact, which is key in tackling grand challenges. Setting up conditions to foster broader organizational impact, instead of individual organizational measures (such as employees flying less), is feasible at an ecosystem level. Using platform ecosystems is thus an intermediate tool for emission reduction across the industry and the ecosystem. Thus, we claim that digitalization is a tool for companies to achieve reduction to become more sustainable, while platforms are tools for reduction and impact for the ecosystems and the industry.

The conditions necessary to drive radical innovation rest on value creation and capture at organizational level and value co-creation and co-capture at an ecosystem level. A radical change for the organizations in the industry could be driven by demand, either from investors or policies and regulations (Tabares et al., 2025).

The governmental and EU policy implications of these insights are to demand not only mandatory GHG emission reports but also budgets from individual companies and entire ecosystems, and to set country- and sector-specific regulations and taxation for breaches of the GHG threshold. Stricter thresholds than the GHG protocol can also be considered for those industries, companies, and ecosystems responsible for upper levels of emissions.

Future research could follow digitalization and platformization, the incremental and radical innovations, and their impact on sustainability in the O&G industry as well as other heavy-asset industries. Other companies, born digital and using data, may have another take on sustainability and may have reached further with digital sustainability than existing incumbents, which should be further explored. Hence, future research could also focus on how established companies can learn from born-digital companies.

## References

Aas, T. H., Breunig, K. J., Hellström, M. M., & Hydle, K. M. (2020). Service-oriented business models in manufacturing in the digital era: Toward a new taxonomy. *International Journal of Innovation Management, 24*(8), 2040002. https://doi.org/10.1142/S1363919620400022

Aas, T. H., Breunig, K. J., Hellström, M. M., & Hydle, K. M. (2023). Organising development processes: The case of product–service systems. *International Journal of Innovation Management, 27*(6), 2350030. https://doi.org/10.1142/S1363919623500305

Adner, R. (2016). Ecosystem as structure. *Journal of Management, 43*(1), 39–58. https://doi.org/10.1177/0149206316678451

Autio, E., & Thomas, L. D. W. (2014). Innovation ecosystems: Implications for innovation management? In M. Dodgson, D. Gann, & N. Phillips (Eds.), *The Oxford handbook of innovation management* (pp. 204–288). Oxford University Press.

Bode, C., Rogan, M., & Singh, J. (2019). Sustainable cross-sector collaboration: Building a global platform for social impact. *Academy of Management Discoveries, 5*(4), 396–414. https://doi.org/10.5465/amd.2018.0112

Christmann, A.-S., Crome, C., Graf-Drasch, V., Oberländer, A. M., & Schmidt, L. (2024). The twin transformation butterfly. *Business & Information Systems Engineering, 66*(4), 489–505. https://doi.org/10.1007/s12599-023-00847-2

Gawer, A., & Cusumano, M. A. (2014). Industry platforms and ecosystem innovation. *Journal of Product Innovation Management, 31*(3), 417–433.

George, G., Merrill, R. K., & Schillebeeckx, S. J. D. (2020). Digital sustainability and entrepreneurship: How digital innovations are helping tackle climate change and sustainable development. *Entrepreneurship Theory and Practice, 45*(5), 999–1027. https://doi.org/10.1177/1042258719899425

Hellemans, I., Porter, A. J., & Diriker, D. (2021). Harnessing digitalization for sustainable development: Understanding how interactions on sustainability-oriented digital platforms manage tensions and paradoxes. *Business Strategy and the Environment, 31*(2), 668–683. https://doi.org/10.1002/bse.2943

Hou, H., & Shi, Y. (2021). Ecosystem-as-structure and ecosystem-as-coevolution: A constructive examination. *Technovation, 100*, 102193. https://doi.org/10.1016/j.technovation.2020.102193

Hydle, K. M., Hellström, M., Aas, T. H., & Breunig, K. J. (2021). Digital servitization: Strategies for handling customization and customer interaction. In M. Kohtamäki, T. Baines, R. Rabetino, A. Z. Bigdeli, C. Kowalkowski, R. Oliva, & V. Parida (Eds.), *The Palgrave handbook of servitization* (pp. 355–372). Springer International Publishing. https://doi.org/10.1007/978-3-030-75771-7_23

Hydle, K. M., Hellström, M., Aas, T. H., & Breunig, K. J. (2022). Entrepreneuring practices: interconnected bundles for digital servitization. In N. A. Thompson, O. Byrne, A. Jenkins, & B. T. Teague (Eds.), *Research handbook on entrepreneurship as practice*. Edward Elgar Publishing. https://doi.org/10.4337/9781788976831.00027

IPCC. (2014). *Climate Change 2014: Synthesis Report* (Contribution of Working Groups I, II and III to the Fifth Assessment Report of the Intergovernmental Panel on Climate Change, Issue).

Jacobides, M. G., Cennamo, C., & Gawer, A. (2018). Towards a theory of ecosystems. *Strategic Management Journal, 39*(8), 2255–2276. https://doi.org/10.1002/smj.2904

John, N., Wesseling, J. H., Worrell, E., & Hekkert, M. (2022). How key-enabling technologies' regimes influence sociotechnical transitions: The impact of artificial intelligence on decarbonization in the steel industry. *Journal of Cleaner Production, 370*, 133624. https://doi.org/10.1016/j.jclepro.2022.133624

Jonsson, K., Mathiassen, L., & Holmström, J. (2018). Representation and mediation in digitalized work: Evidence from maintenance of mining machinery. *Journal of Information Technology, 33*(3), 216–232. https://doi.org/10.1057/s41265-017-0050-x

Jovanovic, M., Sjödin, D., Parida, V. (2022). Co-evolution of platform architecture platform services and platform governance: Expanding the platform value of industrial digital platforms. *Technovation, 118*. https://doi.org/10.1016/j.technovation.2020.102218

Li, T.-T., Wang, K., Sueyoshi, T., & Wang, D. D. (2021). ESG: Research progress and future prospects. *Sustainability, 13*(21), 11663.

Mäkitie, T., Hanson, J., Damman, S., & Wardeberg, M. (2023). Digital innovation's contribution to sustainability transitions. *Technology in Society, 73*, 102255. https://doi.org/10.1016/j.techsoc.2023.102255

McIntyre, D., Srinivasan, A., Afuah, A., Gawer, A., & Kretschmer, T. (2020). Multisided platforms as new organizational forms. *Academy of Management Perspectives, 35*(4), 566–583. https://doi.org/10.5465/amp.2018.0018

McIntyre, D., Srinivasan, A., Afuah, A., Gawer, A., & Kretschmer, T. (2021). Multisided platforms as new organizational forms. *Academy of Management Perspectives, 35*(4), 566–583. https://doi.org/10.5465/amp.2018.0018

Monteiro, E. (2022). *Digital oil: Machineries of knowing*. The MIT Press.

Mohamed, M., Ahokangas, P., Pikkarainen, M. (2023). Complementors' coopetition-based business models in multiplatform ecosystems. *Journal of Business Models, 11*(1), 68–77. https://doi.org/10.54337/jbm.v11i1.7199

Muench, S., Stoermer, E., Jensen, K., Asikainen, T., Salvi, M., & Scapolo, F. (2022). *Towards a green and digital future* (JRC129319). Luxembourg, 2022: Publications Office of the European Union.

Parida, V., Sjödin, D., & Reim, W. (2019). Reviewing literature on digitalization, business model innovation, and sustainable industry: Past achievements and future promises. *Sustainability, 11*(2), 391.

Rolland, K. H., Mathiassen, L., & Rai, A. (2018). Managing digital platforms in user organizations: The interactions between digital options and digital debt. *Information Systems Research, 29*(2), 419–443. https://doi.org/10.1287/isre.2018.0788

Tabares, S., Parida, V., & Chirumalla, K. (2025). Twin transition in industrial organizations: Conceptualization, implementation framework, and research agenda. *Technological Forecasting and Social Change, 213*, 123995. https://doi.org/10.1016/j.techfore.2025.123995

Teece, D. J. (2007). Explicating dynamic capabilities: The nature and microfoundations of (sustainable) enterprise performance. *Strategic Management Journal, 28*(13), 1319–1350.

Törmer, R. L., & Henningsson, S. (2020). Platformization and internationalization in the LEGO Group. In *53rd Hawaii International Conference on System Sciences.*

Yoo, Y., Lyytinen, K., Boland, R., Berente, N., Gaskin, J., Schutz, D., & Srinivasan, N. (2010). *The next wave of digital innovation: Opportunities and challenges: A report on the research workshop "Digital Challenges in Innovation Research".*

Zuboff, S. (1985). Automate/informate: The two faces of intelligent technology. *Organizational Dynamics, 14*(2), 5–18. https://doi.org/10.1016/0090-2616(85)90033-6

# 12

# Making Oil and Gas Smarter, Not Greener: Fossil Fuel Incumbents' Corporate Venture Capital Investments in Digital Entrepreneurs During Sustainability Transitions

Birthe Soppe and Hannah Schupfer

## Introduction

Considerable debate persists among scholars and practitioners regarding the potential and limitations of digital technologies in addressing sustainability challenges. In the face of the grand challenge of global warming, digital innovation developed by forward-looking entrepreneurs are considered particularly promising for advancing the sustainability transition of key economic sectors (George et al., 2021). A sustainability transition involves the diffusion of innovative, transformative

B. Soppe (✉)
Department of Organisation and Learning, University of Innsbruck, Innsbruck, Austria
e-mail: birthe.soppe@uibk.ac.at

TIK Centre for Technology, Innovation and Culture, University of Oslo, Oslo, Norway

H. Schupfer
Department of Strategy, International Management, and Entrepreneurship, King's College London, London, UK



M. Steen et al. (eds.), *Sustainability Transitions and Industrial Transformation*,
https://doi.org/10.1007/978-981-95-4893-4_12

solutions from protected niches—where entrepreneurs typically nurture radical innovations—into the mainstream socio-technical regime, ultimately altering the system's trajectory and the behaviour of its established actors (Geels, 2012; Geels & Schot, 2007).

Large firms operating in carbon-intensive, hard-to-abate industries and sectors could play a crucial role in breaking existing carbon lock-ins and propelling such systemic shifts (Doblinger & Soppe, 2013; Unruh, 2000). However, established firms often struggle with the technological and organizational changes required for transitions to take shape (Geels, 2014; Wright & Nyberg, 2017). Engaging with entrepreneurs—including 'digital entrepreneurs'—has therefore emerged as a promising strategy for regime incumbents to cope with transformative change. Entrepreneurs' propensity to drive innovation makes them attractive investment partners for established firms for whom these partnerships can serve as a pivotal way to intensify niche-regime interactions and respond to growing sustainability concerns (Hockerts & Wüstenhagen, 2010).

However, despite providing the impetus for novel directions, entrepreneurial ventures and their offerings are often marked by significant ambiguity and uncertainty (Santos & Eisenhardt, 2009). Especially digital entrepreneurs, defined as entrepreneurs who heavily rely on digital technology including artificial intelligence, machine learning, big data analytics, and robotics to develop their products and services (Nambisan, 2017), often draw on and recombine diverse technologies and knowledge areas (Pershina et al., 2019). This positions them as multi-sector agents rather than single and well-bounded niche actors. The ambiguity, uncertainty, and complexity of digital entrepreneurship hence raise questions about how fossil fuel incumbents can benefit from engaging with them. Specifically, as research on niche-regime interactions is developing, we have limited insights into how regime actors incorporate the multifaceted entrepreneurial solutions into their business, i.e., whether they are used to enhance existing regime operations or to develop radically novel, sustainable business areas. Hence, we ask the following research question:

How do fossil fuel incumbents incorporate digital entrepreneurial solutions into their business?

Building on a longitudinal case study of a Norwegian oil and gas (O&G) major, we qualitatively investigate the incumbent's corporate venture capital (CVC) investments in entrepreneurial firms between 2007 and 2023. CVC, the practice of established firms making equity investments in entrepreneurial firms (Dushnitsky & Lenox, 2005), is a pivotal way for incumbents to access needed innovation (Pahnke et al., 2015). Drawing on rich archival data and interview-based insights, our findings reveal that the fossil fuel incumbent has made significant CVC investments in digital entrepreneurs over the time span of analysis and considers these investments as an important way to address the transition towards a low-carbon future. Specifically, we find that the incumbent takes a collaborative approach to its CVC investments. Prior to formalizing an investment, a significant period is dedicated to building relationships between the entrepreneurs and the incumbent's employees, with the new ventures' offerings being jointly scoped and developed. This process reflects the considerable resources the incumbent allocates to fostering niche-regime exchanges. Yet, despite the increased alertness to digital entrepreneurial innovation and the corporate narrative of leveraging CVC investments for a low-carbon future, our analysis indicates that a substantial share of the investments in digital entrepreneurs is targeted at enhancing the existing O&G business, i.e., making it more efficient. This reflects an assimilative investment behaviour, where potentially radically new investments are aligned to support conventional business operations, rather than exploring novel pathways.

These findings advance our understanding of fossil fuel firms' investment behaviour, in particular CVC investments, during sustainability transitions. The literature broadly agrees that CVC is crucial for established firms to access knowledge and drive innovation (Dushnitsky & Lenox, 2005, 2006), especially in efforts to reduce fossil fuel dependency (Hegeman & Sørheim, 2021). Our qualitative study offers more nuanced insights and challenges some of these assumptions. It highlights the assimilative nature of incumbents' investment behaviour, showing that while CVC investments provide access to potentially disruptive technologies and facilitate interactions with innovative entrepreneurs, they are primarily leveraged to support and maintain business-as-usual (Wright & Nyberg, 2017).

These insights also contribute to research on niche-regime interactions and digital entrepreneurship in the context of sustainability transitions. Although both concepts are widely recognized as key enablers of such transitions (George et al., 2021; Hockerts & Wüstenhagen, 2010), our findings highlight their underutilized potential when mediated through inert regime actors. By showing how fossil fuel incumbents predominantly engage with digital entrepreneurs and adopt digital innovations to improve their conventional business operations, our study challenges the idea of a seamless sustainability and digital 'twin transition'. Instead, our results indicate that incumbents tend to leverage niche-regime interactions and digital solutions for incremental, rather than transformative, sustainability advancements (Mäkitie et al., 2023). These dynamics may critically limit sustainability developments in hard-to-abate sectors.

We now turn to the theoretical foundation of our study with a focus on niche-regime interactions between incumbents and entrepreneurs, and the controversial scholarly debate on the role of digital innovation for sustainability transitions. After presenting our empirical research setting and methods, we elaborate on and discuss the findings of our study.

## Niche-Regime Interactions Between Fossil Fuel Incumbents and Entrepreneurs During Transitions

While research on niche-regime interactions between incumbents and entrepreneurs is evolving, the profound differences between newcomers and incumbent actors are widely acknowledged within both organization and transitions research. Most accounts therefore focus on the tensions between defensive incumbents and entrepreneurial firms championing innovation (Alvarez & Barney, 2001; Dewald & Bowen, 2010; Gurses & Ozcan, 2015; Wesseling et al., 2014). While incumbents carry with them a legacy organizational identity and purpose deeply linked to conventional operations and practices (Georgallis et al., 2018; Schupfer & Soppe, 2025), entrepreneurs are typically less invested in the status quo, more innovative, and perceived as the drivers of sustainable development—often in opposition to the establishment (Pacheco et al., 2010; York & Venkataraman, 2010). Hence, in both fields of research, incumbents and entrepreneurs are portrayed as highly divergent actors.

However, mounting climate change concerns and the urgency to reduce greenhouse gas emissions have moved incumbents in carbon-intensive industries increasingly towards the centre of negative social attention (Ferns et al., 2022), pressuring them to reconsider their established strategies and business models. With knowledge increasingly residing outside of established firms (Powell & Soppe, 2015), in innovative niches in which entrepreneurial firms develop alternative offerings (Geels et al., 2016), a promising way for fossil fuel incumbents to more seriously engage with sustainability and reduce their fossil fuel dependency are niche-regime interactions with entrepreneurs (Hockerts & Wüstenhagen, 2010).

Niche-regime interactions can take many forms. The innovation literature emphasizes CVC investments in entrepreneurial firms as particularly fruitful means for rigid incumbents to obtain new knowledge and foster innovation (Dushnitsky & Lenox, 2005; Pahnke et al., 2015). CVC has proven especially effective in enabling established firms to transition toward green innovation and contribute to more systemic sustainability shifts (Bendig et al., 2022; Hegeman & Sørheim, 2021). However, existing studies predominantly investigate how innovation and learning affect established firms by relying on large-scale, quantitative data, such as patent applications, as a proxy for innovation and organizational change. While such indicators may offer useful insights into formal innovation outputs, they often provide a limited view of how innovation is actually used and implemented within firms. For example, patents may be filed for strategic or legal reasons and do not necessarily reflect the integration or practical use of new technologies or solutions within a company's operations. As a result, there is little research that explores how incumbents incorporate entrepreneurial innovations into their business operations. Understanding this integration is critical, especially in the context of sustainability transitions, where the access to innovation is not enough; what matters is how those innovations are embedded in incumbent firms' activities.

## The Promise of Digital Entrepreneurship and Digital Innovation for Sustainability Transitions

These issues may be even more relevant in the context of investments in digital entrepreneurs. Digital entrepreneurs are young ventures that rely on digital technology including artificial intelligence and machine learning, big data analytics, cloud computing, robotics, and 3D printing to develop their offerings (Nambisan, 2017). However, they often integrate a complex set of various digital and non-digital technologies to develop their innovative solutions (Pershina et al., 2019). As such, they are multi-sector agents and difficult to categorize into a specific market or business domain (Aversa et al., 2021). This ambiguity, uncertainty, and complexity surrounding digital entrepreneurs can make it particularly difficult for investing firms to clearly assess the scope and value of these ventures, as well as to integrate their solutions into their business.

Indeed, although the potential of digital entrepreneurship and digital innovation to tackle sustainability challenges is widely assumed (George et al., 2021), there are also concerned voices. On the one hand, there is the optimistic assumption that digital technologies will have a mainly positive impact on transitions (e.g., Brynjolfsson & McAfee, 2011; Perez, 2016). Digital technologies are considered an enabler for change (Nambisan, 2017; von Briel et al., 2018; Yoo et al., 2012) with the potential to faciliate a 'twin transition' and help organizations and their sectors to decarbonize. For instance, Perez (2016, p. 191) envisions a 'green global golden age' with the climate crisis being resolved through digital technology. On the other hand, scholars are worried that digital technologies promote unsustainable forms of corporate power, learning myopia, and knowledge depreciation, catalyse growing digital capitalism and a 'new dark age' which deters sustainable development (Balasubramanian et al., 2022; Bridle, 2018; Zuboff, 2019).

This debate suggests that the relationship between digital innovation and sustainability transitions is not a linear one, but much more complex. It seems characterized by various types of combinations, couplings, and possibilities, depending on the social use and construction of this relationship (Mäkitie et al., 2023; Mouthaan et al., 2023).

This complexity highlights the need for research on fossil fuel incumbents and their engagement with digital entrepreneurs. We therefore empirically investigate the ways in which digital entrepreneurial offerings are integrated into incumbent businesses, i.e., whether they are used to enhance conventional, polluting operations or to foster new, sustainable business areas. Such insights seem critical to better understand whether niche-regime interactions between fossil fuel incumbents and digital entrepreneurs have the potential to accelerate the sustainability transition of key economic sectors.

## Research Setting and Methods

An important sustainability transition is the energy transition which challenges particularly established O&G firms to radically reduce their fossil fuel dependency and develop alternative, clean energy solutions based on, for instance, renewable energy technologies. We conducted a qualitative longitudinal study (Yin, 2018) of a large firm in the Norwegian O&G industry over a time span of ca. 15 years, from 2007 until 2023. We specifically chose this incumbent as it holds a lead position in the industry and signalled ambitions to take sustainability challenges seriously by driving the shift to a green energy transition. At the same time, the firm is highly dependent on fossil fuel extraction and sales. Most of its revenues are generated through hydrocarbon operations. The Norwegian O&G industry and this well-entrenched actor present a highly suitable research context for our question of interest given mounting demands to take action on the grand challenge of global warming by investing into alternative energy solutions.

Data were collected as part of a larger research project on incumbents' behaviour during the sustainability transition. For this specific study, we focused our analysis on the incumbent's CVC investments in entrepreneurs between 2007 and 2023. In our case company, collaborations with entrepreneurs are primarily facilitated through the firm's CVC unit and associated investment fund, bringing members of the corporation together with the founders or central developers of the start-ups to jointly develop and commercialize innovation (Dushnitsky &

Lenox, 2006). We primarily investigated CVC investments in digital entrepreneurial ventures. We did so because digital entrepreneurs and their offerings, while characterized by substantial ambiguity, uncertainty, and complexity (Nambisan, 2017; Pershina et al., 2019), are argued to hold widespread potential to reduce established firms' fossil fuel dependency (George et al., 2021).

To systematically collect data on the incumbent's CVC investments over time, we relied on corporate press releases announcing CVC investments in start-ups. We collected and coded all corporate press releases announcing start-up investments between 2007 and 2023, accessed through the incumbent's online news archive. In the press releases, the start-ups were mentioned including a brief overview on the start-ups' technologies and reasons for the investment. Overall, we identified 81 CVC investments during our time frame of analysis. The content of the press releases allowed us to code for whether the investments were in digital or non-digital entrepreneurs, the type of entrepreneurial offering (ranging from oil reservoir recovery technology to battery developments for energy storage), and the various digital (e.g., big data analytics, machine learning, artificial intelligence) and non-digital technologies (i.e., geoscience, chemistry, energy, maritime) the entrepreneurs assembled to develop their offerings. We also coded the press releases on how the incumbent attempted to integrate the entrepreneurial solutions into its operations (i.e., to enhance its O&G or renewable energy business). The collected and coded data comprise ca. 76 pages of content.

To contextualize our data and emerging insights, we collected archival material related to the incumbent's CVC activities and sustainability ambitions. This included publicly available corporate sustainability reports, energy transition plans, and relevant entries from the company's website outlining the overarching goals and rationale behind its corporate investments. We conducted open coding of these documents, focusing on understanding the strategic purpose of CVC investments and identifying key technological areas targeted to support the green energy transition. We also triangulated and enriched the insights from the incumbent's perspective by collecting and coding data on the digital entrepreneurs identified through the CVC portfolio analysis. This allowed us to better understand the multi-sectoral nature of these actors.

Doing so, we gathered information from the entrepreneurs' websites, including their founding year, location, core technologies, targeted commercialization areas, and investor base. To access historical data for earlier investments, we utilized the Wayback Machine to retrieve archived versions of the entrepreneurs' websites. In total, we collected and analysed data on all 34 digital entrepreneurs in which the incumbent invested between 2007 and 2023.

While this rich collection of archival data forms the backbone of our analysis, we also conducted interviews with employees of the incumbent firm who were involved in establishing and managing CVC investments and relationships with entrepreneurs. The incumbent interview partners stem from various departments of the company, including sustainability, renewable energy, and corporate ventures. In total, we conducted nine interviews with employees of our case firm. We also interviewed founders or CEOs of five of the start-ups that had entered investments with the incumbent. The major discussion points in our interviews were the incumbent's attitude to sustainability and approach to start-up investments to facilitate the green energy transition. The interviews also served to gain deeper insights into CVC structures and activities. The interviews were conducted between the years 2020 and 2022, lasted on average 60 minutes, and were transcribed for further analysis.

## Findings

### The Fossil Fuel Incumbent's Focus on CVC Investments in the Context of the Green Energy Transition

Before engaging with the green energy transition, our case company was a dedicated O&G company with only a minor focus on clean energy and sustainability. The organizational culture was very much built upon pride in large-scale projects with a high return on investment in the O&G domain. Profitability, efficiency, and enhancing secure O&G operations were considered important achievements, setting the targets for technological investments. Early on, the incumbent invested in entrepreneurs through its CVC unit to access new knowledge and foster innovation, mainly to enhance the existing O&G business. The CVC unit focused on financial investments in entrepreneurial firms, often following suggestions from other departments that discovered relevant ventures. Investments in entrepreneurs for developing clean energy solutions remained largely unexplored. Indeed, as one of our internal informants shared with us, the company had '*very little consciousness about sustainability; then it was all about growth*'.

Around 2015, an important organizational shift occurred. With a new CEO coming in, tackling global warming and taking action to facilitate the transition towards a low-carbon energy future became a corporate priority. The company changed its name and organizational purpose to reflect a reorientation away from O&G towards a sustainability-oriented energy company, with renewable energy investments considered central as part of this transformation (Schupfer & Soppe, 2025).

These developments had important effects on the incumbent's CVC activities and investment targets. After 2015, the company started expanding its corporate venture activities towards novel business areas involving a wider pool of diverse start-ups. While so far, the CVC investments were aimed at supporting and enhancing the existing O&G business, the unit started to invest in novel and less familiar business areas, including clean and renewable energy. In 2016, the company announced the allocation of a fund worth $200M specifically dedicated to exploring novel business areas and clean energy solutions, presenting its CVC unit as '*an active corporate venture partner dedicated to investing in and supporting innovative companies shaping the future of energy*' (corporate webpage, 2024).

Scaling investments in renewables and low-carbon technologies were now emphasized as a central focus in the transition away from fossil fuels (energy transition plan, 2022). Corporate statements such as '*we believe that the innovation, creativity and agility of entrepreneurs, startups and partners can drive change towards a low-carbon future*' and that '*new and innovative solutions…help meet the world's energy needs while reducing its environmental footprint*' (corporate webpage, 2024) suggest that a core ambition behind the incumbent's CVC investments is to support the sustainability transition of its industry.

As part of these developments, the company turned considerable attention to digital technologies and digitalization as '*a driving force in the energy transition*' and an important enabler for '*a net-zero future*' (corporate webpage, 2024). As stated in a press release from 2022, '*in [our company], digitalisation is a key element to achieve our goals during the energy transition*' (press release, 2022). Digital innovation and digital technologies are frequently emphasized as a central pillar of the corporation's transition strategy, highlighting the critical role of technologies such as artificial intelligence, digital twins, automation and robotics, and 3D printing in enabling the energy transition. Given the incumbent's proclaimed alertness to entrepreneurial investments and digital technology as key components of its green transition strategy, how does the firm approach CVC investments in entrepreneurial ventures?

## The Fossil Fuel Incumbent's Collaborative Approach to CVC Investments

Investigating the underlying organizational processes involved in setting up and managing the CVC investments, our analysis reveals that these are not purely arm's-length equity investments. Before a collaborative investment is formed, there are individual champions or mentors within the incumbent firm involved in building personal ties with the entrepreneurs and following them up over a longer period before an investment is formalized. The incumbent also runs dedicated mentorship and accelerator programmes to work with and scout suitable entrepreneurs. In other words, the incumbent pursues a collaborative approach to its CVC investments, and the entrepreneurial solutions and offerings are often mutually shaped and developed. Although our informants also described the common challenges of incumbent–entrant collaborations, such as balancing the agile, fast-paced mindset of entrepreneurs with the slower, more bureaucratic decision-making processes of established firms, we find that the incumbent's approach to these investments is notably resource-intensive, involving individualized relationships between a corporate representative or mentor and the entrepreneur. This is further illustrated by the following quote from the CEO of one of the interviewed entrepreneurial firms:

> The real story here...innovation between entrepreneurs and big corporates, is not about the relationship between the entrepreneur and the corporate, but it's about the individuals at the start-up and the individuals at the big corporate. There have to be individuals that [have] bought in, or see the vision, or have a need that they see that they are willing to take the personal reputational risk to go to bat for this start-up. [Name of the start-up] doesn't work with [name of the incumbent], we work with a number of individuals within [the incumbent]. (start-up CEO, interviewed in 2021)

These insights are also reflected across the press releases we analysed. For instance, in a press release from 2020, it is stated: '*After following this (...) company for a long time, [the incumbent] is pleased to announce an investment in the company*'. In a press release announcing a start-up investment in 2021, one can read that the incumbent '*has supported the technology development since 2017*'. Likewise, in a press release from 2019, it is stated that '*We are very pleased to have been able to conclude this investment which comes as a direct result of our close long-term cooperation with the highly competent team*'. Thus, fostering these niche-regime interactions entails a significant allocation of effort and resources by the incumbent.

What motivates the incumbent to dedicate substantial financial and human resources to forming and nurturing collaborations with entrepreneurs, including digital entrepreneurs? Are those investments genuinely leveraged to advance the incumbent's stated goal of contributing to the green energy transition? As CVC investments often reflect the broader corporate strategy, analysing the CVC portfolio more closely may yield valuable insights.

## The Fossil Fuel Incumbent's CVC Portfolio and Assimilative Investment Behaviour During Sustainability Transitions

Based on our data, we identified 81 CVC investments in entrepreneurial firms between 2007 and 2023. The number of annual investments increased over the past five years, likely reflecting the expansion of the CVC fund, as discussed above. Analysing the overall distribution of the incumbent's CVC investments in digital vs. non-digital entrepreneurs, we find substantial investments in digital entrepreneurs (i.e., ventures who rely on digital technology to develop their products and services). Out of 81 investments, 34 investments were established with digital entrepreneurs, representing approximately 42% of the overall CVC portfolio. Figure 12.1 illustrates the annual distribution of the incumbent's CVC investments in digital vs. non-digital entrepreneurs over time.

Prominent digital technologies among our sampled entrepreneurs are machine learning, artificial intelligence, and big data analytics. More than 50% of the identified digital entrepreneurs (19 out of the 34 digital entrepreneurs) work with one or a combination of these technologies. The remaining digital entrepreneurs rely on blockchain, robotics, drones, cloud computing, digital imaging and mapping, and 3D computing to develop their specific entrepreneurial solutions. However, all entrepreneurs blend digital technologies with non-digital knowledge stemming from various sectors including energy, physics, chemistry, geoscience, aerospace, and medicine. Thus, examining the technology profile of the digital entrepreneurs more closely, our analysis confirms that they are multi-sector actors relying on a blend of various digital and non-digital technologies to develop their entrepreneurial solutions.

The multi-sector nature of the entrepreneurs was also vividly emphasized in the CVC press announcements which we systematically coded. For instance, in a press release from 2017, one can read:

> [Name of the entrepreneurial firm] delivers substantial operating improvements to customers by combining the best of traditional physics-based methods with modern artificial intelligence and machine learning.

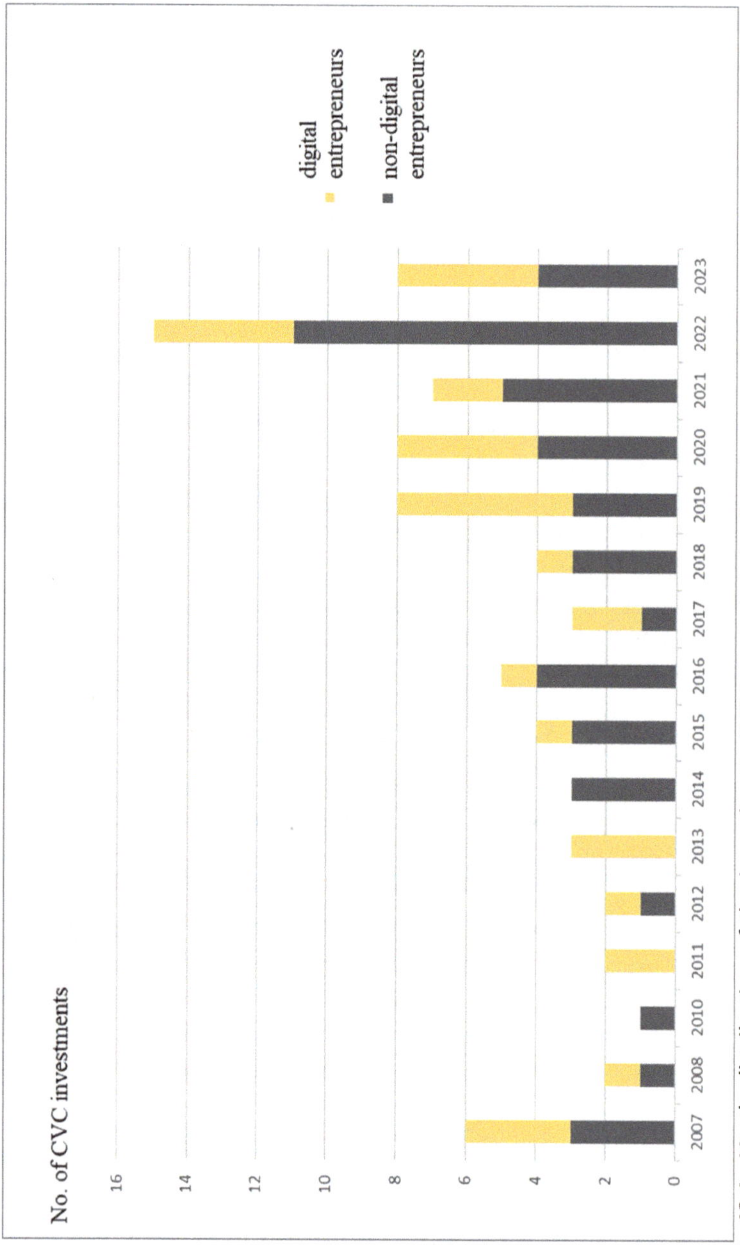


**Fig. 12.1** Yearly distribution of the incumbent's CVC investments in digital vs. non-digital entrepreneurs

Similarly, in an investment announcement from 2020, it is highlighted:

> [Name of the entrepreneurial firm] has developed a software platform that utilises technologies like artificial intelligence, unmanned aerial vehicles and machine learning to assess the status of wind energy installations and predict the most optimal time for proactive maintenance.

Hence, none of the digital entrepreneurs can be clearly categorized into a single business sector or market. This raises the question of how the fossil fuel incumbent incorporates the multifaceted entrepreneurial solutions into its business.

Figure 12.2 illustrates our insights into how the incumbent seeks to integrate investments in digital entrepreneurial ventures. Out of the 34 CVC investments in digital entrepreneurs identified across our observation period, only five investments are clearly aimed at supporting the development of the incumbent's renewable energy business.

The five investments in digital 'renewable' entrepreneurs have been formed between 2019 and 2023. They focus, for instance, on the integration of distributed renewables into the electricity grid and the maintenance of wind power plants. The majority of the investments appear instead focused on enhancing the efficiency and performance of the incumbent's conventional O&G business. They include automated drilling processes, digital subsurface solutions for well and pipeline maintenance, reservoir management, shale drilling, digitally enabled discovery of new O&G fields, and digitally enhanced O&G operations. More recent investments involve digital start-ups working on carbon removal and storage technologies, a technology linked to the incumbent's fossil fuel operations.

Investments in digital O&G ventures often involve very specific technological solutions, emphasizing efficiency gains and ensuring incremental improvements of the incumbent's existing operations and established core business. For instance, one can read: '*The [name of the entrepreneurial firm] technology is a central component within offshore automation and a key to safer and more efficient drilling and well operations*'. In contrast, ambitions related to investments in digital renewable

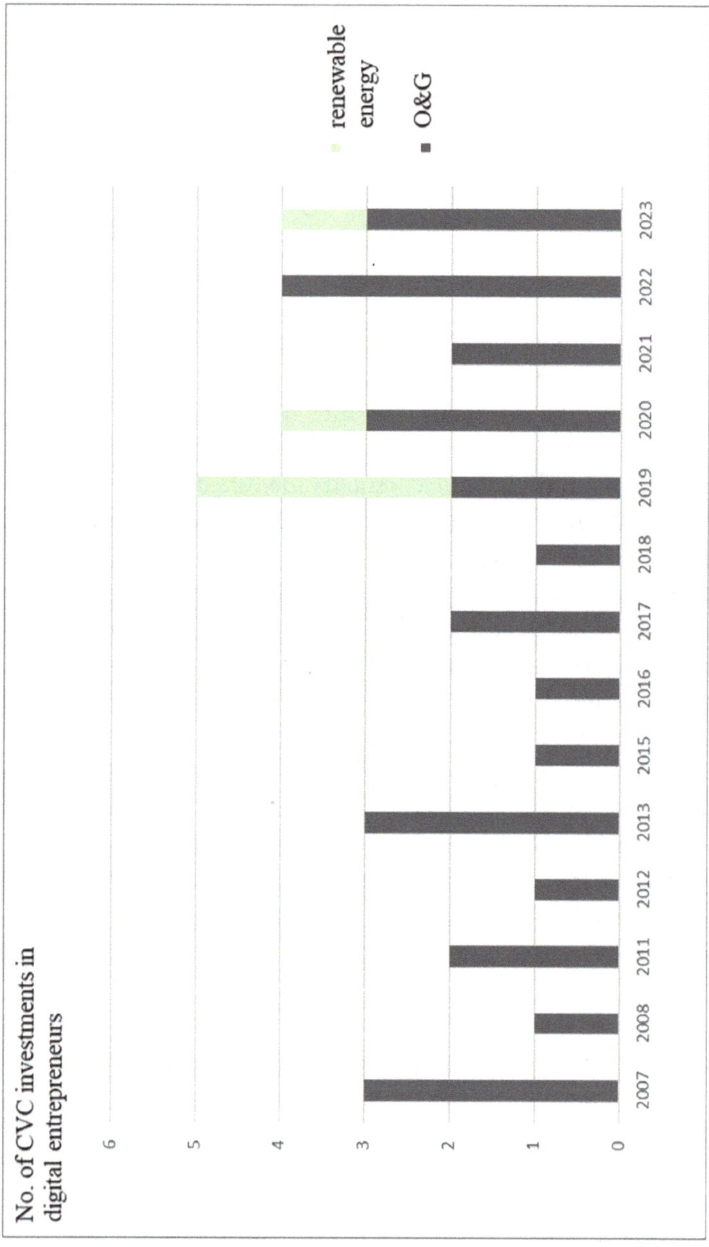


**Fig. 12.2** Integration of digital entrepreneurial solutions into the incumbent's O&G vs. renewable energy business

solutions are often framed in a more open-ended way, broadly emphasizing the '*potential to accelerate the energy transition*' (press releases from 2019).

Taken together, despite the substantive share of investments in digital entrepreneurs and the corporate narrative of leveraging CVC for a low-carbon future, our analysis indicates that the major portion of the incumbent's investments is directed towards enhancing its conventional operations rather than developing radically new business areas. This reflects an assimilative investment behaviour, where investment decisions are predominantly aligned with existing norms and standards to support conventional business operations, rather than being directed towards fostering radically different and transformative pathways. In other words, our analysis suggests that the firm's CVC investments primarily aim at regime maintenance through making conventional fossil fuel operations smarter (i.e., more efficient), rather than making them greener.

## Discussion

Our analysis of a large O&G firm's CVC activities through time was motivated by the desire to better understanding the role of niche-regime interactions among fossil fuel incumbents and entrepreneurs to advancing the sustainability transition of key economic sectors, including energy. Our specific focus on CVC investments in digital entrepreneurs was driven by the research interest to better understand the promise of digital entrepreneurship and digital innovation for sustainability transitions.

First, our findings contribute to our understanding of established firms' investment behaviour during industry transitions. The literature broadly assumes that CVC is crucial for established firms to access novel technologies and drive innovation (Dushnitsky & Lenox, 2005, 2006), especially in the context of established firms seeking to reduce their fossil fuel dependency (Hegeman & Sørheim, 2021; Hockerts & Wüstenhagen, 2010). However, prior work has predominantly relied on quantitative research approaches, such as large panel datasets and patent data, to assess the impact of CVC investments on the 'greening' of invested firms (e.g., Bendig et al., 2022; Dushnitsky & Lenox, 2005). Our qualitative study offers more nuanced insights. It highlights incumbents' assimilative investment behaviour, showing that while CVC investments may indeed provide access to potentially disruptive technologies and foster collaboration with innovative entrepreneurs, they are primarily used to make conventional business operations smarter—such as by improving efficiency—rather than to enable novel, potentially greener pathways. Related, our insights reinforce concerns raised by prior research about the misalignment between green corporate narratives and actual investment behaviour (Li et al., 2022; Patala et al., 2019). The discrepancy between corporate communication and actual investment behaviour appears particularly pronounced in the energy sector, which raises concerns about the sector's ability and willingness to meaningfully address the issue of global warming through adequate corporate investments. In other words, assimilation reinforces business-as-usual and inertia rather than driving systemic change and may critically limit sustainability developments in hard-to-abate sectors.

Second, our study contributes to research on niche-regime interactions and digital entrepreneurship during sustainability transitions. While both concepts are widely considered key drivers of sustainability (George et al., 2021; Hockerts & Wüstenhagen, 2010), our results challenge this assumption by highlighting their limited transformative impact when mediated through incumbent actors. Our findings indicate that fossil fuel incumbents primarily invest in digital entrepreneurs and digital innovation to reduce the environmental impact of their existing fossil fuel operations, rather than to initiate a transition away from conventional fuels. Hence, although niche-regime interactions and digital innovation may offer substantial potential to contribute to sustainability transitions in key socio-economic systems, our analysis suggests that their actual sustainability impact critically depends on how they are utilized by large firms. At present, these technologies appear to be employed primarily for incremental improvements rather than radical sustainability advancements (Mäkitie et al., 2023), which leads us to challenge the idea of a seamless 'twin transition'.

Our study also enriches our understanding of entrepreneurial actors. While it has often been assumed that entrepreneurs stem from a single-niche mobilizing forces to disrupt regimes (Smith & Raven, 2012), we offer a novel conceptualization by showing that entrants often stem from and assemble knowledge from multiple technological domains rather than fitting neatly into a bounded niche. As such, we specify them as multi-sector agents—actors who draw on and recombine knowledge from multiple technological or sectoral knowledge domains. This conceptualization offers a more dynamic account of the cross-sectoral nature of niche-regime interactions within transitions studies. More research is needed to examine how cross-sectoral knowledge recombination can contribute to transitions. Our study also challenges the traditional one-directional view of niche-to-regime influence as we find that many entrepreneurs pursue innovation in a collaborative manner together with established regime actors whose issues they seek to fix. However, this approach may, of course, also facilitate the assimilative deployment of the entrepreneurial offerings into the incumbent business despite the radical potential that lies in the entrepreneurs' multi-sector solutions.

More broadly, our insights complement the traditional macro-level focus of transition research by offering organizational-level insights into niche-regime interactions in the context of sustainability transitions (Köhler et al., 2019). While our study points to persistent organizational barriers in limiting the potential of collective action among diverse actors to achieve sustainability-enhancing outcomes, it highlights the crucial role of individual champions within incumbent firms as boundary spanners between established and novel actors (Smink et al., 2015). These boundary spanners help to form inter-organizational ties to conjointly develop entrepreneurial solutions. Future research is needed to explore the conditions under which such individuals gain legitimacy within their organizations, and the extent to which they can influence strategic decision-making and investment behaviour, thereby driving transformative change.

In this study, we focused on CVC as a specific form of niche-regime interactions. However, such interactions can be established through various other forms including accelerators and incubators or more formal inter-organizational relationships such as learning alliances. One important avenue for future research is to investigate the various forms of niche-regime interactions and their employment for sustainability. Moreover, many entrepreneurs are financially and organizationally supported by more than one incumbent investor. Future research could examine the role and effect of multiple investors on entrepreneurs and their technological developments. We hope that the insights of our study can spark more research on how niche-regime interactions may accelerate or decrease the pace of sustainability transitions.

## References

Alvarez, S. A., & Barney, J. B. (2001). How entrepreneurial firms can benefit from alliances with large partners. *Academy of Management Perspectives, 15*(1), 139–148. https://doi.org/10.5465/ame.2001.4251563

Amit, R., & Han, X. (2017). Value creation through novel resource configurations in a digitally enabled world. *Strategic Entrepreneurship Journal, 11*(3), 228–242. https://doi.org/10.1002/sej.1256

Aversa, P., Huyghe, A., & Bonadio, G. (2021). First impressions stick: Market entry strategies and category priming in the digital domain. *Journal of Management Studies, 58*(7), 1721–1760. https://doi.org/10.1111/joms.12712

Balasubramanian, N., Ye, Y., & Xu, M. (2022). Substituting human decision-making with Machine Learning: Implications for organizational learning. *Academy of Management Review, 47*(3), 448–465. https://doi.org/10.5465/amr.2019.0470

Bendig, D., Kleine-Stegemann, L., Schulz, C., & Eckardt, D. (2022). The effect of green startup investments on incumbents' green innovation output. *Journal of Cleaner Production, 376*. https://doi.org/10.1016/j.jclepro.2022.134316

Bridle, J. (2018). *New dark age: Technology and the end of the future*. Verso Books.

Brynjolfsson, E., & McAfee, A. (2011). *Race against the machine: How the digital revolution is accelerating innovation, driving productivity, and irreversibly transforming employment and the economy*. Digital Frontier Press.

Dewald, J., & Bowen, F. (2010). Storm clouds and silver linings: Responding to disruptive innovations through cognitive resilience. *Entrepreneurship Theory and Practice, 34*(1), 197–218. https://doi.org/10.1111/j.1540-6520.2009.00312.x

Doblinger, C., & Soppe, B. (2013). Change-actors in the U.S. electric energy system: The role of environmental groups in utility adoption and diffusion of wind power. *Energy Policy, 61*, 274–284. https://doi.org/10.1016/j.enpol.2013.07.028

Dushnitsky, G., & Lenox, M. J. (2005). When do incumbents learn from entrepreneurial ventures?: Corporate venture capital and investing firm innovation rates. *Research Policy, 34*(5), 615–639. https://doi.org/10.1016/j.respol.2005.01.017

Dushnitsky, G., & Lenox, M. J. (2006). When does corporate venture capital investment create firm value? *Journal of Business Venturing, 21*(6), 753–772. https://doi.org/10.1016/j.jbusvent.2005.04.012

Ferns, G., Lambert, A., & Günther, M. (2022). The analogical construction of stigma as a moral dualism: The case of the fossil fuel divestment movement. *Academy of Management Journal, 65*(4), 1383–1415. https://doi.org/10.5465/amj.2018.0615

Geels, F. W. (2012). A socio-technical analysis of low-carbon transitions: Introducing the multi-level perspective into transport studies. *Journal of Transport Geography, 24*, 471–482. https://doi.org/10.1016/j.jtrangeo.2012.01.021

Geels, F. W. (2014). Reconceptualising the co-evolution of firms-in-industries and their environments: Developing an inter-disciplinary Triple Embeddedness Framework. *Research Policy, 43*(2), 261–277. https://doi.org/10.1016/j.respol.2013.10.006

Geels, F. W., & Schot, J. (2007). Typology of sociotechnical transition pathways. *Research Policy, 36*(3), 399–417. https://doi.org/10.1016/j.respol.2007.01.003

Geels, F. W., Kern, F., Fuchs, G., Hinderer, N., Kungl, G., Mylan, J., et al. (2016). The enactment of socio-technical transition pathways: A reformulated typology and a comparative multi-level analysis of the German and UK low-carbon electricity transitions (1990–2014). *Research Policy, 45*(4), 896–913. https://doi.org/10.1016/j.respol.2016.01.015

Georgallis, P., Dowell, G., & Durand, R. (2018). Shine on me: Policy support for the emergent solar photovoltaic industry in Europe. *Administrative Science Quarterly, 64*(3), 503–541.

George, G., Merrill, R. K., & Schillebeeckx, S. J. D. (2021). Digital sustainability and entrepreneurship: How digital innovations are helping tackle climate change and sustainable development. *Entrepreneurship Theory and Practice, 45*(5), 999–1027. https://doi.org/10.1177/1042258719899425

Gurses, K., & Ozcan, P. (2015). Entrepreneurship in regulated markets: Framing contests and collective action to introduce Pay TV in the U.S. *Academy of Management Journal, 58*(6), 1709–1739. https://doi.org/10.5465/amj.2013.0775

Hegeman, P. D., & Sørheim, R. (2021). Why do they do it? Corporate venture capital investments in cleantech startups. *Journal of Cleaner Production, 294*. https://doi.org/10.1016/j.jclepro.2021.126315

Hockerts, K., & Wüstenhagen, R. (2010). Greening Goliaths versus emerging Davids—Theorizing about the role of incumbents and new entrants in sustainable entrepreneurship. *Journal of Business Venturing, 25*(5), 481–492. https://doi.org/10.1016/j.jbusvent.2009.07.005

Köhler, J., Geels, F. W., Kern, F., Markard, J., Onsongo, E., Wieczorek, A., & Wells, P. (2019). An agenda for sustainability transitions research: State of the art and future directions. *Environmental Innovation and Societal Transitions, 31*, 1–32. https://doi.org/10.1016/j.eist.2019.01.004

Li, M., Trencher, G., & Asuka, J. (2022). The clean energy claims of BP, Chevron, ExxonMobil and Shell: A mismatch between discourse, actions and investments. *PLOS ONE, 17*(2), Article e0263596. https://doi.org/10.1371/journal.pone.0263596

Mäkitie, T., Hanson, J., Damman, S., & Wardeberg, M. (2023). Digital innovation's contribution to sustainability transitions. *Technology in Society, 73*. https://doi.org/10.1016/j.techsoc.2023.102255

Mouthaan, M., Frenken, K., Piscicelli, L., & Vaskelainen, T. (2023). Systemic sustainability effects of contemporary digitalization: A scoping review and research agenda. *Futures, 149*. https://doi.org/10.1016/j.futures.2023.103142

Nambisan, S. (2017). Digital entrepreneurship: Toward a digital technology perspective of entrepreneurship. *Entrepreneurship Theory and Practice, 41*(6), 1029–1055. https://doi.org/10.1111/etap.12254

Pacheco, D. F., Dean, T. J., & Payne, D. S. (2010). Escaping the green prison: Entrepreneurship and the creation of opportunities for sustainable development. *Journal of Business Venturing, 25*(5), 464–480.

Pahnke, E. C., Katila, R., & Eisenhardt, K. M. (2015). Who Takes You to the Dance? How partners' institutional logics influence innovation in young firms. *Administrative Science Quarterly, 60*(4), 596–633. https://doi.org/10.1177/0001839215592913

Patala, S., Korpivaara, I., Jalkala, A., Kuitunen, A., & Soppe, B. (2019). Legitimacy under institutional change: How incumbents appropriate clean rhetoric for dirty technologies. *Organization Studies, 40*(3), 395–419. https://doi.org/10.1177/0170840617736938

Perez, C. (2016). Capitalism, technology and a green global golden age: The role of history in helping to shape the future. *Rethinking Capitalism: Economics and Policy for Sustainable and Inclusive Growth, 1*, 191–217.

Pershina, R., Soppe, B., & Thune, T. M. (2019). Bridging analog and digital expertise: Cross-domain collaboration and boundary-spanning tools in the creation of digital innovation. *Research Policy, 48*(9). https://doi.org/10.1016/j.respol.2019.103819

Powell, W. W., & Soppe, B. (2015). Boundaries and new organization forms. In J. D. Wright (Ed.), *International encyclopedia of the social & behavioral sciences: Second edition* (2nd ed., pp. 768–772). Elsevier.

Santos, F. M., & Eisenhardt, K. M. (2009). Constructing Markets and Shaping Boundaries: Entrepreneurial Power in Nascent Fields. *Academy of Management Journal, 52*(4), 643–671. https://doi.org/10.5465/amj.2009.43669892

Schupfer, H., & Soppe, B. (2025). Greening from within: The role of organisational purpose shift in building internal legitimacy for fossil fuel incumbents' green innovation. *Industry and Innovation, 32(1)*, 84–107. https://doi.org/10.1080/13662716.2024.2379995

Smink, M., Negro, S. O., Niesten, E., & Hekkert, M. P. (2015). How mismatching institutional logics hinder niche–regime interaction and how boundary spanners intervene. *Technological Forecasting and Social Change, 100*, 225–237. https://doi.org/10.1016/j.techfore.2015.07.004

Smith, A., & Raven, R. (2012). What is protective space? Reconsidering niches in transitions to sustainability. *Research Policy, 41*(6), 1025–1036. https://doi.org/10.1016/j.respol.2011.12.012

Unruh, G. C. (2000). Understanding carbon lock-in. *Energy Policy, 28*(12), 817–830. https://doi.org/10.1016/S0301-4215(00)00070-7

von Briel, F., Recker, J., & Davidsson, P. (2018). Not all digital venture ideas are created equal: Implications for venture creation processes. *The Journal of Strategic Information Systems, 27*(4), 278–295. https://doi.org/10.1016/j.jsis.2018.06.002

Wesseling, J. H., Farla, J. C. M., Sperling, D., & Hekkert, M. P. (2014). Car manufacturers' changing political strategies on the ZEV mandate. *Transportation Research Part D: Transportation and Environment, 33*, 196–209. https://doi.org/10.1016/j.trd.2014.06.006

Wright, C., & Nyberg, D. (2017). An inconvenient truth: How organizations translate climate change into business as usual. *Academy of Management Journal, 60*(5), 1633–1661. https://doi.org/10.5465/amj.2015.0718

Yin, R. K. (2018). *Case Study Research and Applications: Design and Methods* (6th ed.). Sage Publications.

Yoo, Y., Boland, R. J., Lyytinen, K., & Majchrzak, A. (2012). Organizing for innovation in the digitized world. *Organization Science, 23*(5), 1398–1408. https://doi.org/10.1287/orsc.1120.0771

York, J. G., & Venkataraman, S. (2010). The entrepreneur–environment nexus: Uncertainty, innovation, and allocation. *Journal of Business Venturing, 25*(5), 449–463.

Zuboff, S. (2019). *The age of surveillance capitalism: The fight for a human future at the new frontier of power*. PublicAffairs.

# 13

# The Scope and Outcomes of Transformative Policy Missions: A Conceptual Framework and Empirical Exploration of Swedish Industry Decarbonization

John Andersson and Hans Hellsmark

## Introduction

Global warming, biodiversity loss, rising inequality, increasing polarization, and other grand challenges compel society to achieve numerous interrelated transitions to more equitable, resilient, and environmentally benign socio-technical systems. This is not merely about accelerating the pace of innovation, but also involves shaping its directionality towards what can be perceived as desirable outcomes from a local and global context (Andersson et al., 2021). Indeed, the outcomes of transitions will

J. Andersson (✉)
Division Built Environment, Department System Transition, RISE Research Institutes of Sweden, Gothenburg, Sweden
e-mail: johnn.andersson@ri.se

H. Hellsmark
Division of Environmental Systems Analysis, Department of Technology Management and Economics, Chalmers University of Technology, Gothenburg, Sweden


M. Steen et al. (eds.), *Sustainability Transitions and Industrial Transformation*,
https://doi.org/10.1007/978-981-95-4893-4_13

both determine if grand challenges are met and shape the characteristics and distribution of future welfare as they involve industrial creation, transformation, and decline.

In this context, transformative innovation policy has emerged as an alternative to traditional growth-oriented policy approaches (Schot & Steinmueller, 2018). The general idea is to design policy mixes that can mobilize, support, incentivize, and guide a broad range of actors in a collective effort to achieve socio-technical transitions that address societal problems, rather than merely fixing market and system failures (Rogge & Reichardt, 2016; Weber & Rohracher, 2012). However, transitions are cumulative and path dependent, yet complex and uncertain processes, characterized by tensions between established and emerging structures (Geels, 2002; Rip & Kemp, 1998). This makes it genuinely difficult to influence their direction and outcomes, and gives rise to new challenges for policymakers, industrial firms, and other societal stakeholders.

With climate change, biodiversity loss, and other pressing grand challenges, transformative innovation policy increasingly departs from concrete and ambitious goals. These are often set by governments at local, regional, national, and supra-national levels, but may also emerge from, or be further specified within, collaborative stakeholder processes (Haddad et al., 2022). It is also common to frame transformative innovation policy in terms of missions, a notion that draws inspiration from large-scale and government-led scientific programmes such as the twentieth-century mission to put a man on the moon. However, today's missions may not only promote scientific and technological progress, but also focus on shaping systemic change or act as long-term policy frame (Wittmann et al., 2021).

In this chapter, we use the term *transformative policy missions* to refer to major government initiatives that involve goals and interventions oriented towards challenge-driven transformation of socio-technical systems. Previous research has highlighted how the wicked nature of concurrent sustainability challenges gives rise to complexity, uncertainty, and contestation (Levin et al., 2012; Scoones, 2016), and mission-oriented policymaking can be seen as a way to address this wickedness (Janssen et al., 2021). Scholars, not the least in transition management, have also developed conceptual frameworks and related policy recommendations, which emphasize the need for participatory approaches that involve a broad range of actors in collective prioritization among relevant problems and solutions (Voß et al., 2009). However, while the existing literature offers a few analytical frameworks that highlight important aspects (e.g., Andersen et al., 2024; Elzinga et al., 2023; Wanzenböck et al., 2020; Wittmann et al., 2021), there is still a need to bring these insights together in a comprehensive conceptual overview of how the scope and outcomes of transformative policy missions may differ. A shared understanding of these dimensions is important for both analytical and practical purposes.

Our ambition with this chapter is therefore to suggest and illustrate a conceptual framework that can be used for comparing the intention, design, implementation, and outcomes of transformative policy missions. We begin with a discussion based on the scientific literature and introduce the framework, which is then used to explore the ongoing decarbonization of Swedish industry. The empirical account specifically highlights the Industry Leap, which is a major policy programme that departs from the Swedish climate goals to foster collaborative innovation projects oriented towards reaching net-zero emissions from industrial production in Sweden before 2045 (Swedish Government, 2016, 2017).

## Framing the Scope and Outcomes of Transformative Policy Missions

Socio-technical systems may develop along different trajectories, which bring vastly different implications for human and non-human stakeholders (Andersson et al., 2021; Stirling, 2011). A defining feature of transformative innovation policy is that technological development is approached not as a means to stimulate economic growth, but as a way to address climate change, biodiversity loss, inequality, poverty, and other societal problems. Innovation is thus understood in terms of the transformation rather than expansion of production and consumption systems, and the focus is on multi-dimensional change through technical, organizational, social, and behavioural solutions. Transformative policy missions acknowledge this directionality of innovation processes, depart from the idea that some development trajectories are more desirable than others, and give governments a central role in providing goals and guidance.

To design, implement, analyse, and evaluate transformative policy missions, there is a need to explore variety in numerous dimensions. An appropriate starting point is the conceptual framework developed by Andersson et al. (2021), which relates the directionality of innovation processes to a basic morphology of socio-technical systems. In addition, a recent stream of publications that engage with transformative innovation policy offer complementary insights. Wanzenböck et al. (2020) highlight that missions involve contestation and convergence in relation to both problems and solutions. Wittmann et al. (2021) move beyond the distinction between accelerator and transformer missions, as they introduce a more elaborate typology that highlights differences in how missions are defined, designed, and implemented. Elzinga et al. (2023) adopt an innovation systems perspective, proposing a framework for analysing structural and functional change processes generated by missions. In addition, Andersen et al. (2024) demonstrate the importance of a multi-sectoral perspective that accounts for links and dynamics across different socio-technical systems. In the following, we bring together insights from this literature as we frame the scope and outcomes of transformative policy missions.

## Transformative Orientation, Policy Interventions, and System Focus

The scope of a transformative policy mission describes its ambition and delineation in multiple dimensions. To begin with, there is a transformative orientation that departs from an interest in addressing societal problems, which can have a social or environmental character, relating to issues such as poverty, inequality, climate change, and biodiversity loss. These problems can be defined in specific or general terms and be more or less open to different interpretations. For example, some missions may have a narrow focus on the reduction of greenhouse gas emissions, while others promote the global sustainability goals.

Depending on the characteristics of the problem in focus, different solutions become relevant. These may involve a broad spectrum of technical, organizational, and behavioural changes that amount to the development and diffusion of novel products and processes. Although not always associated with the current discourse on missions, solutions may, and often should, also amount to the dismantling and decline of existing structures. However, missions often involve implicit or explicit boundary conditions that exclude some solutions and highlight others. For example, addressing problems within a certain timeframe often requires the scaling up of solutions that already have been tested at a smaller scale rather than pursuing completely new innovations from basic research (Azar & Sandén, 2011). Efforts to address the societal problem in focus may also be coupled with ambitions to realize economic opportunities associated with the development and diffusion of solutions.

Another central aspect that defines transformative policy missions is the policy interventions made. Missions concretize the problem in focus by setting goals (Mazzucato, 2018; Robinson & Mazzucato, 2019), which introduces more or less specific, measurable, and time-bound objectives that serve to guide innovation activities. A defining feature is that overarching goals are set by governments at local, regional, national, or supra-national levels (Haddad et al., 2022). However, these goals may be further specified through collaborative stakeholder processes in which shared problem definitions and objectives emerge. An important difference between missions is accordingly the specificity of overarching goals set by policymakers, which determine the extent to which stakeholders can influence the direction of transformative change. Furthermore, missions involve specific policy instruments oriented towards the set goals. Missions are often linked to major programmes that channel public resources to the development, demonstration, and implementation of new solutions, but they may also employ other tools, such as market subsidies, tax incentives, and regulations, that further catalyse progress in relation to the transformative focus (Sovacool, 2011). The policy instruments associated with missions are parts of broader policy mixes that differ in numerous dimensions, including the underlying policymaking and implementation processes, the objectives and strategies, the design of individual instruments, as well as the overarching consistency, coherence, credibility, and comprehensiveness (Rogge & Reichardt, 2016).

In addition, transformative policy missions focus on different socio-technical systems, which consist of interlinked actors, institutions, and technologies that govern how societal functions related to production and consumption are fulfilled (Andersson et al., 2021). The system focus has a sectoral dimension in which a mission can be broad, with an ambition to transform the entire economy, or focused on a specific industry or value chain defined at some technological scale (e.g., energy, heat and power production, or waste combustion in combined heat and power plants). Similarly, there is a geographical dimension where a mission can be global or focused on a specific region defined at some geographical scale (e.g., nation, county, or municipality).

Notably, different sectors and geographies become relevant depending on whether a mission is approached from the perspective of problems, solutions, policy goals, or policy instruments. For example, solving a problem defined in relation to a specific industry and region is often contingent on innovation processes that transcend pre-conceived boundaries in both sectoral and geographical dimensions (Coenen et al., 2012). Also, policy goals and instruments introduced in a given jurisdiction may be oriented towards transformation in a different geographical region (Sovacool, 2011).

Taken together, this highlights that missions are not only associated with different systems, but also create new multi-system interactions (Andersen et al., 2024; Andersen & Markard, 2020; Rosenbloom, 2020). In addition, the system focus has a temporal dimension, which relates to the time scales at which causes and consequences of problems appear as well as to the short as opposed to long-term viability of solutions. Notably, the temporal orientation is reflected in how time-bound policy goals are defined to create a sense of urgency, stimulate action, ensure consistent progress monitoring, and thereby enable adaptive responses (Mazzucato, 2018).

## Structural and Functional Transformation Processes

Whereas transformative orientation, policy interventions, and system focus together define the scope of a transformative policy mission, its basic purpose is to enact outcomes in terms of dynamic processes that transform real-world socio-technical systems. As shown by the literature on innovation systems (Bergek et al., 2008; Edquist, 1997; Elzinga et al., 2023), these outcomes can be approached from a structural or functional perspective. The former emphasizes how missions give rise to new constellations of actors, institutions, and technologies that engage with activities oriented towards the transformative focus. Indeed, developing and diffusing solutions that address today's grand challenges require cross-sectoral collaborations among both old and new actors, institutions that dislodge lock-ins to unsustainable pathways, technological capacities that allow experimentation and demonstration, and other structural changes.

The functional perspective emphasizes how missions influence transformation processes. It has traditionally been common to distinguish innovation processes, such as knowledge development, entrepreneurial experimentation, market formation, resource mobilization, and legitimation, that govern the emergence of new technologies and industries (Bergek et al., 2008; Hekkert et al., 2007). However, recent literature has shown that when applied to missions, the functional perspective can be extended to also highlight the emergence of directionality and coordination as well as the destabilization of existing systems (Elzinga et al., 2023). Drawing on these insights, we conceptualize the functional outcomes of missions as processes of directionality & coordination, innovation, and destabilization.

Importantly, structural and functional transformation processes initiated and shaped by missions continuously influence their context. Indeed, achieving change in socio-technical systems is why missions are imagined, promoted, and implemented in the first place. However, transformation processes may also shift the transformative focus of a mission and lead to new or re-designed policy interventions. There is accordingly important feedback between the scope and outcomes of transformative policy missions, which suggests that policy interventions should be multifaceted and adaptive to effectively address the resulting complexity and dynamism (Edmondson et al., 2018; Rogge & Reichardt, 2016).

## Analytical Framework

In Figure 13.1, we summarize the discussion with a framework that highlights the scope and outcomes of transformative policy missions. With regard to the scope, missions may have different sectoral, geographical, and temporal delimitation, address different problems and solutions, and involve different policy goals and instruments.[1] On the one hand, missions differ in their focus; some may address climate change while others engage with biodiversity, some may strive for a transformation in Sweden while others target Germany, and so forth. On the other hand, missions differ in their breadth; some initiatives focused on climate change may engage with any means to solving this problem while others only engage with a specific set of solutions, some may have an ambition to promote global change while others have a narrower interest in a specific country, and so forth. Our framework makes it possible to highlight these differences and monitor how the scope of a mission evolves in time. Similarly, as is common in the innovation system literature, analysing structural and functional outcomes reveals whether, and how, a mission promotes dynamic change processes in line with its transformative orientation and system focus. In the next section, we illustrate the merits of this framework as we explore the mission to decarbonize Swedish industry.

[1] As noted above, the policy goals and instruments associated with missions are parts of broader policy mixes, but it is beyond the scope of our framework to capture this complexity.

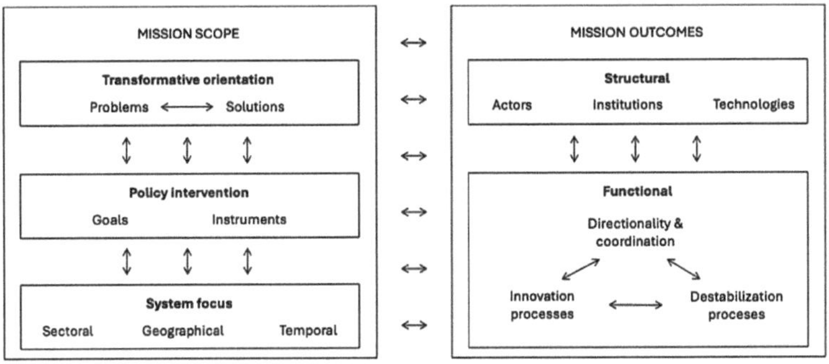


**Fig. 13.1** A conceptual framework that highlights the scope and outcomes of transformative policy missions

## Exploring the Mission to Decarbonize Swedish Industry

An interesting case to explore with the framework introduced in the previous section in the decarbonization of Swedish industry. Although this sector's climate footprint has decreased, slowly but steadily, since the 1990s, recent years have seen an increased focus on net-zero emissions. The development and diffusion of low-carbon industrial solutions have co-evolved with more focused and forceful policymaking, both in Sweden and at the EU level. In 2016, the Swedish climate policy framework introduced the goal to reach net-zero emissions in 2045 (Swedish Government, 2016). To reach this goal several policy instruments have been introduced. A particularly salient instrument at the national level is the Industry Leap (see below for more details), which outlines concrete decarbonization objectives and distributes public funding to research, development, demonstration, and implementation of solutions (Swedish Government, 2017).

Although policymakers in Sweden do not explicitly refer to the decarbonization of Swedish industry as a mission, we here take note of the fact that the government has introduced a bold and time-bound objective, together with forceful and specific policy instruments, to tackle a grand challenge. This is why we conceptualize the decarbonization of Swedish industry as a mission (in an analytical sense). We begin by characterizing the scope of this mission and then use data about projects (co-) funded through the Industry Leap to shine light on some of its structural and functional outcomes. Although this data does not provide a complete picture, it gives a strong indication of general developments oriented towards the mission. This is partly because the Industry Leap is by far the largest national funding scheme for industrial decarbonization, but also since the resources it provides enable Swedish actors to apply for EU-funding, private/public loans, and support through other policy instruments (Andersson & Hellsmark, 2024; Swedish Climate Policy Council, 2024).

## Mission Scope

### System Focus

The mission to decarbonize Swedish industry focuses on socio-technical systems that govern the production of goods in the economy (i.e., not including energy, transportation, agriculture, etc.). This sector can be decomposed into several more specific industries centred around products such as steel, cement, and chemicals. In addition, the Industry Leap and some other policy instruments oriented towards decarbonization associate the production of heat and power in plants fuelled by waste and biomass with the industry sector, since they share many potential solutions (Andersson & Hellsmark, 2024). Note also that reducing industrial emissions is contingent on enabling developments in adjacent sectors (e.g., increased supply of electricity to reduce industrial emissions), while industrial development is necessary to decarbonize yet other activities (e.g., increased supply of electric vehicles to reduce transport emissions).

In the geographical dimension, the mission primarily addresses industrial emissions caused by production activities in Sweden. This means that emissions embedded in imported inputs, as well as emissions arising when exported products are consumed, are excluded (even though they are included when firms disclose their Scope 3 emissions[2]). At the same time, the mission has strong links to developments in other regions. Reducing industrial emissions in Sweden depends on foreign firms, markets, and policies. Developments in Sweden may also play an important role in the decarbonization of economic activities in other regions.

In the temporal dimension, lastly, the mission starts at some point in recent years when the decarbonization of Swedish industry began to take shape as a transformative policy mission. Although efforts to reduce industrial emissions have been in place for decades, it was not until the Swedish climate policy framework was established in 2016 that legally binding goals were put in place. Around the same time, the Industry Leap was launched as a specific policy instrument oriented towards industrial decarbonization. We therefore see 2016 as a starting point for the mission, even though it is certainly contingent on prior developments as well. Furthermore, the temporal scope ends at some point when the mission can be considered complete, which given the current climate goals is around 2045.

## Transformative Orientation

The mission has a transformative orientation that revolves around climate change and the need to rapidly reduce anthropogenic emissions of greenhouse gases (IPCC, 2018). Although this is challenging for most parts of the global economy, it is considered especially difficult for the industry sector (Energy Transitions Commission, 2018; Löfgren & Rootzén, 2021). In particular, emissions from energy-intensive process industries, which produce products such as steel and cement, result not only from the need for heating, but also process-related emissions from the chemical conversion of raw materials such as iron ore and limestone (Energy Transitions Commission, 2018). This means that a shift to renewable energy is not sufficient and highlights the need for a more fundamental transformation. Moreover, process industries are based on

[2] This is similar to the consumption of international air travel and imported goods and services that give rise to emissions in other countries.

continuous and large-scale operations, face long investment cycles, and are exposed to global markets (Löfgren & Rootzén, 2021; Nilsson et al., 2021; Wesseling et al., 2017). As a result, there are high barriers to entry for entrepreneurs and innovative firms as well as a strong lock-in effect to existing technologies.

Not surprisingly, the decarbonization of Swedish industry is slow compared to other sectors. In 2022, industry (including heat & power) was responsible for 18 Mtons CO2eq, which corresponds to 38 per cent of the total emissions in Sweden (Swedish Environmental Protection Agency, 2023a). Almost all fossil emissions stem from energy-intensive process industries, and they are concentrated in a few firms and industrial facilities that produce steel, cement, chemicals, and refined fuels. This can be compared with the heat & power sector, where the emissions are distributed over many organizations and facilities (Fig. 13.2).

Furthermore, reaching the global climate goal of staying within 1.5 degrees will likely require the production of negative emissions through the removal of greenhouse gases from the atmosphere (IPCC, 2018). While this can be achieved in different ways, one of the main options is to capture and store emissions of biogenic carbon. This particularly relevant from a Swedish perspective since the production of pulp and paper, as well as the combustion of biomass and waste in heat and power plants, give rise to very large biogenic emissions (Swedish Environmental Protection Agency, 2023b). Although these emissions are commonly allocated to the land use, land-use change, and forestry (LULUCF) sector, they can accordingly be used by Swedish industrial firms in the production of negative emissions.

Similarly, the industry sector plays an important role in the decarbonization of other sectors. For example, reducing emissions in the transport sector is dependent on the creation of a new industry for the production of batteries in electric vehicles and/or the development of existing industries to produce renewable fuels (Hellsmark & Hansen, 2020). Here as well, Swedish firms may be able to draw on domestic resources, such as renewable electricity and biomass, to meet this demand.

Swedish industry thus plays a dual role in relation to the problem in focus of the mission. On the one hand, there is an urgent need to reduce

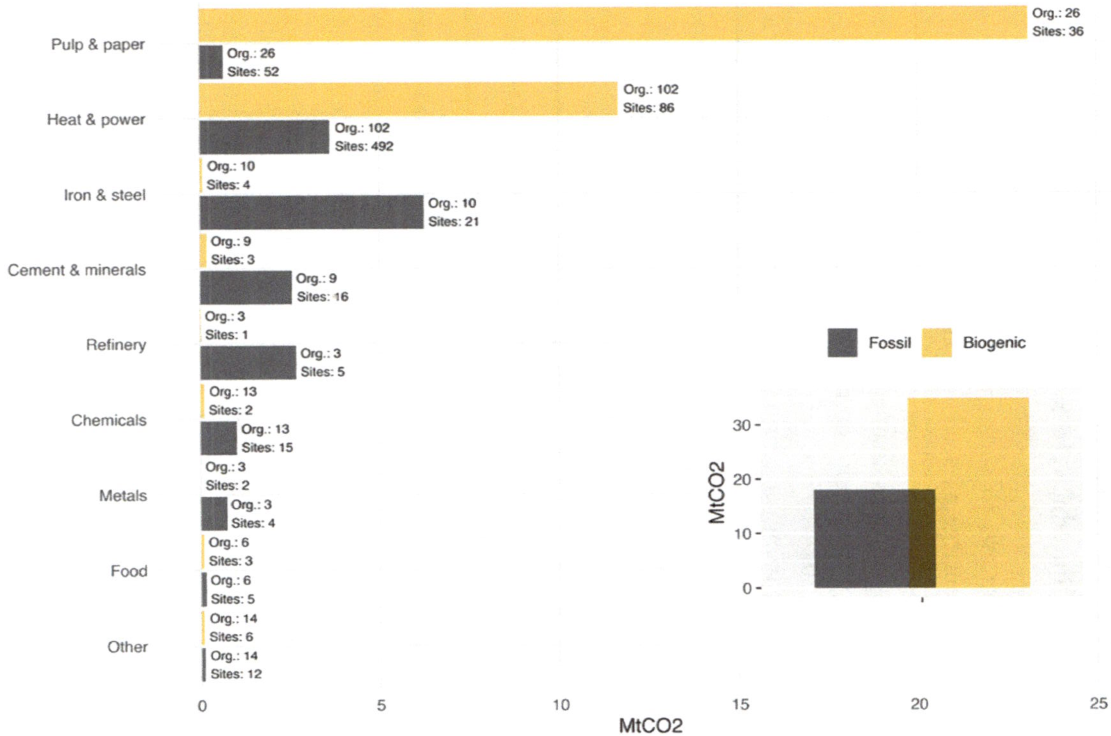


**Fig. 13.2** The distribution of biogenic and fossil emissions (MtCO2) between industry sectors, including heat & power, as well as the number of organizations and industry sites responsible for the emissions within each sector. *Source* Swedish Environmental Protection Agency (2023a)

emissions by transforming existing production activities. On the other hand, the demand for negative emissions, and products that are key for the decarbonization of other sectors, call for the development of new industrial value chains. As we will see in the next section, this is also reflected in the goals introduced by Swedish policymakers.

The mission is accordingly directed towards problems related to climate change. Solutions to these problems have not been specified a priori but rather emerge as actors engage with innovation activities oriented towards the mission. While addressing climate change calls for multiple interlinked solutions that have both technological and social characteristics, a few generic solution pathways can be identified (Andersson & Hellsmark, 2024). First, 'carbon capture' refers to capturing, transporting, and utilizing or storing emissions. As noted above, this pathway can lead to reduced emissions from existing production activities or to the production of negative emissions, depending on whether the technology is applied to emissions with fossil or biogenic origin. Second, 'electricity & hydrogen' involves replacing production processes based on fossil fuels and materials with processes that use renewable electricity and hydrogen as key inputs. Third, 'biofuels & biomass' rather involves replacing existing production processes with processes based on biomass, either as a fuel or as a raw material. Fourth, 'process efficiency' is about increasing the energy or materials efficiency of existing production processes, in order to reduce (but not eliminate) the need for fossil fuels and material. Fifth, 'recycling' involves replacing virgin raw materials with recycled alternatives or circular flows and thereby reducing and possibly eliminating the need for fossil fuels and materials. Lastly, there are numerous 'other solutions' that are not captured by the mentioned pathways, such as radically different production processes, alternative products, and downstream transformations that reduce or eliminate the demand for industrial products.

## Policy Interventions

The mission departs from the overarching Swedish climate policy framework, which stipulates that domestic net emissions, from all economic sectors, shall be zero by 2045 and that negative emissions shall be realized thereafter (Swedish Government, 2016).[3] The Industry Leap decomposes this goal into three interrelated objectives associated with the industry sector:

A. To **decarbonize existing industries** that give rise to fossil emissions.
B. To create new industries that **replace existing industries** with fossil emissions or that offer solutions for decarbonizing other sectors (e.g., transport, agriculture, and energy).
C. To create new industries that **produce negative emissions** (e.g., using biogenic emissions from existing industries).

These objectives reflect the transformative orientation discussed in the previous section and thereby foreground different industries and value chains. When it comes to the decarbonization of existing industrial facilities (A), the focus becomes industries with large fossil emissions (Fig. 13. 2). For the creation of new industries (B), a broad variety of low-carbon products and processes become relevant. Some of these constitute alternatives to existing value chains with large fossil emissions, while others are key resources for the decarbonization of other sectors. Lastly, the creation of new industries that produce negative emissions (C), is strongly associated with production processes that give rise to biogenic emissions.

As mentioned, the Industry Leap is the most salient policy instrument associated with the mission. Between 2018 and 2022 it distributed approximately 200MEUR (2 billion SEK) to over 100 different research, development, demonstration, and implementation projects. The Industry Leap has recently been complemented by the Impact Innovation programme, where two focus areas, developed by stakeholders in a bottom-up process, are highly relevant for industrial decarbonization. Besides these two major funding programmes, the government offers credit guarantees to large investments in low-carbon production plants

[3] The climate policy framework also establishes the Swedish Climate Policy Council as an advisory body, providing guidance on how climate goals can be met and ensuring that measures taken are consistent and effective in reducing emissions.

(Swedish Environmental Protection Agency, 2022) and promotes bio-CCS through a proposed funding mechanism based on reversed auctions and the facilitation of a value chain for carbon transport and storage (Swedish Energy Agency, 2021). There are also a number of national policy instruments that support decarbonization in general or in other economic sectors, such as a carbon tax (from which most industrial emissions are exempted) and various funding programmes. In addition, Swedish policymakers have established the platform Fossil Free Sweden (Fossil Free Sweden, 2021), which facilitates decarbonization dialogues among firms and other actors. This has resulted in several interrelated roadmaps towards industry decarbonization, often supported by broad networks of actors in different sectors.

Lastly, it should be noted that the mission is governed not only by national policy instruments, but also by EU-level climate policy. In particular, the long-standing Emissions Trading System (ETS) allows industrial firms to buy and sell emission allowances within a continuously shrinking emissions budget. This provides a financial incentive for decarbonization, while ensuring that emissions are reduced over time. More recently, the ETS has been complemented by policy instruments that offer additional support and incentives for decarbonization, including the Carbon Border Adjustment Mechanism (CBAM), which aims to prevent carbon leakage and promote a level playing field for European producers, and the Innovation Fund, which supports the development and deployment of low-carbon technologies.

## Mission Outcomes

The outcomes of the mission to decarbonize Swedish industry will be described in two parts focusing, respectively, on structural and functional change. First, we show how the Industry Leap has mobilized actors to engage in networks that cut across sectoral systems. Second, we uncover the formation of directionality by analysing how decarbonization projects (co-)funded by the programme relate to different policy goals, technological solutions, and sectoral systems. Structural and functional outcomes are thus highlighted in relation to some but not all dimensions of the mission scope (Fig. 13. 1). The analysis draws on several interrelated research projects that use emissions statistics, information about collaborative innovation projects, and stakeholder interviews (see e.g., Andersson & Hellsmark, 2024; Hellsmark et al., forthcoming).

## The Emergence of Actor Networks

Industrial decarbonization relies heavily on collaborative networks that involve multiple stakeholders to realize and integrate new solutions in established industries with significant emissions. The Industry Leap has contributed to mobilizing and connecting actors by (co-)funding projects that often involve broad consortia of public and private organizations. In the following section, we examine this structural outcome by identifying and analysing the network of actors that forms through project collaborations. Fig. 13.3 illustrates this network by linking actors that participate in the same project as well as to target industries in focus of decarbonization efforts.[4]

By the end of 2022, 136 actors had received funding from the Industry Leap, including private and public firms, research institutes, universities, and governmental bodies, and 62 of these had received more than 0.1 MEUR. The total budget of all 103 (co-)funded projects amounted to approximately 700 MEUR, out of which 200 MEUR is public funding from the Swedish Energy Agency (Andersson & Hellsmark, 2024). Together, actors involved in the Industry Leap thus make significant investments in development, demonstration, and implementation of low-carbon solutions. Furthermore, actors from all parts of the industry sector are actively engaged, except for the food industry which only accounts for a smaller share (1.5%) of the total industrial emissions (Swedish Environmental Protection Agency, 2023). In terms of reach, the participating firms are responsible for 86% of fossil emissions and

[4] We only consider actors connected to the project leader, not assuming all partners collaborate directly. This approach is essential for large projects, as it highlights central hubs and potential fragmentation, offering a clearer view of collaboration dynamics and actor roles. Note also that for clarity Fig. 13.3 only includes actors that receive more than 0.1 MEUR.

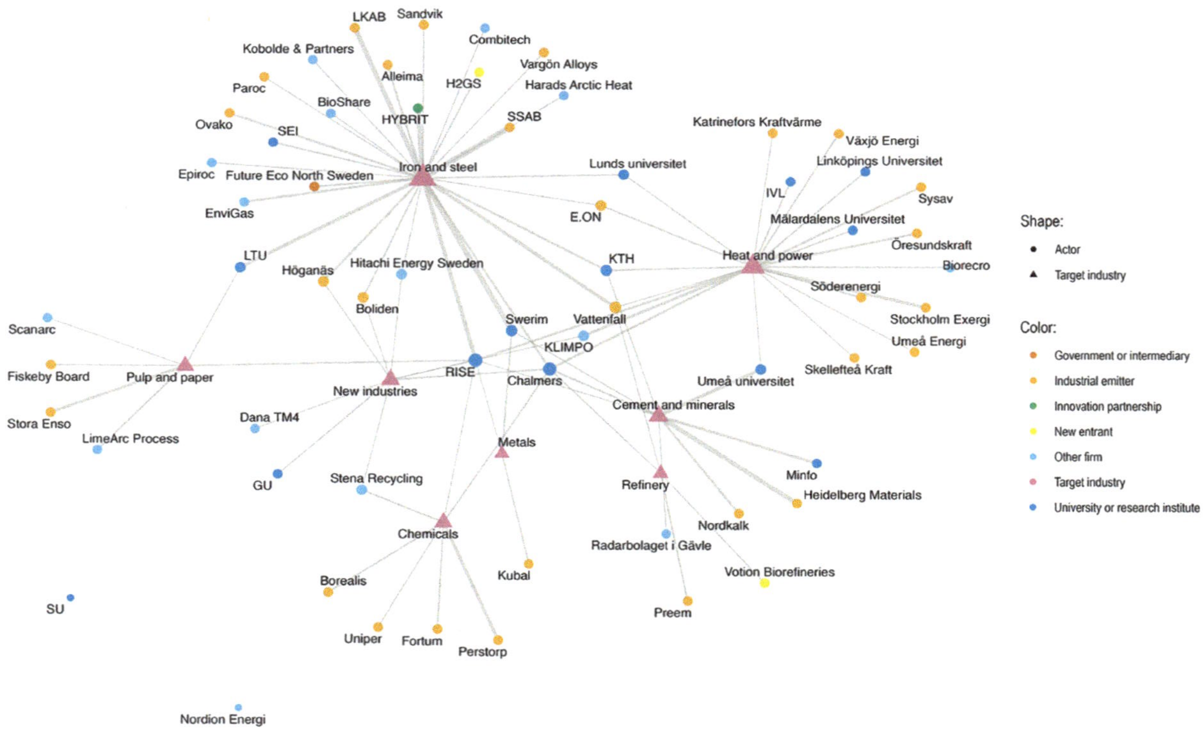


**Fig. 13.3** A network of 62 actors (circle nodes) that receive more than 0.1 MEUR in funding from the Industry Leap to engage in the decarbonization of 9 target industries (triangle nodes). Actors are connected to each other if they participate in the same project and to target industries if they participate in projects with that sectoral system focus. *Source* Hellsmark et al. (forthcoming)

49% of biogenic emissions from industry. Indeed, 19 of the 20 largest emitters are active in the programme. An important mission outcome is thus a high degree of involvement from the incumbent firms.

When examining more closely which actors are engaged in the Industry Leap network, we only find two new entrants. The first one, H2 Green Steel (H2GS), aims to produce steel with reduced carbon emissions using green hydrogen, and the second, Votion Biorefineries, attempts to convert agricultural and forestry residues into biochemicals and biofuels. Both firms are supported by large private investors, including potential customers. Interestingly, the new entrants do not collaborate with other actors in projects (co-)funded by the Industry Leap. This does certainly not mean that they lack such collaborations, but further underlines how the main policy instrument associated with the mission has given rise to a network with a strong focus on incumbents.

The network also includes several strategic collaborations between energy utilities and industrial firms, focusing on reducing emissions in the Iron & steel, Cement, Chemical, and Refinery industry. Utilities, such as Vattenfall, Fortum, Uniper, play an especially important role as "system entanglers" (Löhr & Chlebna, 2023), connecting industrial decarbonization with the energy sector. This is a role that includes facilitating learning and building crucial infrastructure for electricity and hydrogen production (Hedeler and Hellsmark, forthcoming).

In addition, we find that many actors bridge decarbonization efforts across sectoral systems by participating in projects that target different industries. Research institutes and universities often take on this role, with Chalmers University of Technology having the most central position in the network. However, interconnections between the decarbonization of different industries are also created by industrial emitters and other firms, including Borealis, Boliden, Stora Enso, Höganäs, KLIMPO, and Stena Recycling.

What sets the Swedish industry sector apart compared to equivalent actors in other European countries is that almost all incumbent firms have adopted net-zero-emission targets that are more ambitious than, or at least in line with, Swedish and European climate goals (Hansen et al.,

2024). Only one of the 20 largest emitters has not declared an ambition to become fossil-free by mid-century or earlier (Hellsmark et al., forthcoming). However, even though the main emitters have high ambitions and almost all industries are active in the network, we do find significant differences in the level of engagement across individual actors and industries. This becomes particularly evident when focusing on large emitters and relating their fossil emissions to both their co-funding and the number of collaborations with other actors in the network. There are a few main emitters with very focused and strategic collaborations with a handful of actors to realize their technology strategies. In particular, the collaboration between SSAB, LKAB, and Vattenfall (i.e., the HYBRIT innovation partnership), which aims to produce fossil-free steel using hydrogen, stands out in terms of their co-funding in relation to governmental funding. This does not come as a surprise since the technology is being scaled up and implemented at a commercial scale. Other salient large emitters are the oil refinery Preem and the cement producer Heidelberg Materials. Preem has the most collaborations of all the main emitters, searching and exploring different solutions that could be scaled up in the short as well as medium term. Heidelberg Materials is primarily focused on collaborating with actors that could enable CCS on the medium to long term. On the other side of the spectrum (i.e., large emitters investing very little in relation to their ambitions and emissions), we find another oil refinery, St1, as well as municipal firms such as Renova and Tekniska Verken i Linköping with large emissions that originate from waste incineration. Quite surprisingly, these actors have no significant participation in decarbonization projects (co-)funded by the Industry Leap, even though their ambitions in terms of reaching net-zero emissions are high.

## The Formation of Directionality

We now shift focus to the functional perspective on mission outcomes by exploring the formation of directionality as indicated by the orientation of decarbonization projects (co-)funded by the Industry Leap. The analysis decomposes the total budget (public and private) in relation to three aspects of the mission scope: sectoral system focus, policy goals, and solutions (Fig. 13.4).

Our results show that the iron & steel industry is targeted by 48% of the total project funding, followed by the chemicals industry with 35%. There is, accordingly, a significant concentration of resources to projects that target these two industries. At the same time, projects that target pulp & paper, heat & power, cement & minerals, metals, and other [5]industries receive a significant but much smaller share of the total funding, while the refinery and food industries gather minimal attention. Also, projects that target multiple industries at the same time are

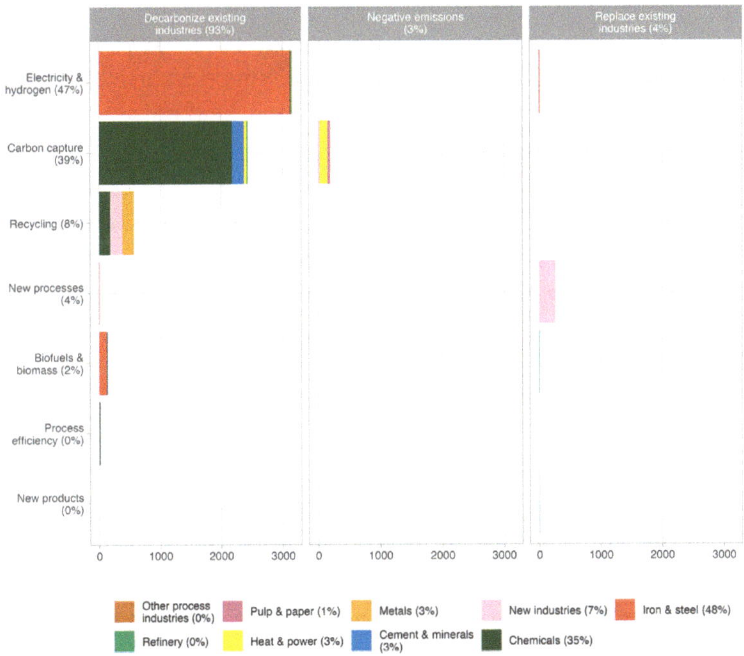


**Fig. 13.4** A decomposition of the total budget of projects (co-)funded by the Industry Leap (MSEK and per cent) in relation to the policy goal, industry, and solution in focus

[5] 1 EUR ≈ 10 SEK.

relatively rare and mainly focus on early-stage knowledge development and policy studies, rather than concrete technological innovations.

When it comes to policy goals, the majority of decarbonization projects focus on reducing emissions from existing industries, which accounts for 93% of the total Industry Leap budget. Less emphasis is placed on efforts to achieve negative emissions through capturing and storing biogenic emissions (bio-CCS), representing only 3% of the total budget. The remaining four per cent is allocated to developing industries and technologies that have the potential to replace existing high-emission industries with new ones that have significantly lower emissions. This clearly aligns with the essence of the Swedish climate goals, which is to rapidly reduce fossil emissions. However, it stands in contrast to the likely need for negative emissions to compensate for residual emissions that are hard to abate (e.g., from the agricultural sector). The relatively low investment in new industries may indicate high entry barriers and a focus on established industries.

In terms of solutions, there is a strong emphasis on electricity & hydrogen and carbon capture, which together make up 86% of the total Industry Leap budget. Hydrogen-based solutions dominate the broad range of projects targeting the iron & steel industry, which reflects significant and focused investment in transitioning to fossil-free steel production. Carbon capture projects are numerous but generally smaller in scale, targeting both the reduction of fossil emissions and the production of negative emissions across various industries. Besides these dominating technological pathways, recycling, biofuels & biomass, and process efficiency receive markedly less attention. Recycling projects mainly target the chemicals industry, aiming to reduce dependence on fossil feedstock through thermochemical plastics recycling. Biofuels & biomass projects are mostly associated with the iron and steel industry, where they explore the use of biomass as a process input and biogenic coal production. Process efficiency projects are scarce but target energy efficiency improvements in the chemicals industry and material efficiency in the cement industry. In addition, there are a few projects oriented towards other solutions that are not easily categorized into generic technological pathways.

Taken together, our results indicate that the formation of directionality is somewhat imbalanced in relation to the mission scope. Decarbonization projects have a strong focus on the iron & steel and chemicals industries, mainly explore technological pathways based on electricity & hydrogen and carbon capture, and pay relatively little attention to the production of negative emissions. However, a broad range of solutions must be implemented across all industrial sectors, albeit in different combinations, to reach net-zero emissions and beyond. This further highlights the salient dominance of incumbent industry in the actor network mobilized by the Industry Leap and suggests a need to stimulate broader experimentation that involves and initiates links between stakeholders from all industry sectors.

## Discussion and Concluding Remarks

In this chapter, we have introduced and illustrated a basic framework for analysing the scope and outcomes of transformative policy missions. The framework decomposes the mission scope into a transformative orientation towards certain problems and solutions, policy interventions including both goals and instruments, as well as a system focus in sectoral, geographical, and temporal dimensions. It also distinguishes structural and functional outcomes, specifically highlighting that the latter involve the provision of directionality and coordination, innovation processes that promote the emergence of new solutions, as well as destabilization processes that govern the dismantling of established systems.

Together with our illustrative case study of industrial decarbonization in Sweden, the framework brings several contributions to the literature. To begin with, it adds depth to the literature on multi-system dynamics by highlighting how missions give rise to interconnectedness and interdependencies across various dimensions. As noted by scholars such as Kanger et al. (2021) and Andersen and Geels (2023), understanding the speed and pathways of net-zero transitions requires a nuanced analysis of how technologies, actors, and institutions interact across multiple systems. Our approach captures these interactions, highlighting the role of cross-sectoral collaborations and system entanglements in driving innovation. This is crucial for understanding how policy interventions in one sector or region can influence and be influenced by developments in other sectors or regions, as seen in the Swedish case where industrial decarbonization is closely linked to the energy and transportation sectors.

Second, the framework advances the literature on transformative innovation policy by supporting the design, implementation, and evaluation of missions. This complements the work of Rogge and Reichardt (2016) and Edmondson et al. (2018), who highlight the need for adaptive and multifaceted policy mixes to address the complexity of socio-technical systems. By departing from a clear idea of the key dimensions of transformative policy missions, policymakers may develop a better understanding of the directionality of mission-driven innovation processes, ensure that interventions are targeted and coherent, and anticipate and manage the complexities and uncertainties associated with transformations that span multiple systems. For instance, in the decarbonization of Swedish industry, understanding the geographical links and dependencies helps in designing policies that leverage international collaborations and market dynamics. Similarly, recognizing the temporal aspects of innovation processes enables the setting of realistic and ambitious timelines for achieving net-zero emissions, ensuring that short-term actions align with long-term goals. This holistic perspective is vital for creating policy mixes that are not only adaptive but also synergistic, addressing multiple barriers and drivers of change simultaneously.

Third, the framework echoes the findings of Andersen and Markard (2020) in emphasizing the importance of context-specific solutions and the need for tailored policy interventions. In the Swedish context, the focus on decarbonizing existing industries, developing new low-carbon industries, and producing negative emissions demonstrates how different goals require different technological and social innovations. Our approach may here help identify the most effective policy instruments for each specific goal and context, enhancing the strategic coherence and effectiveness of transformative innovation policies.

In addition, the framework facilitates a more inclusive and participatory approach to policy design and implementation. By explicitly considering the diverse perspectives and needs of various stakeholders, as highlighted by Heiberg et al. (2022) and Ohlendorf et al. (2023), policymakers can foster broader acceptance and support for transformative initiatives. This is particularly important in addressing the contestation that often arises in multi-system transitions, ensuring that the benefits of innovation are widely shared and that the burdens are equitably distributed.

At the same time, the chapter has limitations that suggest promising avenues for future research. The framework remains formulated at a quite general level and would clearly benefit from more detailed conceptualizations of its constituent parts. In particular, continued conceptual development should engage with the parallel development of theoretical ideas related to mission-oriented innovation systems (Elzinga et al., 2023), which is likely to bring additional insights in the years to come. Future research is also needed to test the framework in broader empirical studies that investigate different types of transformative policy missions. This is the only way to validate that the concepts used to describe and decompose the scope and outcomes of missions are useful for comparative analyses and cumulative knowledge development. In addition, the illustrative case study presented in this chapter should be complemented with investigations that offer a comprehensive account of structural and functional outcomes. This involves moving beyond the limited perspective highlighted here and analysing how missions (i) result in organizational, institutional, and technological change at the structural level, (ii) propel processes of coordination, innovation, and destabilization at the functional level, and (iii) involve feedback between these outcomes and the underlying mission scope (i.e., system focus, transformative orientation, and policy interventions). As with other types of research on sustainability transitions and related policymaking, identifying data sources that can reveal the characteristics of these processes is challenging. Although resource demanding, comprehensive analyses are likely to require a combination of stakeholder interviews and information from a wide range of secondary sources. In this chapter, we relied largely on project data from the Industry Leap, which can be considered a good indicator of broader mission outcomes. However, further explorations in the Swedish context should also assess the Impact Innovation programme, which can be expected to play an important role for future missions related to most sectors and challenges.

In the end, we hope that the analytical framework and illustrative case study presented in this chapter will help bridge the gap between theoretical concepts and practical policy applications. After all, clearly delineating the space in which transformative policy missions unfold is essential for designing effective and coherent policy interventions that drive sustainable and equitable transitions towards net-zero emissions and other societal objectives.

## References

Andersen, A. D., & Geels, F. W. (2023). Multi-system dynamics and the speed of net-zero transitions: Identifying causal processes related to technologies, actors, and institutions. *Energy Research & Social Science, 102*, 103178. https://doi.org/10.1016/j.erss.2023.103178

Andersen, A. D., Geels, F. W., Coenen, L., Hanson, J., Korsnes, M., Linnerud, K., Makitie, T., Nordholm, A., Ryghaug, M., Skjolsvold, T., Steen, M., & Wiebe, K. (2024). Faster, broader, and deeper! Suggested directions for research on net-zero transitions. *Oxford open* Energy, 2, oiad007. doi:https://doi.org/10.1093/ooenergy/oiad007.

Andersen, A. D., & Markard, J. (2020). Multi-technology interaction in socio-technical transitions: How recent dynamics in HVDC technology can inform transition theories. *Technological Forecasting and Social Change, 151*, 119802. https://doi.org/10.1016/j.techfore.2019.119802

Andersson, J., & Hellsmark, H. (2024). Directionality in transformative policy missions: The case of reaching net zero emissions in the Swedish process industry. *Journal of Cleaner Production, 437*, 140664. https://doi.org/10.1016/j.jclepro.2024.140664

Andersson, J., Hellsmark, H., & Sandén, B. (2021). The outcomes of directionality: Towards a morphology of sociotechnical systems. *Environmental Innovation and Societal Transitions, 40*, 108–131. https://doi.org/10.1016/j.eist.2021.06.008

Azar, C., & Sandén, B. A. (2011). The elusive quest for technology-neutral policies. *Environmental Innovation and Societal Transitions, 1*, 135–139.

Bergek, A., Jacobsson, S., Carlsson, B., Lindmark, S., & Rickne, A. (2008). Analyzing the functional dynamics of technological innovation systems: A scheme of analysis. *Research Policy, 37*, 407–429. https://doi.org/10.1016/j.respol.2007.12.003

Coenen, L., Benneworth, P., & Truffer, B. (2012). Toward a spatial perspective on sustainability transitions. *Research Policy, 41*(6), 968–979. https://doi.org/10.1016/j.respol.2012.02.014

Edmondson, D. L., Kern, F., & Rogge, K. S. (2018). The co-evolution of policy mixes and socio-technical systems: Towards a conceptual framework of policy mix feedback in sustainability transitions. *Research Policy, 48*(10), 103555. https://doi.org/10.1016/J.RESPOL.2018.03.010

Edquist, C. (1997). *Systems of Innovation: Technologies, institutions and organizations.* Pinter Publishers.

Elzinga, R., Janssen, M. J., Wesseling, J., Negro, S. O., & Hekkert, M. P. (2023). Assessing mission-specific innovation systems: Towards an analytical framework. *Environmental Innovation and Societal Transitions, 48*, 100745. https://doi.org/10.1016/j.eist.2023.100745

Energy Transitions Commission (2018). Mission Possible: Reaching net-zero carbon emissions from garder-to-abate sectors by mid-century.

Geels, F. W. (2002). Technological transitions as evolutionary reconfiguration processes: A multi-level perspective and a case-study. *Research Policy, 31*, 1257–1274. https://doi.org/10.1016/S0048-7333(02)00062-8

Haddad, C. R., Nakić, V., Bergek, A., & Hellsmark, H. (2022). Transformative innovation policy: A systematic review. *Environmental Innovation and Societal Transitions, 43*, 14–40. https://doi.org/10.1016/j.eist.2022.03.002

Hansen, T., Andersson, J., Finstad, J., Hanson, J., Hellsmark, H., Mäkitie, T., Nordholm, A., & Steen, M. (2024). How aligned are industry strategy and government policy for the decarbonization of energy-intensive process industries? *Climate Policy, 24*(9), 1149–1162. https://doi.org/10.1080/14693062.2024.2363490

Hedeler, B., & Hellsmark, H. Forthcoming. *The Role of System Entanglers in Mission-Oriented Innovation: Facilitating Cross-Sectoral Industrial Decarbonization.*

Heiberg, J., Truffer, B., & Binz, C. (2022). Assessing transitions through socio-technical configuration analysis—A methodological framework and a case study in the water sector. *Research Policy, 51*(1), 104363. https://doi.org/10.1016/j.respol.2021.104363

Hekkert, M. P., Suurs, R. A. A., Negro, S. O., Kuhlmann, S., & Smits, R. E. H. M. (2007). Functions of innovation systems: A new approach for analysing technological change. *Technological Forecasting and Social Change, 74*, 413–432. https://doi.org/10.1016/j.techfore.2006.03.002

Hellsmark, H., Andersson, J., & Hedeler, B. (forthcoming). Leaders and laggards: The role of incumbents in transformative policy missions. *Under Review.*

Hellsmark, H., & Hansen, T. (2020). A new dawn for (oil) incumbents within the bioeconomy? Trade-offs and lessons for policy. *Energy Policy, 145*, 111763. https://doi.org/10.1016/j.enpol.2020.111763

IPCC (2018). Global warming of 1.5°C. An IPCC special report on the impacts of global warming of 1.5°C above pre-industrial levels and related global greenhouse gas emission pathways, in the context of strengthening the global response to the threat of climate change.

Janssen, M. J., Torrens, J., Wesseling, J. H., & Wanzenböck, I. (2021). The promises and premises of mission-oriented innovation policy - a reflection and ways forward. *Science and Public Policy, 48*, 438–444. https://doi.org/10.1093/scipol/scaa072

Kanger, L., Schot, J., Sovacool, B. K., van der Vleuten, E., Ghosh, B., Keller, M., Kivimaa, P., Pahker, A.-K., & Steinmueller, W. E. (2021). Research frontiers for multi-system dynamics and deep transitions. *Environmental Innovation and Societal Transitions, 41*, 52–56. https://doi.org/10.1016/j.eist.2021.10.025

Levin, K., Cashore, B., Bernstein, S., & Auld, G. (2012). Overcoming the tragedy of super wicked problems: Constraining our future selves to ameliorate global climate change. *Policy Sciences, 45*, 123–152. https://doi.org/10.1007/s11077-012-9151-0

Löfgren, Å., & Rootzén, J. (2021). Brick by brick: Governing industry decarbonization in the face of uncertainty and risk. *Environmental Innovation and Societal Transitions, 40*, 189–202. https://doi.org/10.1016/j.eist.2021.07.002

Löhr, M., & Chlebna, C. (2023). Multi-system interactions in hydrogen-based sector coupling projects: System entanglers as key actors. *Energy Research & Social Science, 105*, 103282. https://doi.org/10.1016/j.erss.2023.103282

Mazzucato, M. (2018). Mission-oriented innovation policies: Challenges and opportunities. *Industrial and Corporate Change, 27*, 803–815. https://doi.org/10.1093/icc/dty034

Nilsson, L. J., Bauer, F., Åhman, M., Andersson, F. N. G., Bataille, C., De, S., Ericsson, K., Hansen, T., Johansson, B., Lechtenböhmer, S., Sluisveld, M. V., & Vogl, V. (2021). An industrial policy framework for transforming energy and emissions intensive industries towards zero emissions.

*Climate Policy, 21*(8), 1053–1065. https://doi.org/10.1080/14693062.2021.1957665

Ohlendorf, N., Löhr, M., & Markard, J. (2023). Actors in multi-sector transitions - discourse analysis on hydrogen in Germany. *Environmental Innovation and Societal Transitions, 47*, 100692. https://doi.org/10.1016/j.eist.2023.100692

Rip, A., & Kemp, R. (1998). Technological change. In S. Rayner & E. L. Malone (Eds.), *Human choice and climate change* (pp. 327–399). Battelle Press. https://doi.org/10.1007/BF02887432

Robinson, D. K. R. R., & Mazzucato, M. (2019). The evolution of mission-oriented policies: Exploring changing market creating policies in the US and European space sector. *Research Policy, 48*, 936–948. https://doi.org/10.1016/j.respol.2018.10.005

Rogge, K. S., & Reichardt, K. (2016). Policy mixes for sustainability transitions: An extended concept and framework for analysis. *Research Policy, 45*(8), 1620–1635. https://doi.org/10.1016/j.respol.2016.04.004

Rosenbloom, D. (2020). Engaging with multi-system interactions in sustainability transitions: A comment on the transitions research agenda. *Environmental Innovation and Societal Transitions, 34*, 336–340. https://doi.org/10.1016/j.eist.2019.10.003

Schot, J., & Steinmueller, W. E. (2018). New directions for innovation studies: Missions and transformations. *Research Policy, 47*(9), 1583–1584. https://doi.org/10.1016/J.RESPOL.2018.08.014

Scoones, I. (2016). The politics of sustainability and development. *Annual Review of Environment and Resources, 41*, 293–319. https://doi.org/10.1146/annurev-environ-110615-090039

Sovacool, B. K. (2011). An international comparison of four polycentric approaches to climate and energy governance. *Energy Policy, 39*, 3832–3844. https://doi.org/10.1016/j.enpol.2011.04.014

Stirling, A. (2011). Pluralising progress: From integrative transitions to transformative diversity. *Environmental Innovation and Societal Transitions, 1*, 82–88. https://doi.org/10.1016/j.eist.2011.03.005

Swedish Climate Policy Council (2024). Klimatpolitiska rådets rapport 2024.

Swedish Energy Agency (2021). Första, andra, tredje ... Förslag på utformning av ett stödsystem för bio-CCS ER 2021: 31.

Swedish Environmental Protection Agency. (2022). *Industrins klimatomställning Underlagsrapport till regeringsuppdraget om Näringslivets klimatomställning*. Swedish Environmental Protection Agency.

Swedish Environmental Protection Agency (2023a). Industri, utsläpp av växthusgaser [WWW Document].Retrieved December 20, 2023, from URL https://www.naturvardsverket.se/data-och-statistik/klimat/vaxthusgaser-utslapp-fran-industrin/

Swedish Environmental Protection Agency (2023b). Biogena koldioxidutsläpp och klimatpåverkan [WWW Document].Retrieved January 31, 2023, from URL https://www.naturvardsverket.se/amnesomraden/klimatomstallningen/omraden/klimatet-och-skogen/biogena-koldioxidutslapp-och-klimatpaverkan/

Swedish Government (2016). Regeringens proposition 2016/17:146. Ett klimatpolitisk ramverk för Sverige.

Swedish Government (2017). Förordning (2017:1319) om statligt stöd till åtgärder som bidrar till industrins klimatomställning.

Voß, J.-P., Smith, A., & Grin, J. (2009). Designing long-term policy: Rethinking transition management. *Policy Sciences, 42*, 275–302. https://doi.org/10.1007/s11077-009-9103-5

Wanzenböck, I., Wesseling, J. H., Frenken, K., Hekkert, M. P., & Weber, K. M. (2020). A framework for mission-oriented innovation policy: Alternative pathways through the problem—Solution space. *Science and Public Policy, 47*(4), 474–489. https://doi.org/10.1093/scipol/scaa027

Weber, K. M., & Rohracher, H. (2012). Legitimizing research, technology and innovation policies for transformative change: Combining insights from innovation systems and multi-level perspective in a comprehensive "failures" framework. *Research Policy, 41*(6), 1037–1047. https://doi.org/10.1016/j.respol.2011.10.015

Wesseling, J. H., Lechtenböhmer, S., Åhman, M., Nilsson, L. J., Worrell, E., & Coenen, L. (2017). The transition of energy intensive processing industries towards deep decarbonization: Characteristics and implications for future research. *Renewable and Sustainable Energy Reviews, 79*, 1303–1313. https://doi.org/10.1016/J.RSER.2017.05.156

Wittmann, F., Hufnagl, M., Lindner, R., Roth, F., & Edler, J. (2021). Governing varieties of mission-oriented innovation policies: A new typology. *Science and Public Policy, 48*(5), 727–738. https://doi.org/10.1093/scipol/scab044

# 14

# The More the Merrier? A Study on Multiple Mission-Specific Innovation Systems in the Dutch Energy Sector

Remi Elzinga, Matthijs J. Janssen, and Simona O. Negro

## Introduction

There is increasing global emphasis on using innovation policy to tackle complex societal challenges such as climate change and resource scarcity (Schot & Steinmueller, 2018). A key approach in this context is mission-oriented innovation policy (MIP), which aims to provide directionality through ambitious, clear, and time-bound goals (Mazzucato, 2018). These goals help guide search and negotiation processes that align innovation efforts (Janssen et al., 2021). Directionality arises from and shapes

R. Elzinga · M. J. Janssen · S. O. Negro
Copernicus Institute of Sustainable Development, Utrecht University, Utrecht, The Netherlands

R. Elzinga (✉)
PBL Netherlands Environmental Assessment Agency NL, Hague, The Netherlands
e-mail: r.elzinga@uu.nl

M. J. Janssen
Dialogic Innovation & Interaction, Utrecht, The Netherlands



M. Steen et al. (eds.), *Sustainability Transitions and Industrial Transformation*,
https://doi.org/10.1007/978-981-95-4893-4_14

societal actors' perspectives on which problems to address and which solutions to pursue (Janssen et al., 2023; Klerkx et al., 2025). MIP thus aims to aligning these perspectives, enabling targeted R&D and diffusion policies to overcome the inertia underlying persistent societal issues, to transform socio-technical systems and facilitate broader sustainability transitions (Hekkert et al., 2020; Wittmann et al., 2020).

Although scholars have spent much attention to understanding a mission's potential for facilitating transformation, the origins and nature of mission directionality remain a topic of fierce debate (Parks, 2022; Andersson et al., 2026; De Graaff et al., 2025). Especially in the case of wicked societal challenges, the clarity and justification of a mission's scope can be ambiguous and contested (Wanzenböck et al., 2020). Complicating matters further, the rising popularity of mission-oriented thinking has led policymakers to introduce a multitude of parallel missions. In the OECD's analysis of national MIPs, Larrue (2021) uses the notion of 'umbrella frameworks' to classify policies comprising multiple missions. A striking example is the Dutch take on MIP, which consists of 25 missions addressing a diverse array of complex societal challenges, such as public health, clean water, sustainable energy, and greenhouse gas emissions (Janssen, 2020).

An important but largely overlooked question concerns the co-existence of these parallel missions. Each individual mission is supposed to provide guidance, but when existing next to each other this guidance might become weaker rather than stronger. This especially holds when multiple missions consider the same problems and solutions, with associated actors and resources. It remains unclear how actors engaging with or affected by multiple missions should navigate potential conflicts among competing mission goals or solution directions (Andersen & Geels, 2023; Andersen et al., 2023; Kanger et al., 2021; Löhr & Chlebna, 2023). From research on the development and interaction of technology pathways we know that prioritized directions can be reinforcing but also weakening (Sandén & Hilman, 2011). How this works out for co-existing missions, each with their own scope and influence on innovation system dynamics, has hitherto not been studied in depth.

This study investigates the effects of pursuing multiple missions, the nature of their interactions, their influence on shaping directionality, and their impact on desired transitions. To explore mission interaction, we analyse the mission-specific innovation systems (MIS) around two co-existing missions (Elzinga et al., 2023; Hekkert et al., 2020; Wesseling & Meijerhof, 2023). A MIS can be understood as the collection of actors, networks, and institutions involved in pursuing and realizing a mission goal—be it through creating new structures, processes, and innovations or through the transformation or dismantling of existing (and hindering) structures and processes. We aim to uncover potential alignments or tensions among the respective innovation systems supporting each mission, identifying factors that could either facilitate or hinder mission success (Rosenbloom, 2020). Recognizing opportunities for inter-mission alignment may reveal intervention points that could be leveraged to accelerate change.

The remainder of this chapter is structured as follows. The next section introduces the MIS framework and outlines the analytical steps for studying mission-specific innovation systems in a multi-system context. After this, the methodology is detailed and the results are presented, examining problem and solution framings, key actors, and drivers and barriers influencing the development of technologies within two Dutch missions linked to renewable energy and circularity. The final section concludes the book chapter with a discussion on directionality in interacting missions, its influence on solution directions, and offers insights for accelerating transitions shaped by multiple missions, aiming to inform both research and policymaking.

## Interactions of Mission-Specific Innovation Systems

This section explores how multiple missions shape directionality and influence innovation systems. Directionality guides priorities in research, investment, and collaboration, impacting all major system components. The MIS framework puts the concept of missions central to assess how innovation systems are characterized and driven by high and continuous directionality (Fig. 14.1) (Elzinga et al., 2023). Since missions aimed at societal challenges require restructuring of actors, networks, and institutions (Wittmann et al., 2020), evaluating mission progress benefits from an innovation systems perspective. We discuss how the often-contested directionality of different missions influences the development and diffusion of solutions, and how innovation systems can be assessed accordingly.

## MIS in a Multi-System Context

The MIS framework provides a basis for analysing how structures, actors, institutions, and processes support or hinder the development and diffusion of solutions to societal challenges (Elzinga et al., 2023; Hekkert et al., 2020). This aligns with recent calls to broaden innovation system analysis in sustainability transitions beyond single, technology-focused systems (Andersen & Geels, 2023; Berkhout et al., 2004; McMeekin et al., 2019; Papachristos et al., 2013; Raven, 2007). Studies of interacting systems often take a socio-technical perspective, focusing on systems that fulfil societal functions like transport or food (Geels, 2018; Rosenbloom, 2020). Rosenbloom (2020) argues that understanding sustainability transitions requires examining overlapping structures and system functions—such as the flow of resources and information—across multiple innovation systems. This approach is essential for understanding system interactions and their implications for sustainability transitions.

Bergek et al. (2015) examine how multiple technological innovation systems interact by analysing their structures and key processes. They show systems may connect through sectoral or geographic proximity, or via political and institutional structures. They distinguish between vertically overlapping systems—linked through value chains—and horizontally overlapping systems that share inputs, assets, or outputs. Innovations in such systems can compete or complement one another, influencing the dynamics within each system. The current study begins with two horizontally overlapping mission-specific innovation systems (MISs), each tied to a distinct mission. Their partial overlap includes actors, networks, institutions, and shared solution directions. When these solutions are technological, they could be analysed as separate

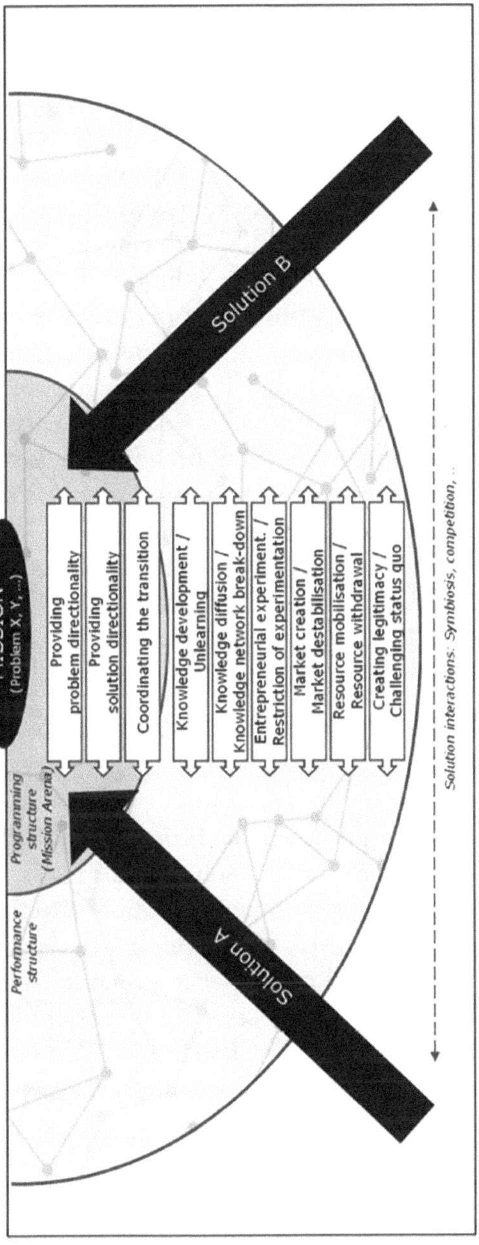


**Fig. 14.1** The MIS framework as proposed by Elzinga et al. (2023)

but interacting technological innovation systems. Thus, a single MIS involving multiple solutions inherently spans several interconnected innovation systems. This implies that multiple missions and their corresponding MISs constitute a complex, multi-system context.

Figure 14.2 illustrates the relationship between innovation system components across two missions, each targeting partially overlapping societal problems. These problems are addressed through solution directions (SDn)—strategic approaches like reducing vs. recycling waste—which guide the development of specific solutions (Sn), such as chemical or mechanical recycling. Competition may arise when different solutions require similar resources. Alternatively, solution directions can also reinforce each other, especially when they tackle the same problems and rely on shared (technological or institutional) developments. For example, both reducing and recycling waste could benefit from waste incineration fees, which create financial incentives to minimize waste and promote recycling by making incineration more expensive.

In the subsequent explanation of the different analytical steps of the MIS framework, we discuss what considerations apply when facing a multi-mission (and thus multi-system) context.

## Assessing Multiple Mission-Specific Innovation Systems

The MIS framework, as proposed by Elzinga et al. (2023), consists of three analytical steps: a problem-solution diagnosis (for analysing a mission's problem framings, associated solution directions, and their interactions), a structural analysis (for examining the actors and institutions in the programming and performance structure), and a functional analysis (for assessing how key processes influence the acceleration of solutions and the destabilization of regime barriers). This section explores how a multi-system context affects these analytical steps.

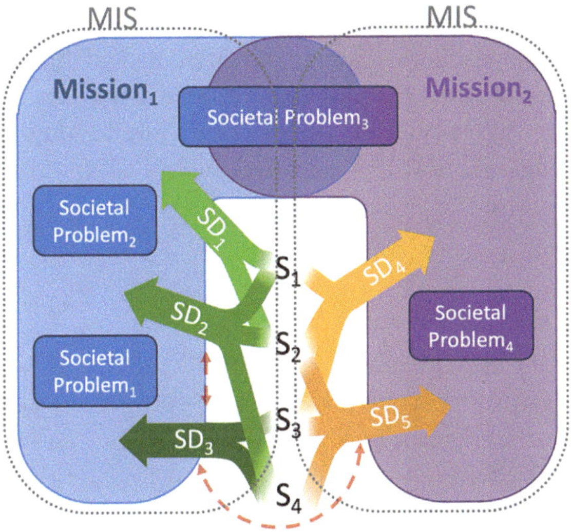


**Fig. 14.2** Visualization of interacting mission-specific innovation systems. The mission-specific innovation systems can include different or overlapping Societal Problems and Solution Directions (SD). Solutions (S) can be part of different SDs for different missions. Red arrows indicate interactions among solution directions

## Problem-Solution Diagnose

The first step in assessing a MIS is to identify the problem framings tied to the mission and the corresponding solution directions (black arrows in Fig. 14.1). A solution direction consists of technological and social innovations aimed at addressing specific problems related to the mission. Mapping these directions—both those prioritized or overlooked—lays the groundwork for further structural and functional analysis. This step also includes assessing how solution directions interact (positively, neutrally, or negatively) and their maturity (Sandén & Hilman, 2011; Elzinga et al., 2023). The dominance of particular solution trajectories over others can be evaluated by market share, actor engagement, and speed of development.

In contexts with multiple missions, each mission might prioritize different problems, increasing the diversity of both problem framings and proposed solutions. This leads to competing rationales and logics that can cause contestation between innovation systems. As a result, stakeholder negotiations to frame the missions might become fiercer and polarized compared to innovation systems affected by a single mission. This study pays close attention to the dominance of specific mission framings and how they may overshadow alternative missions and solution directions.

## Structural Analysis

The second step in MIS analysis involves identifying the actors, infrastructures, and institutions involved in developing and diffusing of solution directions. Understanding which actors shape the mission's directionality and the tensions this creates is vital for assessing the system's ability to mobilize and align stakeholders around a shared goal. The MIS framework distinguishes two structures: programming and performance (Elzinga et al., 2023). The programming structure—or mission arena—involves actors coordinate, steering, and shaping the mission, while performative actors are involved in implementing and developing solution. Actors can be part of both structures.

Analysing the number, type, and composition of actors involved in both structures of the mission arena is essential for conducting a thorough structural analysis (Elzinga et al., 2023). When analysing systems shaped by multiple missions, overlapping or interacting arenas may emerge (Sandén & Hilman, 2011; Elzinga et al., 2023). This study explores the division and influence of actors across different mission arenas. The dynamics between innovation systems—and the dominance of certain actors—can be understood by examining how actors are positioned, distributed, and overlap within these arenas.

## Functional Analysis

The third and final step is a functional analysis that evaluates the performance of various solution directions, recognizing that their development is shaped by interactions with complementary and competing options. Elzinga et al. (2023) proposed a set of system functions that aim to reflect the distinct features of MIS, allowing better monitoring of transformational processes (Table 14.1). This approach acknowledges that mission success depends on both innovation and destabilization. The analysis also helps identify functional bottlenecks, offering potential intervention points for policymakers to improve system performance.

A functional analysis should be conducted for each solution direction to identify key processes hindering development, interactions causing systemic issues, and which direction may become dominant. While a full analysis was completed for all solutions under both missions (see De Koning et al., 2024), this article focuses on selected functions per electricity technology to explore how specific processes shape transition dynamics in multi-mission contexts. Strong performance of one solution direction—through access to resources, legitimacy, or supportive legislation—can unintentionally obstruct the development of competing solutions within the same or overlapping missions.

**Table 14.1** MIS functions from Elzinga et al., 2023

| Programming functions |
|---|
| Providing problem directionality: Efforts to converge the selection, prioritization, and framing of societal challenges. |
| Providing solution directionality: Efforts to converge the selection, prioritization, and visions for solutions and solution directions. |
| Coordinating the transition: Orchestration and monitoring of collective efforts for transition routes through, e.g., intermediaries, transition teams, and platforms. |
| **Performance functions** |
| *Innovation side* |
| Entrepreneurial experimentation: Actions to tests, trials, and experiments with solutions to promote learning. |
| Knowledge development: Learning about societal challenges and (potential) innovations through research, development, learning-by-doing, etc. |
| Knowledge diffusion: Dissemination of knowledge through, e.g., conferences, reports, consultations, publications, etc. |
| Market creation: Establishing, shaping, and protecting (niche) markets. |
| Resource mobilization: Acquiring and distributing resources (e.g., financial, human) to help develop and diffuse innovations. |
| Creating legitimacy: Efforts directed to earning socio-political and institutional support from stakeholders (e.g., citizens) for the mission-oriented transition. |

# Methodology

## Case Description

The Netherlands has introduced multiple missions aimed at achieving its climate goals, including the Net-Zero Act and the mission for a fully circular economy. To mitigate the high environmental impact of modern society on the climate, the Dutch government introduced a mission on achieving a fully circular economy, and a mission on advancing the electrification of society. The electrification or net-zero mission, inspired by the Paris Climate Agreement of 2015, and operationalised with specific goals and targets in the Netherlands four years later, aims to push innovation to reduce costs and increase renewable electricity production.[1] Moreover, the Net-Zero Act focuses on decarbonising the industrial sector through electrification and reducing greenhouse gas emissions to net-zero by 2050. This legislative framework supports industries in transitioning to renewable energy sources and boosting energy efficiency. Additionally, the Circular Economy (CE) mission, initiated by the "Raw Materials Agreement" in 2017, seeks to minimize waste and optimize material use, with the mission to be fully circular by 2050 providing a long-term vision (Janssen, 2020).[2] In the Dutch policy context, circular R-strategies—such as refuse, rethink, reduce, reuse, repair, refurbish, remanufacture, repurpose, recycle, and recover—play a key role in advancing the transition towards a circular economy.

To study the effect of having multiple missions in the development and diffusion of solutions, we study how the MIS's around those missions comprise and influence the innovation systems of different energy technologies, namely wind turbines, solar panels, and electric vehicle batteries. How these energy technologies develop—whether through policy incentives or market-driven innovation—must align with the two missions' specific objectives and might therefore create

[1] https://www.rijksoverheid.nl/onderwerpen/klimaatverandering/klimaatakkoord/wat-is-het-klimaatakkoord.

[2] https://www.rijksoverheid.nl/documenten/rapporten/2017/01/24/grondstoffenakkoord-intentieovereenkomst-om-te-komen-tot-transitieagenda-s-voor-de-circulaire-economie.

contestation and conflicts among missions and their envisioned solution directions.

## Data Gathering and Analysis

Within the study, various methods were used to collect data. Firstly, desk research involved analysing news sites, policy documents, R&D project documentation, and other publicly available sources reporting on the development of the technologies under study. This analysis focused on innovation activities related to different solution directions within the missions. These documents provided a foundational understanding of the innovation system‘s functioning, which was further explored through stakeholder interviews.

A total of 33 interviews was conducted to map out the organization and functioning of the innovation system, with a focus on the barriers hindering the further development of the solution directions. The analytical steps as outlined above will be applied to the different electricity technologies and compared among each other. Interview data was used to identify the dominant problem and solutions to tackle these problems based on the perceptions by stakeholders. In the functional analysis we will provide scores for each function based on interview data and describe exemplary innovation dynamics. To warrant a complete representation of the sector, a diverse set of stakeholders was interviewed, including research institutions, NGOs, producers, trade associations, governments, recyclers, investors, advocacy organizations, and entrepreneurs. An overview is provided in Table 14.2. In the remainder of this research, references to an interview will specify the respective technology and type of actor.

Finally, intermediate results were tested and validated in workshops with expert from the sector. The data gathered from desk research and interviews was validated by academic researchers, policymakers, and sector representatives. Their collective expertise helped identify and correct any inconsistencies. Two workshops were hosted, one at the start and one at the end of the research process, involving 16 participants.

**Table 14.2** List of interviews and workshops per studied technology. In-text references consist of first letter of type of actor, first letter of studied technology, and the number of interviewed actors. For example, the code for the second interviewed branch organization for PV panels is B-S-2

| Type of actor | Number of interviewed actors | | |
|---|---|---|---|
| | Wind turbines | Solar PV panels | Batteries EV |
| Branch organization | 1 | 2 | 3 |
| Circular entrepreneur | 2 | 2 | 6 |
| Grid operator | 1 | | |
| Producer | 1 | | |
| Governmental organization | 3 | 1 | 1 |
| Knowledge institute | 3 | 1 | 3 |
| Investment agency | | 1 | 1 |
| Sectoral organization | | 1 | |
| All technologies | | | |
| Validation workshop | 16 participants | | |

# Results

This section discusses the results of the MIS analysis and illustrates the dynamics caused by the interaction of the two mission-specific innovation systems and the technologies they impact upon. This section is structured according to the analytical steps of the MIS framework. For each analytical step, the overlapping system dynamics of the technologies are first addressed, followed by specific examples from the technological case studies.

## Problem-Solution Diagnoses

### Problems

To start, the missions for electrification and a circular economy address overlapping societal challenges but differ in their prioritization of these problems. Both aim to reduce $CO_2$-emissions and decrease the reliance on fossil resources (Fig. 14.3). These goals are institutionally supported by agreements such as the Dutch climate agreement and the Dutch coalition agreement. A significant target within these missions is to reduce Dutch $CO_2$-emissions by 49% by 2050.[3]

Additionally, interviewed actors identify supply security as a problem relevant within both the mission for electrification and circularity. The rapid transfer or relocation of the production chain for the various electricity technologies, particularly to China, is causing stress and uncertainty in Europe. Alternatively, supply security is a material-oriented issue as well. Better material management, keeping materials in the loop and therewith decrease dependency on import is also addressed in the circular economy mission. However, actors perceive this issue predominantly as a significant uncertainty that could hinder the energy transition; a shortage of materials and products might prevent the achievement of climate goals, thus framing supply risk as part of the mission for electrification. Consequently, various efforts are being made within both the PV sector and the electric vehicle batteries-sector to establish a local supply chain.

Lastly, specifically within the circular economy mission, the growing volume of discarded wind turbine blades, solar panels, and electric vehicle batteries is emerging as a significant issue. Effective and high-quality processing methods to recover the materials after use of these technologies and their components are still lacking or in its infancy[4,5] leading to substantial material loss. This problem is also receiving considerable media attention, with reports highlighting the dumping and burial of turbine blades in the desert,[6] the shredding of solar panels for use as road fill, and the poor material recovery rates from electric vehicle batteries (Interview P-W-1, K-W-1, K-W-2).

3 https://www.government.nl/topics/climate-change/climate-policy#:~:text=To%20combat%20climate%20change%2C%20the,a%2095%25%20reduction%20by%202050.

4 https://www.siemensgamesa.com/global/en/home/explore/journal/recyclable-blade.html.

5 https://www.vestas.com/en/media/company-news/2023/vestas-unveils-circularity-solution-to-end-landfill-for-c3710818.

6 https://www.nrc.nl/nieuws/2020/10/30/wat-zijn-de-wieken-van-oude-windturbines-waard-a4018058.

## Solutions

### *Cost Reduction*

A first solution direction considered relevant by actors addressing the dominant problems is "*cost reductions*" through improving the efficiency of renewable electricity technologies, scaling them, and improving their performance. Interviewed actors are convinced that, in order to reduce $CO_2$-emissions on a large scale, renewable electricity technologies must competitively price against electricity generated from fossil fuel-based technologies to replace them. Achieving this necessitates the various electricity technologies to be scaled up, starting with an increase in the number of wind turbines, solar panels, and electric vehicle batteries in use (Fig. 14.3). Additionally, these technologies must become more efficient and effective. Significant investments have been made to improve solar cell technology and wind turbines have rapidly become larger and more efficient, all aimed at reducing the Levelized Cost of Energy as much as possible.

### *Recycling*

Secondly, the solution direction '*recycling*' is believed to address the dominant problems in the sector, like the increasing volume of wasted products, $CO_2$-emissions, and the supply security of rare earth metals. By recycling, the disposal or incineration of electricity technologies, which leads to substantial $CO_2$-emissions, can be prevented. Additionally, recycling can help retain rare materials within the local economy, reducing dependency on non-European suppliers. Stakeholders emphasize that building local recycling capacity is essential to really contribute to the outlined societal problems (Interview B-W-1, G-W-3, K-W-3, B-S-1, B-S-2, G-S-1, B-B-2, B-B-3). Local production facilities will however be required to incorporate these recycled raw materials into the manufacturing of new products.

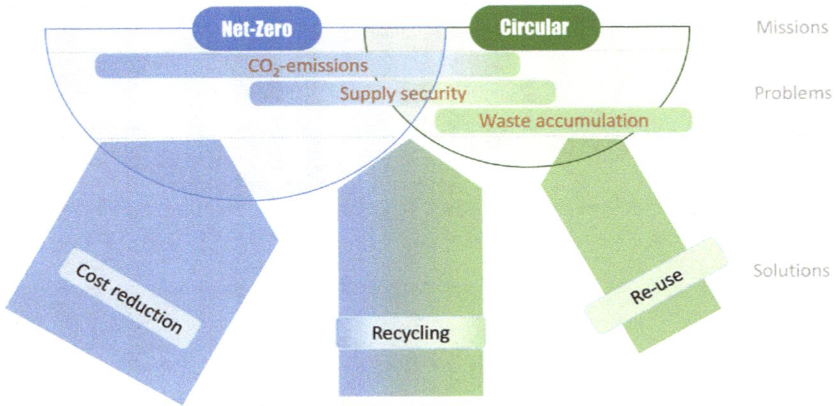


**Fig. 14.3** Problem-solution Diagnose of the interacting mission-specific innovation systems for electrification (net-zero mission) and circular economy (circular mission). Problems and solutions are coded by colour to match the mission they are associated with. A gradient indicates they are associated with both missions. The width of a solution direction indicates the level of support from actors

*Lifetime Extension*

A final identified solution direction is '*lifetime extension*'. *Lifetime extension* encompasses several circular or innovation strategies aimed at keeping products in use for as long as possible after they have been brought to market. This includes reusing, repairing, and refurbishing products. According to stakeholders, lifetime extension is a good strategy to promote supply security. When materials and products remain in use longer, the demand for new materials decreases, thereby reducing dependence on the countries from which these materials and products are imported.

## Interactions

The goals associated with the electrification mission are perceived by interviewed stakeholders to provide stronger directionality than those set for circularity. As it comes with a more detailed and convincing vision on what the future state of the socio-technical system might look like, the electrification mission is seen as dominant over the relatively vague (and thus less activating) circularity mission. This dominance significantly impacts which solution directions are supported in the innovation systems, as these solutions must primarily contribute to achieving the electrification mission's goals, namely reducing $CO_2$-emissions. Solutions aimed at addressing other problems receive less legitimacy, resources, and support. Institutions are primarily focused on scaling up electricity technologies, business models are designed for scaling up and resources are invested in enhancing performance (electricity generation). This focus hinders the circularity of the analysed electricity technologies as circular innovations lack resources, institutional support, and market demand (Interviews P-W-1, G-W-2, G-W-3, I-S-1, S-S-1, G-B-1, C-B-5, C-B-6).

According to several interviewees, alignment between solution directions from different missions can create synergistic effects (Interviews G-W-1, G-W-2, C-W-2, K-W-3, G-S-1, C-S-1, C-S-2, G-B-1, B-B-1, validation workshop). Both missions help to establish legitimacy and allocate resources, encouraging actors to innovate along the solution direction that both missions have in common. For instance, in the solution direction *cost reduction*, consider the interaction among the aim of enhancing the efficiency of electricity technologies to contribute to the net-zero mission, and the goal of reducing material usage through more efficient designs. The solution direction *cost reduction* stimulates innovations that maintain or improve a technology's functionality while using fewer materials. This aligns with both the circular economy mission, by reducing material usage, and the climate mission, by improving the performance and efficiency of the technology. An empirical example of this synergy is the recent development of car batteries. Innovations in the design of Li-ion batteries have reduced costs and material usage by using fewer expensive metals, while simultaneously improving battery performance and halving the required amount of lithium (Interview B-B-3). The amount and widespread support for these cost reducing innovations has been made possible by the legitimacy and directionality of both missions and, according to interviewed actors, could not have been achieved through support from just one of the two missions (Interview G-B-1, B-B-1, I-B-1, C-B-1).

Alternatively, when solutions do not align with the goals of both missions, the objectives of the circular economy are often overlooked. A prime example is the trend of increasing the size of wind turbines. Larger turbines convert wind into electricity more effectively, thereby reducing the Levelized Cost of Energy. Designers and manufacturers have successfully scaled up wind turbines to such an extent that those responsible for their installation and decommissioning have struggled to keep up. As a result, there is a continuous need for expansion and adaptation of infrastructure (such as installation ships and cranes) and production capacity (including larger production halls and machinery) (Interview W-B-1, C-W-2, N-W-1, P-W-1, G-W-2, K-W-1). These investments are challenging to amortize due to the rapid pace of new innovations and the substantial quantities of materials needed for the ongoing replacement of these capacities. Consequently, circularity in the design of wind turbines is largely neglected as resources are diverted elsewhere.

Another example is the competitive interaction between the solution direction *reuse* and *cost reduction* for PV panels. The Dutch government views solar PV as a crucial future energy source, with the aim that 30% of the Dutch energy mix will come from solar energy by 2035[7],[8] To achieve this goal, the available space for installing PV panels must be used as efficiently as possible. As a result, panels are often prematurely discarded and replaced with newer, more efficient models. This practice does not fully utilize the remaining economic and technological lifespan of the existing panels, which contradicts the circular solution direction focused on lifetime extension.

## Structural Analysis—Mission Arenas

To illustrate the potential tensions among the mission arenas of the focal missions, we examine the distribution of actors across various structures and assess which actors have the capacity to create directionality within these structures. Notably, many influential actors are situated in the arena of the net-zero mission, like steering committees and policy roundtables (Fig. 14.4). For instance, energy companies play a substantial role in the transition due to their significant investments and stakes in the sector. According to interviewed actors, these companies invest heavily in the development and installation of PV panels with the aim of scaling up electricity production (Interview I-S-1, S-S-1). Moreover, energy companies often hold considerable shares in wind and solar parks, thus exerting a strong directional influence in the sector. Additionally, the Netherlands Enterprise Agency (RVO.nl), which manages public investments, focuses more on scaling up renewable electricity technologies than on enhancing their circularity. The number of subsidy programmes and the resources allocated to energy, such as the SDE++ and HER+ programmes, surpass those dedicated to circularity (Interview G-W-2, G-W-3, G-S-1, I-B-1).

When examining the type distribution of actors in the arenas, it becomes evident that the net-zero mission arena features a more diverse

[7] https://www.google.com/url?sa=t&source=web&rct=j&opi=89978449&url=https://www.rijksoverheid.nl/binaries/rijksoverheid/documenten/rapporten/2023/10/09/monitor-zon-pv-2023-in-nederland/Monitor%2BzonPV%2B2023%2Bdefinitief%2Bcorrectie%2B2023%2B10%2B06.pdf&ved=2ahUKEwjC7aH2l_mFAxXuhv0HHfqODH0QFnoECBIQAw&usg=AOvVaw01w1cmve5XLVHUtFMIwNSJ.

[8] https://www.netbeheernederland.nl/publicatie/rapport-ii3050-scenarios.

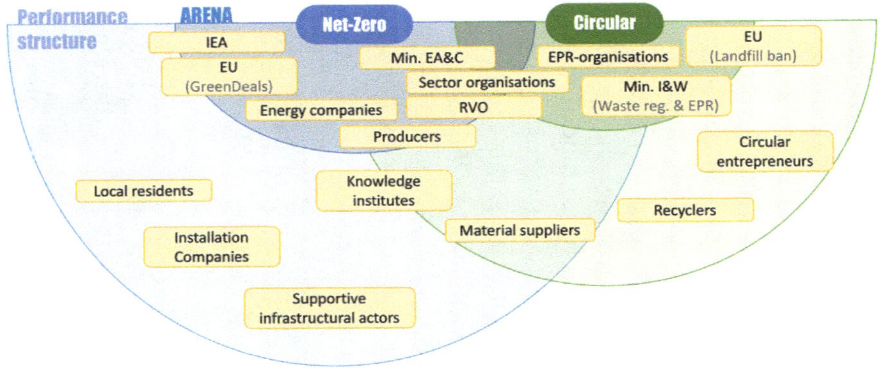


**Fig. 14.4** Visualization of the mission arena (or programming structure) and performance structure of both studied mission-specific innovation systems. Yellow boxes represent actors or groups of actors. Abbreviations used in the figure include the International Energy Agency (IEA), the European Union (EU), the Ministry of Economic Affairs and Climate (Min. EA&C), the Ministry of Infrastructure & Water Management (Min. I&W), Extended Producer Responsibility organizations (EPR), and the Netherlands Enterprise Agency (RVO). The structures for the circular mission are smaller as less actors are represented

range of actors than the circular mission arena. The net-zero mission includes stakeholders such as the International Energy Agency, energy companies, producers, and governments, creating a heterogeneous mix of actors involved in shaping the mission. This diversity ensures broad support for the net-zero mission across the supply chain. Alternatively, the circular arena is primarily composed of government bodies, particularly the Ministry of Infrastructure and Water Management (I&W). The extended producer responsibility (EPR) policy from I&W, shaped by the European EPR policy for batteries and PV panels, indirectly draws producers into the arena. As a result, it is unsurprising that, according to interviewed actors, the circular economy mission is less well-known, more contested in its implementation, and appears to have less support (Interview B-S-1, B-S-2, S-S-1).

Regarding the performance structures, the actors operating in the net-zero mission are more developed and better funded than those focused on the circular mission. Examples include large established organizations such as port authorities involved in wind turbine placements, a rapidly growing number of companies installing solar panels, and car leasing companies promoting e-mobility. Additionally, knowledge institutes are perceived to be more focused on improving energy efficiency and performance rather than, e.g., recyclability (Interview K-W-2, K-W-3, C-S-2, K-S-1, C-B-5). This well-supported set of operational actors ensures that the performance structure for the net-zero mission functions effectively. In contrast, the circular economy mission, where predominantly SMEs operate the performance structure, faces significant challenges. Circular entrepreneurs focused on the reuse of batteries, PV panels, and wind turbines report increasing difficulties in maintaining profitability and fear that such business models may disappear entirely (Interview C-W-2, C-S-2, C-B-4, C-B-5, C-B-6). The business model for recyclers is also not profitable (in the case of PV recycling and wind turbine blades) or is under significant pressure (for batteries).

## Functional Analysis

In this section, we describe the functional analysis of the different solution directions. The functional analysis is executed for each solution direction, to better understand the specific processes undermining the solutions development; the interactions among functions creating systemic problems; and ultimately to track which direction is most likely to emerge as dominant and steer further development of the technologies. At the top of each subsection, we present an overview of all functions and their scoring. Following this, we delve deeper into the functions most significant to the dynamics observed, highlighting those that are particularly revealing for understanding the performance and development of the different solution directions.

### Wind Turbines

As shown in Figure 14.5, the *cost reduction* solution direction outperforms the other solution directions *recycling* and *lifetime extension. Lifetime extension* faces significant barriers, including challenges in steering and coordinating the transition, limited access to necessary knowledge, and the absence of a viable business model. While recycling is regarded as a feasible approach to address the dominant problems, its functioning is constrained by insufficient resources and unfavourable market conditions.

Zooming in, a key innovation system function that significantly impacts the performance of various wind turbine solution directions is *market creation.* The government's procurement policy plays a pivotal role in shaping the wind sector in the Netherlands, exerting significant influence over its development and direction. The awarding criteria within the tenders are predominantly financially oriented, requiring potential operators to maximize electricity production at the lowest possible cost (Interview N-W-1, G-W-2, G-W-3). This approach drives the market towards scaling up to reduce the Levelized Cost of Energy, thereby greatly stimulating actors involved in the *efficiency* solution direction. Conversely, the current policy strongly impedes the *lifetime extension*

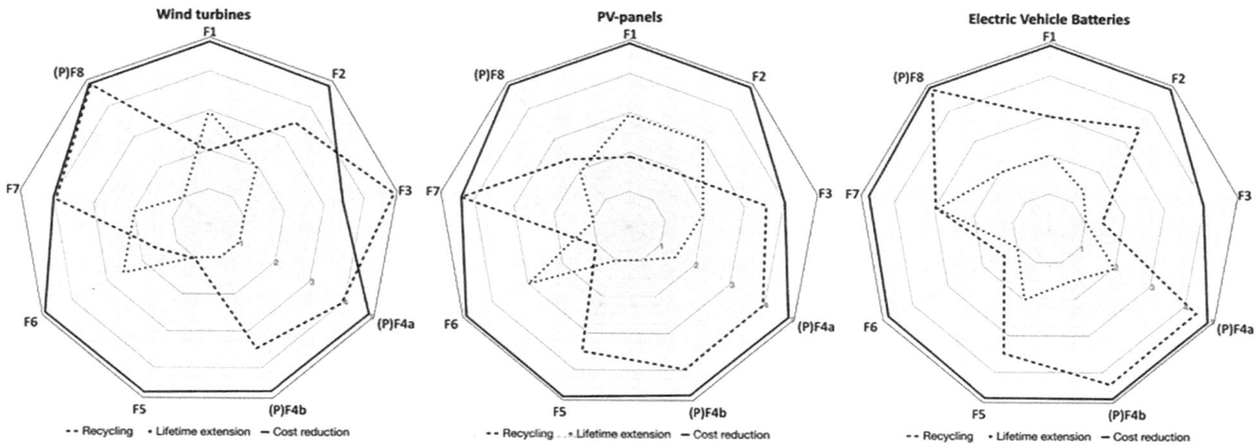


**Fig. 14.5** Spider diagram showing the scores for the MIS functions for wind turbines, PV panels and EVB. Programming functions are labelled PF. A score of 5 indicates a well performing function accelerating the transition, whereas a 1 indicates a poorly performing function hampering development.Functions labels: F1 Entrepreneurial experimentation, F2 Knowledge development, F3 Knowledge diffusion, (P)F4a Problem-directionality, (P)F4b Solution-directionality,F5 Market creation, F6 Resource mobilization, F7 Creating legitimacy, (P)F8 Coordination

solution direction. Wind parks typically receive operational permits for 25 to 30 years, even though turbines have a longer economic and technological lifespan of up to 45 years (Interview W-B-1, K-W-3). As a result, these turbines are often replaced prematurely because subsidies and permits expire, making way for newer and more efficient turbines.

The *recycling* solution direction also faces significant obstacles due to current (procurement) policies. Although circularity was recently included in a tender for the first time in the Netherlands, actors believe it needs to remain a stable criterion over a longer period to effectively influence the sector's direction (Interview W-B-1, P-W-1). Furthermore, there is no additional policy to promote the circular economy; interviewed actors even describe the waste policy as confusing, inconsistent, and obstructive to recycling turbine blades. The National Waste Management Plan (in Dutch: LAP3) prohibits the landfilling and incineration of composites of which these blades are made unless alternative processing costs exceed €200/ton (Interview G-W-1, G-W-2, G-W-3). Given that recycling composites from turbine blades averages €1400/ton, blades are often incinerated, mixed into asphalt, or left in storage awaiting alternative processing methods.

The continuous drive to make turbines larger and more efficient has diverted R&D interest away from circular design, explaining the immaturity of innovations for more easily recyclable blades. This dynamic shows that current market-forming instruments promote the *efficiency* solution direction but severely hinder other (circular) solution directions. Consequently, resources are not invested in making turbines more circular (function *resource mobilization*), and knowledge development does not focus on circularity (function *knowledge development*). This, in turn, has negatively affected other functions and the overall performance of the innovation system for circularity.

### PV Panels

The dominance of the *cost reduction* solution direction over *recycling* and *lifetime extension* is also evident in the functional analysis of PV panels

(Fig. 14.5). While all functions related to *cost reduction* score exceptionally well, the other solution directions show inconsistent performance. Large differences can be seen in functions like *directionality, knowledge development, knowledge diffusion,* and *resource mobilization.*

For the PV sector, a critical determinant for the functioning of the different solution directions is the function *knowledge development. Knowledge development* shows strong differences among the solution directions. As the Dutch government set the goal for solar power to constitute 30% of the energy mix by 2035 (Interview G-S-1), the innovation policy for solar panels primarily focuses on scaling up the number of solar panels in the Netherlands and improving their efficiency. Several knowledge institutes in the Netherlands are dedicated to improving or developing new technologies to enhance efficiency, holding a robust knowledge position with institutions such as TNO, TU Delft, University of Groningen, and Eindhoven University of Technology (Interview K-S-1). Additionally, actors believe that the strategy for optimizing knowledge development might prove beneficial for reducing the $CO_2$-emissions of the energy sector by developing improved technologies, but as well for the problem of supply security. Developing innovations which significantly improve the performance of technologies makes these innovations attractive to other countries and would therefore create mutual dependence (Interview B-S-1, B-S-2, I-S-1, S-S-1).

Contrary, interviewed actors indicate a hampering lack of knowledge regarding the recycling and reuse of solar panels. Current technologies for recycling solar panels are not well-developed, leading to low-quality recycling processes and limited material recovery (Interview C-S-1, C-S-2). Alternative recycling technologies remain prohibitively expensive, and their development is challenging and unattractive due to an unprofitable business case and a lack of investments (Interview I-S-1, K-S-1). Moreover, there is uncertainty about which panels on the market are suitable for reuse, and there is insufficient knowledge to develop standards for this purpose. The focus on improving efficiency creates gaps in the knowledge base and capabilities of actors to achieve circularity.

## Electric Vehicle Batteries

While the solution direction *cost reduction* demonstrates dominance over the other solution directions once again, *recycling* performs comparably well in areas such as *directionality, coordination, entrepreneurial experimentation,* and *knowledge development* (Fig. 14.5). In contrast, the lifetime extension solution lags in these functions, as well as in several other key areas.

In contrast to the other renewable electricity technologies, the solution direction *cost reduction* yields benefits for the net-zero and the circularity mission for electric vehicle batteries. The function *coordination* is a key function in the electric vehicle batteries-case and exemplifies this dynamic. The sector is strongly shaped by the European Battery Directive (established in 2006 and updated in 2023), which aligns goals of the mission for electrification and circularity. For instance, the directive aims to phase out hazardous and scarce materials, making batteries more efficient and safer while striving to enhance performance. Additionally, the directive imposes various circular requirements on battery design and disposal, such as the mandatory inclusion of a specified percentage of recycled materials in the production of new batteries and the compulsory recovery of materials at the end of their life cycle. In the Netherlands, these obligations are implemented using an Extended Producer Responsibility scheme (Interviews G-B-1, B-B-1, B-B-2). This regulatory framework showcasing how well-coordinated policies can create a supportive environment for achieving multiple mission goals.

Moreover, the key actors in the sector have managed to unite by establishing a growth fund that focuses on sustainable material supply, development, and scaling up of electric vehicle batteries technology. The National Growth Fund project, called 'Material Independence & Circular Batteries', is led by several influential organizations, including the EPR-organization, governmental agencies such as the Ministry of Economic Affairs and Climate Policy, and the Battery Competence Cluster, a network organization aimed at making battery competencies and knowledge available to the Dutch industry. The growth fund project comprises multiple smaller projects—examples include recycling, material refining, battery innovations and production, and reuse—in which both resources and attention are devoted to the objectives of the net-zero and CE missions.

## Discussion and Conclusion

The aim of this study was to better understand the interaction between two mission-specific innovation systems, namely those around the net-zero mission for the electrification of the Netherlands and its industry and around the mission for a fully circular economy. Our findings suggest that, overall, the net-zero mission is building more traction and in various respects dominates the circular mission. The nature and strength of the interaction between the two mission-specific systems varies for the different analytical steps and the system elements they pertain to (such as the structure, problem framings, and the development of solutions).

The first step in our multi-mission MIS analysis, i.e., the problem-solution diagnosis, highlights that reducing $CO_2$-emissions is a central issue within both missions. In the perception of actors, solutions for tackling this problem are seen as more valuable and urgent compared to other solutions. This dynamic could provide an initial explanation for the dominance of the solution direction *cost reduction* over others, as it is mainly focused on reducing $CO_2$-emissions. A second societal issue in the sector is supply security, which is resource-oriented and thus more aligned with the CE-mission. Nevertheless, actors engaging with the net-zero mission have adeptly reframed the supply security problem to fit within the net-zero scope: the sector must better manage its resources to safeguard the capability to produce and install the necessary electricity technologies to achieve climate goals. These actors demonstrate their ability to amplify their momentum by incorporating new societal problems into the problem-solution constellation of the mission and strengthen its directionality.

The second analytical step, the structural analysis, reveals various interactions between the two mission-specific innovation systems. In comparison, the arena for the net-zero mission is more densely populated, with a broader and more balanced representation of the sector.

Additionally, these actors possess more (political) power and resources, which further explains the dominance of the net-zero mission. It is also noteworthy that different ministries within the Dutch government drive the respective missions. While the former Ministry of Economic Affairs and Climate Policy plays a central role in the net-zero mission, the Ministry of Infrastructure and Water Management leads the CE-mission. Future research could investigate whether the differences in leadership and available policy instruments of these ministries affect the promotion of their respective missions.

For the third analytical step, this research demonstrates that the configuration of key processes for developing and diffusing solutions—the innovation system functions—can simultaneously stimulate and hinder different solution directions. These innovation system functions often provide reference points for policymakers considering potential interventions. Therefore, it is essential to gain a better understanding of how these processes affect the promotion and development of solution directions or technology, especially for their impact on interacting solution directions. Efforts should focus on stimulating key processes that do not support solution directions at the cost of other directions, or alternatively designing interventions that enable these functions to benefit several solution directions simultaneously.

In sum, we can conclude the co-existence of multiple missions and corresponding innovation system necessitates holistic mission management to ensure balanced progress across the diverse objectives. When multiple missions affect systems simultaneously, one mission often becomes dominant, naturally attracting a disproportionate share of resources, support, and legitimacy. This dominance can lead to an imbalance, with the prevailing mission shaping agendas, policy attention, and funding distribution, often at the expense of the other mission. Effective mission management is therefore essential to mediate these dynamics, potentially fostering alignment. In the studied case, successful circular economy policies in renewable electricity technology will require a coordinating, cross-sectoral vision to create and leverage synergies between circularity and the energy transition. This vision should establish clear expectations, set goals, provide direction, and most importantly highlight the added value of circularity in relation to the net-zero mission.

A potential starting point to create synergies between the two studied missions for the Dutch energy sector could be the shared focus on the societal issue of supply security (Kivimaa et al., 2022). Addressing this challenge has brought together actors from both innovation systems and offers opportunities for solutions focused on efficiency, recycling, and extending the lifespan of technologies. Policymakers can use this shared perspective to build broader support for innovation across both systems, thereby optimizing the use of allocated resources, enhancing legitimacy, and tapping into the capabilities of a wider range of stakeholders.

In conclusion, the current research shows that setting multiple missions does not necessarily result in mutually reinforcing interactions or an increase in the legitimacy and momentum of the missions. Introducing multiple missions or goals can lead to confusion among actors, fragmentation of focus and resources, contestation regarding the directionality driving the systems, and ultimately the paralysis of both transitions. It is crucial to map out the systems driven by these missions to fully understand how the missions interact across different system components. When missions are better aligned—in terms of formulation, perception, and interpretation—with the systems they aim to guide, synergies can emerge that may significantly boost innovation and transformation.

## References

Andersen, A. D., & Geels, F. W. (2023). Multi-system dynamics and the speed of net-zero transitions: Identifying causal processes related to technologies, actors, and institutions. *Energy Research & Social Science, 102*, 103178.

Andersen, A. D., Geels, F. W., Steen, M., & Bugge, M. M. (2023). Building multi-system nexuses in low-carbon transitions: Conflicts and asymmetric adjustments in Norwegian ferry electrification. *Proceedings of the National Academy of Sciences*, 120(47), e2207746120.

Andersson, J., Hellsmark, H., & Johansson, E. (2026). Varieties of disagreement in transformative policy missions: A Q study on the decarbonization of Swedish industry. *Environmental Innovation and Societal Transitions*, 58, 101069.

Bergek, A., Hekkert, M., Jacobsson, S., Markard, J., Sandén, B., & Truffer, B. (2015). Technological innovation systems in contexts: Conceptualizing contextual structures and interaction dynamics. *Environmental Innovation and Societal Transitions, 16*, 51–64.

Berkhout, F., Smith, A., & Stirling, A. (2004). Socio-technological regimes and transition contexts. *System innovation and the transition to sustainability: Theory, evidence and policy, 44*(106), 48–75.

De Graaff, S., Wanzenböck, I., & Frenken, K. (2025). The politics of directionality in innovation policy through the lens of policy process frameworks. *Science and Public Policy, 52*(3), 418–432.

Elzinga, R., Janssen, M. J., Wesseling, J., Negro, S. O., & Hekkert, M. P. (2023). Assessing mission-specific innovation systems: Towards an analytical framework. *Environmental Innovation and Societal Transitions, 48*, 100745.

Geels, F. W. (2018). Disruption and low-carbon system transformation: Progress and new challenges in socio-technical transitions research and the Multi-Level Perspective. *Energy Research & Social Science, 37*, 224–231.

Hekkert, M. P., Janssen, M. J., Wesseling, J. H., & Negro, S. O. (2020). Mission-oriented innovation systems. *Environmental Innovation and Societal Transitions, 34*, 76–79.

Janssen, M. (2020). Post-commencement analysis of the Dutch 'Mission-oriented Topsector and innovation Policy'strategy. Utrecht University-Copernicus Institute of Sustainable Development. Mission-Oriented Innovation Policy Observatory.

Janssen, M. J., Torrens, J., Wesseling, J. H., & Wanzenböck, I. (2021). The promises and premises of mission-oriented innovation policy—A reflection and ways forward. *Science and Public Policy, 48*(3), 438–444.

Janssen, M. J., Wesseling, J., Torrens, J., Weber, K. M., Penna, C., & Klerkx, L. (2023). Missions as boundary objects for transformative change: Understanding coordination across policy, research, and stakeholder communities. *Science and Public Policy, 50*(3), 398–415.

Kanger, L., Schot, J., Sovacool, B. K., van der Vleuten, E., Ghosh, B., Keller, M., et al. (2021). Research frontiers for multi-system dynamics and deep transitions. *Environmental Innovation and Societal Transitions, 41*, 52–56.

Kivimaa, P., Brisbois, M. C., Jayaram, D., Hakala, E., & Siddi, M. (2022). A socio-technical lens on security in sustainability transitions: Future expectations for positive and negative security. *Futures, 141*, 102971.

Klerkx, L., Begemann, S., & Janssen, M. (2025). Mission co-creation and the balance between exploitative and explorative innovation: Insights from the

Dutch circular agriculture transition. *Science and Public Policy, 52*(1), 128–145.

Koning, A., Elzinga, R., Balkenende, R., van Heusden, F., & Kleijn, R. (2024). *Productgroepanalyse hernieuwbare energietechnologie: Aangrijpingspunten voor beleid*. CML.

Larrue, P. (2021). *The design and implementation of mission-oriented innovation policies: A new systemic policy approach to address societal challenges*. OECD.

Löhr, M., & Chlebna, C. (2023). Multi-system interactions in hydrogen-based sector coupling projects: System entanglers as key actors. *Energy Research & Social Science, 105*, 103282.

Mazzucato, M. (2018). Mission-oriented innovation policies: Challenges and opportunities. *Industrial and Corporate Change, 27*(5), 803–815.

McMeekin, A., Geels, F. W., & Hodson, M. (2019). Mapping the winds of whole system reconfiguration: Analysing low-carbon transformations across production, distribution and consumption in the UK electricity system (1990–2016). *Research Policy, 48*(5), 1216–1231.

Papachristos, G., Sofianos, A., & Adamides, E. (2013). System interactions in socio-technical transitions: Extending the multi-level perspective. *Environmental Innovation and Societal Transitions, 7*, 53–69.

Parks, D. (2022). Directionality in transformative innovation policy: Who is giving directions? *Environmental Innovation and Societal Transitions, 43*, 1–13.

Raven, R. (2007). Co-evolution of waste and electricity regimes: Multi-regime dynamics in The Netherlands (1969–2003). *Energy Policy, 35*(4), 2197–2208.

Rosenbloom, D. (2020). Engaging with multi-system interactions in sustainability transitions: A comment on the transitions research agenda. *Environmental Innovation and Societal Transitions, 34*, 336–340.

Sandén, B. A., & Hillman, K. M. (2011). A framework for analysis of multi-mode interaction among technologies with examples from the history of alternative transport fuels in Sweden. *Research Policy, 40*(3), 403–414.

Schot, J., & Steinmueller, W. E. (2018). Three frames for innovation policy: R&D, systems of innovation and transformative change. *Research Policy, 47*(9), 1554–1567.

Wanzenböck, I., Wesseling, J. H., Frenken, K., Hekkert, M. P., & Weber, K. M. (2020). A framework for mission-oriented innovation policy: Alternative pathways through the problem–solution space. *Science and Public Policy, 47*(4), 474–489.

Wesseling, J., & Meijerhof, N. (2023). Towards a Mission-oriented innovation systems (MIS) approach, application for Dutch sustainable maritime shipping. *PLOS Sustainability and Transformation, 2*(8), e0000075.
Wittmann, F., Hufnagl, M., Lindner, R., Roth, F., & Edler, J. (2020). Developing a typology for mission-oriented innovation policies (no. 64). Fraunhofer ISI discussion papers-innovation systems and policy analysis.

# 15

# The Transformative Capacity of Green Industrial Policy

Teis Hansen, Ulrich Elmer Hansen, and Markus Steen

## Introduction

Industrial policy is back in the Global North, after having been considered for a few decades an unfortunate policy approach of the past. Thus, the answer to the rhetorical question in the title of the highly influential paper by Pack and Saggi (2006) 'Is There a Case for Industrial Policy?' is today 'yes' in many influential policy circles. However, this new generation of industrial policies, commonly referred to as Green Industrial

T. Hansen (✉)
Department of Food and Resource Economics, University of Copenhagen, Copenhagen, Denmark
e-mail: teis.hansen@ifro.ku.dk

T. Hansen · M. Steen
Department of Technology Management, SINTEF, Trondheim, Norway

U. E. Hansen
DTU Wind and Energy Systems, Technical University of Denmark, Risø, Denmark



M. Steen et al. (eds.), *Sustainability Transitions and Industrial Transformation*,
https://doi.org/10.1007/978-981-95-4893-4_15

Policy (GIP),[1] often comes with an aim to combine industrial development with greening of the economy, particularly through a focus on decarbonisation.

A main backdrop to the emergence of GIP is the Paris Agreement. Given that ambitions to achieve net-zero materialize, decarbonisation and the associated strategies of different countries fundamentally impacts the global economy (Arcand & Meadowcroft, 2025). First, fierce competition is already present around key mitigation solutions (e.g., renewable energy technologies, electrification, carbon capture and storage [CCS]), and associated supply chains. A key question for policymakers is if and how they support industrial development and transformation notably in domestic manufacturing and production capacity, for example related to renewable energy technologies. Second, policymakers also need to consider the industries and technologies that are incompatible with the net-zero target, and the (increasing) economic risks that they pose. They also need to be mindful of the socio-economic impacts of hydrocarbon phase-outs not only in regions that extract/produce fossil energy, but also where hydrocarbons constitute a key input (e.g., process industries). This points, third, to decarbonisation directly or indirectly affecting all sectors, or socio-technical systems as conceptualized in sustainability transitions studies (Truffer et al., 2022), and the need for a multi-system understanding of decarbonisation (Andersen & Geels, 2023). In turn, this raises several questions pertaining to GIP, which have received scant attention, and a need to understand how this novel generation of industrial policy sits within a broader context of policy targets and instruments.

More specifically, there remains a need for critically reflecting on the transformative capacity of this 'green' policy approach, given its emphasis on domestic competitive industries that opens up for tensions with several transformative priorities, including technology phase-out, reduced consumption, and wide-ranging institutional changes. Transition scholars argue that innovation policy needs to be embedded in a broader policy framework (Alkemade et al., 2011; Foxon & Pearson,

[1] Industrial policy is in many instances combined with other societal goals, such as creating positive socio-economic benefits (Arcand & Meadowcroft, 2025). In this chapter, however, we focus on the 'greening ambition' of the new wave of industrial policies in the Global North.

2008). Consequently, this chapter aims to discuss synergies and conflicts between GIP and third generation innovation policy. As such we take our point of departure in the literature on Transformative Innovation Policy (TIP). TIP seeks to stimulate radical changes towards greater sustainability in systems delivering services essential for our societies, and emphasizes the need for policy to address directionality, demand articulation, and policy coordination (Weber & Rohracher, 2012). We elaborate on each of these aspects, considering whether industrial policy is aligned with TIP, and consider possibilities of GIP to spark concurrent decarbonisation both within and across different sectors. However, first, we present the core tenets of TIP and the new wave of industrial policy.

## Transformative Innovation Policy

The field of sustainability transitions is concerned with understanding change processes towards greater sustainability in the systems delivering essential services we all rely on such as energy or transport (Markard et al., 2012). This rapidly expanding literature is accompanied by similar changes in work on innovation policy (e.g., Kuhlmann & Rip, 2018) where notions of sustainability and grand challenges have recently become salient. Schot and Steinmueller (2018) conceptualized these developments as a 3rd generation of innovation policy focused on transformative change, which is clearly distinguishable from previous innovation policy approaches that were focused on innovation as an engine for economic growth, taking a more linear approach to innovation (1st generation), or innovation systems (2nd generation). In addition to TIP, other 3rd generation approaches include mission-oriented innovation policy (Mazzucato, 2021) and challenge-based innovation policy (Boon & Edler, 2018; Coenen et al., 2015), which—despite some differences (Haddad et al., 2022)—all constitute part of the 'normative turn' in innovation policy (Uyarra et al., 2019). While this generational separation of innovation policy may overemphasize aspects of difference rather than similarity, some aspects to TIP are evidently novel[2] (Grillitsch et al., 2021).

First, the aim of TIP is distinct in terms of scope of change, since the ambition of policy is to modify entire socio-technical systems rather than stimulate single innovations. As highlighted by Alkemade et al. (2011, 127), given the fundamentally different aims of transition policy and (1st and 2nd generation) innovation policy, alignment is only *"expected in the case where policy seeks to create new profitable industries that contribute to a more sustainable society"*. Thus, TIP entails that policy consciously seeks to stimulate transformative change through innovations and efforts that contribute to addressing central sustainability challenges.

Following Weber and Rohracher (2012), emphasis on transformative change necessitates attention to policy challenges that are not in focus when the rationale for policymaking is to address market or structural innovation system failures.[3] Specifically, TIP highlights the importance of directionality, demand articulation, policy coordination, and reflexivity (see also Table 15.1). First, ensuring a clear *direction* to innovation policy entails that policymakers define the preferred envisioned future for society, and reserve and prioritize support for research and innovation efforts that contribute positively to this (Weber & Rohracher, 2012). As highlighted by Diercks et al. (2019), directionality in TIP can be considered a layering on priority-setting in previous technology policies, however, with a main focus on societal outcomes rather than promoting specific technologies or sectors. Second, TIP also involves a significant focus on *demand articulation*, i.e., diffusion and uptake of innovations (including knowledge about the practices of technology users) in society to ensure effect, such as actual decarbonisation of activities. While demand articulation is emphasized in TIP, it connects (in particular when considered in the context of industrial policy) to previous work on lead markets that also made its way into national

---

[2] Arguably, the distinction between the 1st and the 2nd generation of innovation policy is clear, with the 2nd conceptualizing innovation as the outcome of systemic processes.

[3] Other conceptualizations of TIP are available in the literature, e.g., Schot and Steinmueller (2018) which place emphasis on processes for inclusive deliberation as a precondition for experimentation (see also Grillitsch et al., 2019; Sengers et al., 2019). However, in this chapter we operationalize TIP through the four dimensions of Weber and Rohracher (2012) that have been widely adopted in the literature.

and European policymaking (Quitzow et al., 2014). In short, the lead markets approach focuses on the role of specific demand characteristics in the development of innovations with potential for international diffusion (Beise, 2004; Quitzow et al., 2014). Third, a greater need for *coordination* of policies across multiple policy fields follows from the complexity and encompassing character of transitions across entire socio-technical systems. The emphasis on coordination follows previous work on innovation system policy (Lundvall & Borrás, 2005), but is considered even more important in the case of TIP, which requires breaking down multiple policy silos. Fourth, the importance of *reflexivity*, i.e., the need for continuous monitoring, anticipation, and inclusion of stakeholders, is underlined to ensure adaptive policymaking that reflects the multiple distributed arenas involved in governance. While this draws on more traditional approaches to policy monitoring and evaluation, it also goes significantly beyond this, both in terms of foci and approaches to evaluation (Mickwitz et al., 2021; Molas-Gallart et al., 2021), but also in engagement of distributed actors in policy processes. In below sections consider these aspects referring to directionality, demand articulation, policy coordination, and reflexivity.

The second novel aspect of TIP that distinguishes it from previous innovation policy approaches is its focus on intentional destabilization of existing competing and less sustainable solutions in addition to innovation support (Turnheim & Geels, 2012). Given the advantages accumulated through decades of existence, in the form of established infrastructures, production facilities, standards, regulations, use patterns, etc., established solutions are unlikely to be outcompeted by greener alternatives simply through policies focused on supporting them (Kivimaa & Kern, 2016; Rogge & Johnstone, 2017). In this way, TIP includes both support for innovation and 'exnovation' through the phase-out of less sustainable alternatives (David, 2017). This is a key difference between TIP and the widely utilized mission-oriented innovation policy approach (Mazzucato, 2018, 2021), which shares many similarities with TIP (e.g., strong role for governments) but is less concerned with phase-out.

**Table 15.1** Overview of policy foci across the three generations of innovation policy

| 1st generation<br>*Market failures* | 2nd generation<br>*Structural system failures* | 3rd generation<br>*Transformational system failures* |
|---|---|---|
| Information asymmetries: Uncertainty about outcomes and short time horizon of private investors lead to undersupply of funding for R&D. | Infrastructural failure: Lack of physical and knowledge infrastructures. | Directionality failure: Lack of shared vision regarding the goal and direction of the transformation process and/or insufficient targeted funding for R&D and demonstration projects and infrastructures. |
| Knowledge spillover: Public good character of knowledge and leakage of knowledge leads to socially sub-optimal investment in (basic) R&D. | Institutional failures: Hard institutional failure: Formal institutions such as laws, regulations, and standards (in particular regarding IPR and investment) that hinder innovation.<br>Soft institutional failure: Informal institutions (e.g., social norms and values, culture) that hinder innovation. | Demand articulation failure: Insufficient spaces for anticipating and learning about user needs to enable the uptake of innovations. Absence of orienting and stimulating signals from public demand. Limited uptake of more sustainable solutions. |

(continued)

**Table 15.1** (continued)

| 1st generation<br>*Market failures* | 2nd generation<br>*Structural system failures* | 3rd generation<br>*Transformational system failures* |
|---|---|---|
| Externalization of costs: The possibility to externalize costs leads to innovations that can damage the environment or other social agents. | Interaction or network failure: Strong network failure: Repeated cooperation in closely tied networks leads to lock-in into established trajectories and a lack of infusion of new ideas.<br>Weak network failure: Limited inter-organizational interaction and knowledge exchange inhibits exploitation of complementary sources of knowledge and processes of interactive learning. | Policy coordination failure: Lack of policy coordination across different systemic levels, including spatial scales (e.g., regional-national-European), policy fields (e.g., research and innovation policy and different sectoral policies), and parts of the policy cycle (e.g., policy formulation in ministries and policy implementation in agencies). |
| Over-exploitation of commons: Public resources are over-used in the absence of institutional rules that limit their exploitation (tragedy of the commons). | Capabilities failure: Lack of appropriate competencies and resources at actor and firm level prevent access to new knowledge, and lead to an inability to adapt to changing circumstances, and to switch from old to new technological trajectories. | Reflexivity failure: Insufficient ability of the system to monitor, anticipate, and involve actors in processes of self-governance; Lack of distributed reflexive arrangements to connect different discursive spheres, provide spaces for experimentation and learning. |

*Source* Slightly modified from Scordato et al. (2022, 326), based on Weber and Rohracher (2012)

## The New Wave of Industrial Policy

Broadly speaking, industrial policies refer to targeted government interventions aimed at changing the structure of the economy by allocating resources to certain domestic industries (Bailey et al., 2019; Cimoli et al., 2008; Rodrik, 2014). Often, debates concerning the usefulness of industrial policies involve two rather contrasting perspectives: one that questions the state's ability to 'pick winners' and one that advocates an active role for the state (see, e.g., Evans, 1998; Schmitz et al., 2015). Such debates, however, greatly oversimplify the role of the state, market forces, and the importance of industrial policies in achieving industrial development historically (Gerschenkron, 1962). Hence, Rodrik (2009) argues that policymakers should not be concerned with the question of whether industrial policy should be used, but rather how it should be designed and implemented.

Historically, two main types of industrial policies have supported domestic industries. 'Import-substitution' policies aim at protecting domestic industries by restricting the import of goods (Cimoli et al., 2008). Industrial policy measures to this end can for example include local content requirements, import quotas and duties, exchange rate regulation, and premiums for locally produced goods. Alternatively, 'export-oriented' industrial policies (Gereffi, 2014) focus on attracting foreign investments by reducing trade barriers and implement measures aimed at enabling domestic industries to link up with foreign companies and export markets. Both types of industrial policies can be related to 1st and 2nd generation innovation policy, as they seek to address market or system failures linked to, e.g., incentives to innovate and establishment of intra-organizational linkages.

Traditionally, industrial policy was motivated mainly by scarcity of resources and a need for focusing policy support on industries with possibilities for significant improvements in productivity. The Norwegian petroleum industry is an example, having been supported by various industrial policy measures since its emergence in the early 1970s (Sæther et al., 2011). Today, industrial policy interventions are to greater extent focused on industries that also contribute to achieving other public policy goals, in particular decarbonisation (but also industries delivering products considered strategically important[4] which is beyond the scope of this chapter). In this context, the concept of 'green' industrial policies has increasingly been used in both academia and policy circles (Arcand & Meadowcroft, 2025). According to Altenburg and Rodrik (2017, 11), GIPs involves "*any government measure aimed to accelerate the structural transformation towards a low-carbon, resource-efficient economy in ways that also enable productivity enhancements in the economy*". This understanding of GIP aims at improving the environmental sustainability of the economy, while supporting one or more domestic industries.

A prominent example of GIP is the Inflation Reduction Act (IRA) in the US, which includes policy measures (e.g., consumer tax credits, rebates, grants) aiming at stimulating demand for various technologies that can contribute to decarbonisation, both on the supply (e.g., wind turbines, solar PV) and demand (e.g., EVs) side, as well as various other solutions such as hydrogen, CCS, and batteries. At the same time, the IRA focuses on encouraging innovation and domestic manufacturing of such technologies, for example through tax incentives, subsidies, and R&D funding. A range of complementary policies and acts have been adopted in the US to protect domestic industries from foreign competition, for example through setting high import tariffs on solar panels and electric vehicles. This competition is essentially China, which by implementing a broad set of industrial policy instruments over many years now dominates the global manufacturing and supply of key low-carbon technologies such as solar PV modules and batteries (Allan & Nahm, 2025; Gong & Hansen, 2023).

A similar GIP 'package' as the IRA has been adopted in the European Union to nurture technology diffusion while supporting domestic industries of member states through the 'EU Green Deal Industrial Plan' and related policies such as the Critical Raw Materials Act. The former encompasses a variety of policy initiatives to ensure climate and environmental targets are met while supporting inclusive and just economic growth, thus reflecting how industrial policies often encompass more than one societal goal. The latter is much more focused and aims to reduce Europe's dependence on material inputs (e.g., rare earth metals)

[4] See, e.g., the US and EU Chips Acts, focused on semiconductor production in the two areas.

that are essential to the manufacturing of low-carbon and high-tech products such as EVs and wind turbines. Taken together, these different policy packages also reflect a recognition of multi-sector transformation across systems of production, distribution, and consumption.

GIPs have been used across many other emerging economies in addition to China, such as India, South Africa, and Brazil, in relation to renewable energy (Millot & Rawdanowicz, 2024). On the demand side, various financial support systems such as auction schemes for large-scale solar PV and wind power have been adopted to create a domestic market, while requirements for local content, job creation, ownership, and other criteria have been included in procurement systems to stimulate supply-side development of domestic industries (Hansen et al., 2020).

Millot and Rawdanowicz (2024) suggest there is clear evidence that GIPs have contributed to cost reductions in key renewable energy technologies. Nevertheless, the impacts or results of GIP policy packages are not necessarily easy to evaluate (Arcand & Meadowcroft, 2025; Harrison et al., 2017; Millot & Rawdanowicz, 2024; Rodrik, 2014). This is firstly due to industrial policy being part of a broader policy mix, making it challenging to discern the effects of industrial policy versus other policies' effects. Also, industrial policies are always highly context dependent, making it challenging to generalize from particular policy experiences (Warwick & Nolan, 2014). Secondly, countries differ considerably with regard to industrial policy having the intended effects, such as when a 'picking winners' strategy fails, as exemplified by the failure of the Swedish shipyard industry to remain competitive in the 1970s, despite significant government support (Carlsson et al., 2018). Bailey et al. (2019) argue that policy attempts to develop a particular technology or industry need to consider the strength of the innovation system associated with that technology/industry. This reflects how countries have very different conditions for implementing effective GIP, leading some scholars to suggest that GIP needs to be attuned to the presence of comparative advantage within a particular technology, industry, or knowledge domain in order to be effective (Harrison et al., 2017).

Malhotra and Schmidt (2020) argue that industrial policy needs to be attuned to the technological characteristics of low-carbon technologies. They identify three main types of technologies, depending on the degree of design complexity and the need for customisation. A first type of technologies (e.g., solar PV modules) are easily replicated (standardized and mass-produced products for global markets) and therefore easily result in first-mover disadvantages. An example in solar PV module manufacturing, which was initially led by the US, Japan, and Germany, but currently dominated by China. A second type of technologies have highly complex designs and need for customisation (e.g., combined cycle gas turbine, nuclear and biomass power plants), and are challenging to result in cost reductions and industrial development. Here, coordination is needed across countries and sectors to learn from application- and project-specific experiences (Binz et al., 2017). Malhotra and Schmidt (2020) argue that the 'sweet-spot' for GIP is for the third type of technologies that sit in the middle of the spectrum on both design complexity and/or customisation need, and include technologies such as EVs, wind turbines, and rooftop solar PV. Here, countries can reap first-mover advantages in developing technologies that can subsequently be exported to global markets. This necessitates a mix of supply- and demand-side policies and may therefore most easily be achieved by large economies with substantial domestic market potential and resource availability. However, the experience of for instance Denmark in wind power tells us that also smaller countries can, if conditions are conducive, take the lead and sustain a prominent position within the development and production of certain low-carbon technologies that are exported globally (Karnøe & Garud, 2012; Skjærseth et al., 2023; Steen et al., 2024).

In summary, while it is only a few years ago that Aiginger and Rodrik (2020, 191) suggested that *"[o]ne can imagine the greening of industrial policy"*, today this is reality. Whether we can expect GIPs to also contribute to wider transformations towards environmental sustainability, is the question we turn to in the following sections.

## Directionality in Industrial Policy?

In the words of Weber and Rohracher (2012, 1042), directionality *"points to the necessity not just to generate innovations as effectively and efficiently as possible, but also to contribute to a particular direction of transformative change"*. This normative aspect of deciding on a preferred direction marks a substantial difference to previous understanding of innovation policy. Rather, the rationale for policy intervention goes beyond creating conducive conditions for innovation in general, to selecting specific areas where innovation is needed and even creating room for such innovations through phase-out of established technologies (Grillitsch et al., 2021).

To some extent, GIPs can be seen to honour the need for greater directionality in policy, as it reflects a stronger intervention of the state beyond correcting market failures, to stimulating the transition towards a decarbonised economy and accompanying domestic industries and jobs. For example, the EU's Green Deal Industrial Plan aims to foster a development whereby European industry leads the green transition (European Commission, 2020). As such it is attuned to the broader EU climate and socio-economic targets to reduce GHG emissions by at least 55% by 2030 and be climate neutral by 2050. Similarly, attention towards investments in regions with a risk of losing from the low-carbon transition due to specializations related to fossil fuel extraction or use, is arguably an important element in considering the potential negative distributional effects of the transition, thereby addressing potential opposition towards progressive climate policies (Flegal, 2023). In this way, the vision of a decarbonised future is clearly providing direction to the focus of GIPs.

At the same time, there are also clear limits to the relative importance of decarbonising the economy for the design and implementation of industrial policies. Most notably, there are important scalar trade-offs between the policy objectives of limiting global warming and growing domestic industries, where industrial polices generally prioritize the latter. Indeed, in many cases, rapid and substantial diffusion of low-carbon technologies can occur on the basis of imported equipment (Gomel & Rogge, 2020; Hansen, Hanson, et al., 2024). The efficiency with which industrial policy addresses global warming is compromised by the ambition of domestic industrial development, most clearly illustrated by the increasing severance of trade disputes between developed and emerging economies in connection to low-carbon technologies (Meckling, 2021). From a climate change mitigation perspective, it is unfortunate if diffusion of renewables and other low-carbon technologies is slowed down by policies that resemble national protectionism. In this way, there appears to be a clear hierarchy between the words where 'industrial' has greater prominence than 'green'.

Furthermore, for GIP to provide a clear direction towards transformative change, it would have to include not just support for emerging low-carbon technologies, but also directly address the phase-out of polluting solutions (David, 2017; Rinscheid et al., 2021). There are examples of such phase-out instruments in the EU policy package, such as the ban on sale of new fossil fuel-powered cars from 2035 (European Parliament, 2023). However, Narassimhan et al. (2024) conceptualize the general absence of such phase-out policies as 'deliberate misalignment', where governments provide incentives for new technologies, while they simultaneously resist stronger measures to discontinue old technologies. Indeed, in a recent overview of industrial policies in nine OECD countries, Criscuolo et al. (2023) suggest that on average, GIPs account for only 13% of policy expenditure. In this way, GIPs may contribute to policy layering (Streeck & Thelen, 2005) where new policies to support emerging (green) technologies and sectors are simply added on top of policies sustaining existing unsustainable forms of production and consumption. In the US in particular, regulations limiting high-emission activities are left out-of-scope for industrial policies (EIU, 2023).

Also in Europe many countries continue to subsidize fossil fuels both directly and indirectly, especially after Russia's invasion of the Ukraine (CAN Europe, 2023) when new and more expensive (such as liquefied natural gas (LNG)) needed to be imported, including from other continents. This reflects that transition ambitions may be difficult to uphold in times of crisis. A telling example is also the temporary petroleum tax relief package that the Norwegian Government introduced in 2020 in response to the economic downturn caused by the Covid-19 pandemic (Rystad Energy, 2023). While this in a twist of fate did result in production ramp-up that subsequently eased Europe's break-away from

Russian gas, critics argue that the tax relief may result in substantial long-term financial losses and, more importantly, postpone the much-needed energy transition as well as an industrial transformation away from petroleum-industry dependence in Norway.

In summary, such policy layering is unlikely to result in a transition due to the substantial advantages of established socio-technical configurations accumulated over decades of development that result in a need for direct phase-out policies (Kivimaa & Kern, 2016). As concluded by EIU (2023, 1), *"on their own, these plans are insufficient for countries to meet their emissions targets"*. It should be noted that this critique is not unique to GIPs but also valid for mission-oriented innovation policy, which largely disregards destabilization efforts.

## Demand Articulation in Industrial Policy?

Demand articulation highlights the importance of complementing the traditional emphasis on technology-push policies with attention to the uptake of innovations, including learning about user needs and preferences. Obviously, for low-carbon innovations to result in a transition, there is a need for rapid diffusion throughout society.

Demand articulation and investment support for green technologies is perhaps the part GIPs where the transformative potential is most clearly articulated. In addition, investment support is also in some cases linked to other policies with demand-side impacts, such as public procurement regulations ensuing an uptake of the low-carbon products resulting from the investment support (Flegal, 2023; Uyarra et al., 2020). Thus, to some extent, this takes into account transformation needs across systems (e.g., of energy production and use), even if systematic attention to co-occurring modifications and the need for investment support across such systems is not always clear in GIPs.

Despite the attention to stimulating investment and demand, the prioritization of industrial development described above implies that diffusion of low-carbon technologies is lower than it would have been without trade tariffs. The hallmark example of this is the import restrictions in the US on Chinese solar PV modules (Meckling, 2021) that has spurred investments by Chinese producers across Southeast Asia in order to serve the American market from in particular Vietnam and Malaysia. Most recently, restrictions on solar PV imports from these countries are becoming implemented, to protect domestic producers in the US. As prices remain at a higher level than they would have been in the absence of tariffs, government funds set aside for industrial policy initiatives result in lower emission reductions than they would otherwise have resulted in (see also EIU, 2023). On the flipside, however, the extensive cost reductions in solar PV modules in China have been enabled at least in part by a multifaceted set of industrial policy instruments and initiatives in China over several years (Millot & Rawdanowicz, 2024), coupled with vast export opportunities due to comprehensive renewable energy demand-side policies in countries like the US and Germany (Harrison et al., 2017). Hence, in an increasingly interconnected global economy, the dynamics of market demand, industrial policy, and trade regulations may interact in complex ways that either inhibit or encourage technology diffusion.

In addition, the demand side of industrial policy is limited to stimulating demand of low-carbon technologies, rather than reducing demand per se, which is unlikely to stimulate increased industrial activity. Of course, GIP may stimulate diffusion of solutions that reduce the need for inputs in the use phase (e.g., energy efficiency programmes limiting energy needed for heating and cooling of buildings), but actual sufficiency measures are out-of-scope for industrial policy, as is often the case for decarbonisation policy (Hansen, Andersson, et al., 2024).

## Policy Coordination in Industrial Policy?

While the need for coordination of policies across scales and policy areas is not in itself a unique feature of policies for transformative change, the need for coordination is arguably unprecedented due to the high complexity and interconnectedness of such transitions, spanning across many parts of society (Grillitsch et al., 2021; Weber & Rohracher, 2012). Indeed, it may be argued that the necessity of coordination across different realms of policymaking is a main characterizing feature of GIPs, which sets it apart from previous types of industrial policies.

Here it is important to distinguish between policies supporting industries (supply side) and policies supporting market creation (demand side) in different sectoral contexts. For example, the market for large-scale renewable energy involves a politically created demand where energy policies are essential. In contrast, the market for organic food involves a primary consumer-based demand, which the government can stimulate by supporting the supply side of organic agricultural production. This means that for GIPs to be effective, policymakers need to ensure that industrial policy and market-creation measures are closely aligned and coordinated with the respective areas of policymaking in the sector in question.

Some GIPs clearly demonstrate attention to such policy coordination. To exemplify, Flegal (2023) mentions the development of new structures spanning existing institutions in the US such as the Joint Office of Energy and Transportation, which is central to expanding the EV charging network across the US. In the EU, the comprehensive policy package introduced in recent years, including its GIP elements, demonstrate how the EU GIP sits within a much broader set of policies and policy instruments to achieve near- and long-term overarching targets related both to sustainability as well as a more equitable society. For example, as part of its funding programme for climate action, the EU's innovation fund supports a host of projects that aim to develop new technologies both for clean energy production and for mitigating end-of-pipe emissions (European Commission, 2024). As part of this, many projects focused on CCS have received funding, but none of these are connected to the continued use of fossil fuels for power production.

It is however evident that much more could be done to ensure policy coordination. Fragmentation across policies, financing instruments, and countries is one of the key challenges outlined in the recent Draghi Report, suggesting that EU industrial policies are designed in ways that limit their impact and effectiveness (Draghi, 2024). Multi-level coordination challenges are also evident, as regional level policies are frequently misaligned with policies at higher levels (Jolly et al., 2023; Matti et al., 2017; Steen et al., 2019), in some cases driven by capture of industrial policy by vested interests, which strengthen regional lock-ins (Hassink & Kiese, 2021). Policy coordination across the policy strategy and instrument levels are also frequently a challenge, as ambitious objectives are not supported by equally ambitious policies (for an example, see Schebesta & Candel, 2020). Further, the 'deliberate misalignment' described above by Narassimhan et al. (2024) highlights that there are limits to policy coordination whereby industrial policy may lack coordination with climate policies, when such coordination would entail worsened prospects for domestic low-carbon industry development. Fundamentally, coordination of climate and industrial policies depends on whether policymakers assess that domestic industries can compete in emerging green technologies, thereby profiting economically from the energy transition. When this appears likely—or policymakers can at least make this argument—industrial policies that are more aligned with climate policies will be more feasible to implement.

## Reflexivity in Industrial Policy?

Considering whether reflexivity is prominent in GIPs requires attention to whether and how industrial policies prioritize a broad inclusion of stakeholders, as well as are associated with continuous monitoring and anticipation functions. Both aspects are central to ensure adaptable policies that do not become locked-in to supporting specific industries. Ensuring broad involvement of private sector actors in governance processes is key to aligning government actions with industry development (Rodrik, 2014). Such 'embeddedness' of industrial polices can be in the form of government bodies overseeing progress and coordinating support schemes with industry associations, as an important channel of policy feedback and distributed policy learning (Evans, 1998).

A broad inclusion of stakeholders is a central element in mitigating the risk that rent-seeking incumbent actors strongly influence the selection of prioritized industries in GIPs (Schmitz et al., 2015). Regime capture of GIP is a likely outcome if only a narrow group of stakeholders are involved and consulted in the decision-making process, potentially leading to fixed portfolios of interventions that inhibit new directions (see also van der Loos et al., 2024). Aiginger and Rodrik (2020, 192) suggest that *"sustained collaboration between the public and private sectors around issues of productivity and social goals"* is central to avoid *"ex ante selection of activities"* that may follow from more traditional top-down selection processes. We would however add that it is critically important to move beyond a narrow and selective involvement of private sector actors. Involvement of independent actors may influence GIPs to ensure that policies strategically target industries with the greatest potential for societal in general, rather than specifically for incumbent industries, as illustrated by the independent intermediary organization Propulsion Québec's influence on industrial policy in Canada (Arcand et al., 2024). According to Allan and Nahm (2025), the important question for policymakers is therefore how they can support their domestic firms and industries to become inserted into the global value chains for green technologies through GIPs.

In terms of monitoring of implemented industrial policies (on both programme and project levels), there is a general agreement regarding the need for implementing more traditional management methods, such as milestone approaches. It is also important to learn from organizations that have successfully implemented these, exemplified by the Advanced Research Projects Agency-Energy (ARPA-E) (Flegal, 2023). However, there is little attention to more reflexive practices, such as formative approaches to evaluation (Molas-Gallart et al., 2021). While GIPs should 'discipline' the private sector by setting clear objectives, rules, and measurable targets, policymakers should adopt a flexible approach allowing changes along the way (Rodrik, 2014). Further, monitoring activities may even have negative implications for the transformative potential of industrial policies, if they are primarily concerned with the industrial outcomes of GIPs. To illustrate, the Draghi Report suggests the importance of monitoring the competitiveness implications of the implementation of the EU's Carbon Border Adjustment Mechanism, and, if negative, postpones the phase-out of free allowances in the EU Emissions Trading System for energy-intensive industries (Draghi, 2024). Overall, we suggest that there is significant room for implementing more reflexive practices in GIPs.

## Conclusions and Reflections

Based on the discussion in this chapter, we conclude that GIP involves two main interrelated dimensions. The first concerns broadening the scope of industrial policies to include governmental aspirations to improve the environmental sustainability of the economy, for example by supporting the diffusion of renewables. The second is related to the ambition of improving capabilities and competitiveness of domestic manufactures of such technologies through industrial policy measures, including demonstration projects and R&D, and in some instances local content requirements.

In our view (see also, e.g., Arcand & Meadowcroft, 2025) industrial policies are a necessary ingredient in policy mixes to transform current unsustainable systems of production and consumption. The task at hand is far too monumental and the time available far too short to rely only on policies correcting market failures. However, when considering whether industrial policy is aligned with Transformative Innovation Policy (TIP) it becomes clear that the two policy approaches are both friends and foes (borrowing the terminology of Alkemade et al., 2011). The latter is reflected most prominently in the industry protectionist measures that are features of GIPs in different countries—both in the Global North and South. These may be detrimental to technological learning and cost reductions that can allow for more rapid diffusion. Nevertheless, the legitimacy of 'green' policies may be strengthened if accompanied by value and job creation (Vona, 2019), which in many instances will require measures misaligned with the free market logic that has prevailed in recent decades. This underscores the challenging balancing acts that policymakers need to achieve in the years to come to ensure a just and inclusive transformation towards sustainability. Indeed, the transformation will create winners and losers, which requires a policy approach that manages to counteract populist resistance and opposition to ambitious climate policy.

Answering the question of whether and how GIP policy has a transformative potential is therefore not straightforward. Much will depend on how GIPs are designed and implemented, as we lack a blueprint guaranteeing success, as industrial policy involves a 'bet' and policy experimentation while being highly context dependent. As argued by Allan et al. (2021, 15) a positive view of GIP emphasizes the opportunity to create positive feedback loops between low-carbon technology manufacturing and other sectors (including, e.g., finance), while a more pessimistic view sees GIP potentially resulting in competitive duplication where countries wanting to support domestic manufactures and stimulate deployment of those are unable to *"replicate cost reductions achieved by competitors"*. In this way, the introduction of fierce GIPs in various countries to achieve dominant positions in the (ibid.) *"green industrial race"* risks producing *"lots of competition but little decarbonization"*.

For GIPs to be effective, therefore, there is a need to ensure policy mix alignment and coordination, including with broader climate and environmental targets and the innovation 'supply side', and to ensure that they are attuned to complex multi-system and multi-sector dynamics across industries and value chains. Achieving legitimacy for ambitious GIPs may also be easier when there is shared understanding of problems and alignment between climate or other environmental targets on multiple governance levels, and when the regional or national innovation system can deliver relevant solutions (Bugge et al., 2022). Considering the actually implemented GIPs, these policies to some extent reflect needs for both clear directionality towards decarbonisation, demand articulation, reflexivity, and coordination between different policy areas, however, attention to these priorities are only pursued when they support domestic industry development. In the absence of such alignment, the 'green' component of GIP becomes downplayed. Thus, industrial policy continues to also support unsustainable solutions, and it may be particularly challenging to maintain the green direction in times of crisis.

**Acknowledgements** The authors acknowledge financial support by the Research Council of Norway through the INTRANSIT centre (grant 295021) and the Ministry of Foreign Affairs of Denmark through the project entitled 'The industrial return on the green transition: Renewable energy in Vietnam (REturn)' (Research Grant no. 23-M05-DTU).

## References

Aiginger, K., & Rodrik, D. (2020). Rebirth of Industrial Policy and an Agenda for the Twenty-First Century. *Journal of Industry, Competition and Trade, 20*(2), 189–207. https://doi.org/10.1007/s10842-019-00322-3

Alkemade, F., Hekkert, M. P., & Negro, S. O. (2011). Transition policy and innovation policy: Friends or foes? *Environmental Innovation and Societal Transitions, 1*(1), 125–129. https://doi.org/10.1016/j.eist.2011.04.009

Allan, B., Lewis, J. I., & Oatley, T. (2021). Green Industrial Policy and the Global Transformation of Climate Politics. *Global Environmental Politics, 21*(4), 1–19. https://doi.org/10.1162/glep_a_00640

Allan, B. B., & Nahm, J. (2025). Strategies of Green Industrial Policy: How States Position Firms in Global Supply Chains. *American Political Science Review, 119*(1), 420–434. https://doi.org/10.1017/S0003055424000364

Altenburg, T., & Rodrik, D. (2017). Green industrial policy: Accelerating structural change towards wealthy green economies. In T. Altenburg & C. Assmann (Eds.), *Green Industrial Policy. Concept, Policies, Country Experiences* (pp. 1–22). UN Environment; German Development Institute/Deutsches Institut für Entwicklungspolitik (DIE).

Andersen, A. D., & Geels, F. W. (2023). Multi-system dynamics and the speed of net-zero transitions: Identifying causal processes related to technologies, actors, and institutions. *Energy Research & Social Science, 102*, 103178. https://doi.org/10.1016/j.erss.2023.103178

Arcand, B., Allan, B., & Eaton, D. (2024). *Assessing the Impact of Propulsion Québec.* The Transition Accelerator. https://transitionaccelerator.ca/wp-content/uploads/2024/09/TA_Assessing_the_Impact_of_PropulsionQuebec_EN_v1.0.pdf

Arcand, B., & Meadowcroft, J. (2025). Net zero by 2050: the case for green industrial policy. *Environmental Politics, 34*(3), 551–571. https://doi.org/10.1080/09644016.2024.2375899

Bailey, D., Glasmeier, A., & Tomlinson, P. R. (2019). Industrial policy back on the agenda: putting industrial policy in its place? *Cambridge Journal of Regions, Economy and Society, 12*(3), 319–326. https://doi.org/10.1093/cjres/rsz018

Beise, M. (2004). Lead markets: country-specific drivers of the global diffusion of innovations. *Research Policy, 33*(6–7), 997–1018. https://doi.org/10.1016/j.respol.2004.03.003

Binz, C., Gosens, J., Hansen, T., & Hansen, U. E. (2017). Toward Technology-Sensitive Catching-Up Policies: Insights from Renewable Energy in China. *World Development, 96*, 418–437. https://doi.org/10.1016/j.worlddev.2017.03.027

Boon, W., & Edler, J. (2018). Demand, challenges, and innovation. Making sense of new trends in innovation policy. *Science and Public Policy, 45*(4), 435–447. https://doi.org/10.1093/scipol/scy014

Bugge, M. M., Andersen, A. D., & Steen, M. (2022). The role of regional innovation systems in mission-oriented innovation policy: exploring the problem-solution space in electrification of maritime transport. *European Planning Studies, 30*(11), 2312–2333. https://doi.org/10.1080/09654313.2021.1988907

CAN Europe. (2023). Fossil Fuel Subsidies in the EU. https://caneurope.org/content/uploads/2023/03/Fossil-Fuels-Subsidies-Report.pdf

Carlsson, B., Eliasson, G., & Sjöö, K. (2018). The Swedish industrial support program of the 1970s revisited. *Journal of Evolutionary Economics, 28*(4), 805–835. https://doi.org/10.1007/s00191-018-0581-5

Cimoli, M., Dosi, G., & Stiglitz, J. E. (Eds.). (2008). *Industrial Policy and Development: The Political Economy of Capabilities Accumulation.* Oxford University Press.

Coenen, L., Hansen, T., & Rekers, J. V. (2015). Innovation Policy for Grand Challenges. *An Economic Geography Perspective. Geography Compass, 9*(9), 483–496. https://doi.org/10.1111/gec3.12231

Criscuolo, C., Díaz, L., Lalanne, G., Guillouet, L., van de Put, C.-E., Weder, C., & Deutsch, H. Z. (2023). *Quantifying industrial strategies across nine OECD countries* (Vol. 150). OECD Publishing. https://doi.org/10.1787/5f2dcc8e-en

David, M. (2017). Moving beyond the heuristic of creative destruction: Targeting exnovation with policy mixes for energy transitions. *Energy Research & Social Science, 33*, 138–146. https://doi.org/10.1016/j.erss.2017.09.023

Diercks, G., Larsen, H., & Steward, F. (2019). Transformative innovation policy: Addressing variety in an emerging policy paradigm. *Research Policy, 48*(4), 880–894. https://doi.org/10.1016/j.respol.2018.10.028

Draghi, M. (2024). The Future of European Competitiveness Part A: A competitiveness strategy for Europe.

EIU. (2023). *The global green subsidy race*. T. E. I. Unit.

European Commission. (2020). Industry and the Green Deal. Retrieved August 16, from https://commission.europa.eu/strategy-and-policy/priorities-2019-2024/european-green-deal/industry-and-green-deal_en

European Commission. (2024). Innovation Fund projects. Retrieved August 16, from https://climate.ec.europa.eu/eu-action/eu-funding-climate-action/innovation-fund/innovation-fund-projects_en

Evans, P. (1998). Transferable lessons? Re-examining the institutional prerequisites of East Asian economic policies. *The Journal of Development Studies, 34*(6), 66–86. https://doi.org/10.1080/00220389808422546

Flegal, J. (2023). Industrial policy + climate policy. In Roosevelt Institute (Ed.), *Industrial Policy Synergies: Reflections from Biden Administration Alumni*. Roosevelt Institute.

Foxon, T., & Pearson, P. (2008). Overcoming barriers to innovation and diffusion of cleaner technologies: some features of a sustainable innovation policy regime. *Journal of Cleaner Production, 16*(1), S148–S161. https://doi.org/10.1016/j.jclepro.2007.10.011

Gereffi, G. (2014). A Global Value Chain Perspective on Industrial Policy and Development in Emerging Markets. *Duke Journal of Comparative and International Law, 24*, 433–458.

Gerschenkron, A. (1962). *Economic backwardness in historical perspective: a book of essays*. Belknap Press of Harvard University Press.

Gomel, D., & Rogge, K. S. (2020). Mere deployment of renewables or industry formation, too? Exploring the role of advocacy communities for the Argentinean energy policy mix. *Environmental Innovation and Societal Transitions, 36*, 345–371. https://doi.org/10.1016/j.eist.2020.02.003

Gong, H., & Hansen, T. (2023). The rise of China's new energy vehicle lithium-ion battery industry: The coevolution of battery technological innovation systems and policies. *Environmental Innovation and Societal Transitions, 46*, 100689. https://doi.org/10.1016/j.eist.2022.100689

Grillitsch, M., Hansen, T., Coenen, L., Miörner, J., & Moodysson, J. (2019). Innovation policy for system wide transformation: The case of Strategic Innovation Programs (SIPs) in Sweden. *Research Policy, 48*(4), 1048–1061. https://doi.org/10.1016/j.respol.2018.10.004

Grillitsch, M., Hansen, T., & Madsen, S. (2021). Transformative Innovation Policy: A novel approach? In B. Godin, D. Vinck, & G. Gaglio (Eds.), *Handbook of Alternative Theories of Innovation (pp. 276–291)*. Edward Elgar. https://doi.org/10.4337/9781789902303.00028

Haddad, C. R., Nakić, V., Bergek, A., & Hellsmark, H. (2022). Transformative innovation policy: A systematic review. *Environmental Innovation and Societal Transitions, 43*, 14–40. https://doi.org/10.1016/j.eist.2022.03.002

Hansen, T., Andersson, J., Finstad, J., Hanson, J., Hellsmark, H., Mäkitie, T., Nordholm, A., & Steen, M. (2024). How aligned are industry strategy and government policy for the decarbonization of energy-intensive process industries? *Climate Policy, 24*(9), 1149–1162. https://doi.org/10.1080/14693062.2024.2363490

Hansen, T., Hanson, J., Mäkitie, T., Normann, H., & Steen, M. (2024). Scalar orientation of policies and technology legitimacy: The case of decarbonization in Norway. *Environmental Innovation and Societal Transitions, 53*, 100919. https://doi.org/10.1016/j.eist.2024.100919

Hansen, U. E., Nygaard, I., Morris, M., & Robbins, G. (2020). The effects of local content requirements in auction schemes for renewable energy in developing countries: A literature review. *Renewable and Sustainable Energy Reviews, 127*, 109843. https://doi.org/10.1016/j.rser.2020.109843

Harrison, A., Martin, L. A., & Nataraj, S. (2017). Green industrial policy in emerging markets. *Annual Review of Resource Economics, 9*(1), 253–274.

Hassink, R., & Kiese, M. (2021). Solving the restructuring problems of (former) old industrial regions with smart specialization? Conceptual thoughts and evidence from the Ruhr. *Review of Regional Research, 41*(2), 131–155. https://doi.org/10.1007/s10037-021-00157-8

Jolly, S., Steen, M., Hansen, T., & Afewerki, S. (2023). Renewable energy and industrial development in pioneering and lagging regions: the offshore wind industry in southern Denmark and Normandy. *Oxford Open Energy, 2*, oiad010. https://doi.org/10.1093/ooenergy/oiad010

Karnøe, P., & Garud, R. (2012). Path Creation: Co-creation of Heterogeneous Resources in the Emergence of the Danish Wind Turbine Cluster. *European Planning Studies, 20*(5), 733–752. https://doi.org/10.1080/09654313.2012.667923

Kivimaa, P., & Kern, F. (2016). Creative destruction or mere niche support? Innovation policy mixes for sustainability transitions. *Research Policy, 45*(1), 205–217. https://doi.org/10.1016/j.respol.2015.09.008

Kuhlmann, S., & Rip, A. (2018). Next-Generation Innovation Policy and Grand Challenges. *Science and Public Policy, 45*(4), 448–454. https://doi.org/10.1093/scipol/scy011

Lundvall, B.-Å., & Borrás, S. (2005). Science, Technology, and Innovation Policy. In J. Fagerberg, D. C. Mowery, & R. R. Nelson (Eds.), *The Oxford Handbook of Innovation* (pp. 599–631). Oxford University Press.

Malhotra, A., & Schmidt, T. S. (2020). Accelerating Low-Carbon Innovation. *Joule, 4*(11), 2259–2267. https://doi.org/10.1016/j.joule.2020.09.004

Markard, J., Raven, R., & Truffer, B. (2012). Sustainability transitions: An emerging field of research and its prospects. *Research Policy, 41*(6), 955–967. https://doi.org/10.1016/j.respol.2012.02.013

Matti, C., Consoli, D., & Uyarra, E. (2017). Multi level policy mixes and industry emergence: The case of wind energy in Spain. *Environment and Planning C: Politics and Space, 35*(4), 661–683. https://doi.org/10.1177/0263774x16663933

Mazzucato, M. (2018). Mission-Oriented Research & Innovation in the European Union.

Mazzucato, M. (2021). *Mission economy: A moonshot guide to changing capitalism*. Penguin UK.

Meckling, J. (2021). Making Industrial Policy Work for Decarbonization. *Global Environmental Politics, 21*(4), 134–147. https://doi.org/10.1162/glep_a_00624

Mickwitz, P., Neij, L., Johansson, M., Benner, M., & Sandin, S. (2021). A theory-based approach to evaluations intended to inform transitions toward sustainability. *Evaluation, 27*(3), 281–306. https://doi.org/10.1177/1356389021997855

Millot, V., & Rawdanowicz, L. (2024). *The return of industrial policies. Policy considerations in the current context*. OECD Publishing. https://doi.org/10.1787/2226583X

Molas-Gallart, J., Boni, A., Giachi, S., & Schot, J. (2021). A formative approach to the evaluation of Transformative Innovation Policies. *Research Evaluation, 30*(4), 431–442. https://doi.org/10.1093/reseval/rvab016

Narassimhan, E., Myslikova, Z., & Gallagher, K. S. (2024). Strategies for green industrial and innovation policy–an analysis of policy alignment, misalignment, and realignment around dominant designs in the EV sector. *Environmental Research Letters, 19*(1), 014029. https://doi.org/10.1088/1748-9326/ad101e

Pack, H., & Saggi, K. (2006). Is There a Case for Industrial Policy? A Critical Survey. *The World Bank Research Observer, 21*(2), 267–297. https://doi.org/10.1093/wbro/lkl001

European Parliament. (2023). EU ban on the sale of new petrol and diesel cars from 2035 explained. Retrieved August 16, from https://www.europarl.europa.eu/topics/en/article/20221019STO44572/eu-ban-on-sale-of-new-petrol-and-diesel-cars-from-2035-explained

Quitzow, R., Walz, R., Köhler, J., & Rennings, K. (2014). The concept of "lead markets" revisited: Contribution to environmental innovation theory. *Environmental Innovation and Societal Transitions, 10*, 4–19. https://doi.org/10.1016/j.eist.2013.11.002

Rinscheid, A., Rosenbloom, D., Markard, J., & Turnheim, B. (2021). From terminating to transforming: The role of phase-out in sustainability transitions. *Environmental Innovation and Societal Transitions, 41*, 27–31. https://doi.org/10.1016/j.eist.2021.10.019

Rodrik, D. (2009). Industrial Policy: Don't Ask Why, Ask How. *Middle East Development Journal, 1*(1), 1–29. https://doi.org/10.1142/S1793812009000024

Rodrik, D. (2014). Green industrial policy. *Oxford Review of Economic Policy, 30*(3), 469–491. https://doi.org/10.1093/oxrep/gru025

Rogge, K. S., & Johnstone, P. (2017). Exploring the role of phase-out policies for low-carbon energy transitions: The case of the German Energiewende. *Energy Research & Social Science, 33*, 128–137. https://doi.org/10.1016/j.erss.2017.10.004

Rystad Energy. (2023). Double relief: Norwegian tax break will see gas and oil supply to Europe increase. Retrieved August 16, from https://www.rystadenergy.com/news/double-relief-norwegian-tax-break-will-see-gas-and-oil-supply-to-europe-increase

Sæther, B., Isaksen, A., & Karlsen, A. (2011). Innovation by co-evolution in natural resource industries: The Norwegian experience. *Geoforum, 42*(3), 373–381. https://doi.org/10.1016/j.geoforum.2011.01.008

Schebesta, H., & Candel, J. J. L. (2020). Game-changing potential of the EU's Farm to Fork Strategy. *Nature Food, 1*(10), 586–588. https://doi.org/10.1038/s43016-020-00166-9

Schmitz, H., Johnson, O., & Altenburg, T. (2015). Rent Management–The Heart of Green Industrial Policy. New Political Economy, 20(6), 812–831. doi:https://doi.org/10.1080/13563467.2015.1079170

Schot, J., & Steinmueller, W. E. (2018). Three frames for innovation policy: R &D, systems of innovation and transformative change. Research Policy, 47(9), 1554–1567. doi:https://doi.org/10.1016/j.respol.2018.08.011

Scordato, L., Bugge, M. M., Hansen, T., Tanner, A., & Wicken, O. (2022). Walking the talk? Innovation policy approaches to unleash the transformative potentials of the Nordic bioeconomy. *Science and Public Policy, 49*(2), 324–346. https://doi.org/10.1093/scipol/scab083

Sengers, F., Wieczorek, A. J., & Raven, R. (2019). Experimenting for sustainability transitions: A systematic literature review. *Technological Forecasting and Social Change, 145*, 153–164. https://doi.org/10.1016/j.techfore.2016.08.031

Skjærseth, J. B., Hansen, T., Donner-Amnell, J., Hanson, J., Inderberg, T. H. J., Nielsen, H. Ø., Nygaard, B., & Steen, M. (2023). *Wind Power Policies and Diffusion in the Nordic Countries*. Comparative Patterns. Palgrave Macmillan.

Steen, M., Faller, F., & Fyhn Ullern, E. (2019). Fostering renewable energy with smart specialisation? Insights into European innovation policy. *Norsk Geografisk Tidsskrift-Norwegian Journal of Geography, 73*(1), 39–52. https://doi.org/10.1080/00291951.2018.1554696

Steen, M., Mäkitie, T., Hanson, J., & Normann, H. E. (2024). Developing the industrial capacity for energy transitions: Resource formation for offshore wind in Europe. *Environmental Innovation and Societal Transitions, 53*, 100925. https://doi.org/10.1016/j.eist.2024.100925

Streeck, W., & Thelen, K. (2005). *Beyond continuity: Institutional change in advanced political economies*. Oxford University Press.

Truffer, B., Rohracher, H., Kivimaa, P., Raven, R., Alkemade, F., Carvalho, L., & Feola, G. (2022). A perspective on the future of sustainability transitions research. *Environmental Innovation and Societal Transitions, 42*, 331–339. https://doi.org/10.1016/j.eist.2022.01.006

Turnheim, B., & Geels, F. W. (2012). Regime destabilisation as the flipside of energy transitions: Lessons from the history of the British coal industry (1913–1997). *Energy Policy, 50*, 35–49. https://doi.org/10.1016/j.enpol.2012.04.060

Uyarra, E., Barbara, R., & Dale-Clough, L. (2019). Exploring the normative turn in regional innovation policy: responsibility and the quest for public value. *European Planning Studies, 27*(12), 2359–2375. https://doi.org/10.1080/09654313.2019.1609425

Uyarra, E., Zabala-Iturriagagoitia, J. M., Flanagan, K., & Magro, E. (2020). Public procurement, innovation and industrial policy: Rationales, roles, capabilities and implementation. *Research Policy, 49*(1), 103844. https://doi.org/10.1016/j.respol.2019.103844

van der Loos, A., Frenken, K., Hekkert, M., & Negro, S. (2024). On the resilience of innovation systems. *Industry and Innovation, 31*(1), 42–74. https://doi.org/10.1080/13662716.2023.2269110

Vona, F. (2019). Job losses and political acceptability of climate policies: why the 'job-killing' argument is so persistent and how to overturn it. *Climate Policy, 19*(4), 524–532. https://doi.org/10.1080/14693062.2018.1532871

Warwick, K., & Nolan, A. (2014). *Evaluation of industrial policy: methodological issues and policy lessons.* OECD Publishing. https://doi.org/10.1787/5jz181jh0j5k-en

Weber, K. M., & Rohracher, H. (2012). Legitimizing research, technology and innovation policies for transformative change: Combining insights from innovation systems and multi-level perspective in a comprehensive 'failures' framework. *Research Policy, 41*(6), 1037–1047. https://doi.org/10.1016/j.respol.2011.10.015

# 16

# Navigating Twin Transitions: A Policy Framework to Align Digital Innovation with Sustainability, Applied to the Dutch Processing Industry

Nikhil John, Joeri Wesseling, and Koen Frenken

## Introduction

The twin transition concept broadly aims to combine green and digital transitions. The European Commission introduced this idea around 2019, highlighting digital technologies as "key" enablers for achieving sustainability goals in its Green Deal (European Commission, 2019). Subsequent policies, like the *2020 Industrial Strategy*, however lacked focus on interlinkages between sustainability and digital transitions. To

**Supplementary Information** The online version contains supplementary material available at https://doi.org/10.1007/978-981-95-4893-4_16.

N. John · J. Wesseling · K. Frenken
Section of Innovation Studies, Copernicus Institute of Sustainable Development, Faculty of Geosciences, Utrecht University, Utrecht, The Netherlands

N. John (✉)
Process Systems Engineering, Department of Chemical Engineering, Faculty of Applied Sciences, Delft University of Technology, Delft, Netherlands
e-mail: N.John@tudelft.nl



M. Steen et al. (eds.), *Sustainability Transitions and Industrial Transformation*,
https://doi.org/10.1007/978-981-95-4893-4_16

help realize twin transitions, more effective policy is needed not only at the European, but also national and subnational levels (European Commission, 2020; Wurm & Wittmann, 2023).

To inform twin transition policy, sustainability scholars have critically examined the interplay between digitalization and sustainability, including potential negative environmental effects (Berkhout & Hertin, 2004; Erlinghagen & Markard, 2012; Hankel, 2014). Transition scholars have recently highlighted additional relevant aspects of the twin transition such as the broader socio-technical system dynamics (John et al., 2022; Mäkitie et al., 2023; Meijer, 2024; Mouthaan et al., 2023), the misalignment between digital and low-carbon trajectories (Fouquet & Hippe, 2022; John et al., 2022), unintended consequences (Bohnsack et al., 2022), the reinforcement of problematic institutional structures (Bennich, 2024; John et al., 2022; Klerkx et al., 2019), and techno-optimism (Kovacic et al., 2024; Mouthaan et al., 2023). Such critical perspectives and deeper diagnoses of digital and sustainability dynamics need to be fed back into policymaking processes.

We take a closer look at the policymaking challenges related to twin transitions by focusing on the energy-intensive processing industries (EPIs) in the Netherlands. This sector is particularly relevant due to its high carbon emissions and significant potential for decarbonization (Bataille et al., 2018; Wesseling et al., 2017). Advanced digital innovations are emerging in EPIs (John et al., 2022; Ometto et al., 2019; Siemons et al., 2020) and although many digital-related policymakers don't yet focus on transformative sustainability policy, policymakers and other stakeholders increasingly aim to achieve twin transition goals in this sector (Aazami, 2021; European Commission, 2020). To this end, more research is needed on how to align digital innovation policy with the decarbonization transition. Therefore, our research question is: "How can specific policy goals, intervention logics and instruments be formulated towards sustainable twin transitions in the processing industry?"

This study is the first (exploratory) attempt to construct policy goals from the ground up for a "whole" twin transition context. This is a challenging task because single transition goals are difficult to manage, let alone synergizing two transitions. In our case, the deep decarbonization industry challenge is compounded by the need to navigate the complex dynamics of digital innovation. Therefore, we construct a framework to answer the research question using a pragmatic philosophy. Pragmatic transitions research emphasizes the idea that the value of knowledge lies in its usefulness and applicability to real-world situations, particularly in informing policy (Zolfagharian et al., 2019). We attempt to knowledge broker between policymakers and sustainability science (Wittmayer & Schäpke, 2014) to provide pragmatic direction on twin transitions. This also involves balancing acknowledging the potential negative effects of digitalization for sustainability, as pointed out by various scholars, while recognizing evidence within the nature of digital innovation as a "technical" solution for societal challenges. This approach is inherently value-laden; however, if done reflexively, it can help explain the dynamics of twin transitions better while also contributing to guiding them towards sustainability.

We build on various innovation and transition perspectives along with recent case study research in digital innovation and EPIs, synthesizing their key insights to inform a twin transition policy framework. The framework is organized in two generalizable policy approaches: the technological approach and the socio-institutional approach. These are not mutually exclusive, but rather complementary, as they highlight different yet crucial aspects of navigating twin transitions. Through these two approaches, we subsequently identify four policy intervention logics and underlying instruments.

The rest of the chapter is structured as follows. The next section outlines the theoretical basis for both technological and socio-institutional approaches, based on which we present a twin transition policy framework. We then apply this framework to an EPI case study, describing the methods and demonstrating how to derive subsequent policy insights. We conclude with limitations and additional discussions.

## Twin Transitions: How Do we Conceptualize Them and their Policy Focus?

### Framing Twin Transitions

This section discusses how twin transitions have been conceptualized, focusing on analytical depth and relevance for policymaking. We broadly see digitalization as the integration of digital technologies into economic and social activities, involving the adoption of artificial intelligence (AI), data-driven systems, automation, internet of things, etc. Such adoption has the potential to transform how industries, organizations, and societies operate, enabling new production methods, organizational models, and value chains. While transition scholars using the Multi-Level Perspective have positioned digitalization across all of its three conceptual levels: landscape, regime, and niche (John et al., 2022), we frame digitalization as a "meta-regime" (Schot & Kanger, 2018). The digital meta-regime, in addition to fostering new digital technologies, has been empirically shown to create and modify socio-institutional structures within the industries it interacts with (Erlinghagen & Markard, 2012; John et al., 2022; Klerkx et al., 2019; Meijer, 2024).

We conceptualize "Digital Innovation Systems" (DIS)—which capture digital technologies and their surrounding socio-technical system—as a manifestation of the digital meta-regime within specific industrial contexts (Andersen et al., 2024; John et al. 2022, 2024; Schot & Kanger, 2018). Hence, multiple DIS can be nested within the digital meta-regime, corresponding to the different sectors impacted by digitalization; see Fig. 16.1.

### Considerations for Twin Transition Policy

Transitions are defined as fundamental shifts in the structures of socio-technical systems. Transitions literature recognizes that incremental innovations help reinforce the established regime without coping with the root sustainability challenges, and that radical innovations have more potential to create structural change (Geels & Schot, 2007; Kemp,

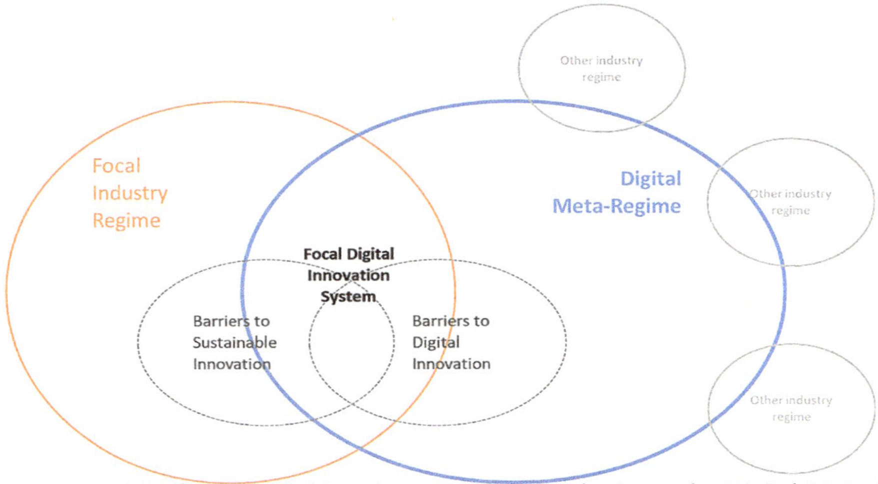


**Fig. 16.1** A multi-system model of twin transition dynamics with emphasis on the Digital Meta-Regime and a focal Digital Innovation System

1994). Mäkitie et al. (2023) conceptualize interactions between digital and sustainable technologies using the notions of incremental and radical innovation, categorizing twin innovations into four types: incremental twin innovation, digitally supported sustainable innovation, sustainability-supported digital innovation, and radical twin innovation. While this typology covers the ideal types of innovation categories, the pattern of twin technologies that emerge in industries may vary. For example, previous research on hard-to-abate sectors has shown that incremental digital innovation in fossil-based application areas threatens to increase lock-in and reinforce path dependencies, leading to a misaligned twin transition (John et al., 2022). On the other hand, scholars have also shown that digital innovation aligns more closely with circular economy goals in manufacturing than with the energy system transition (Wurm & Wittmann, 2023). Overall, incremental innovations with minimal sustainability benefits often entrench existing systems, undermining sustainable twin transitions.

The critical issue from this twin transition discourse is the contrasting rationales underpinning digital and sustainable innovations. Digital innovation is driven by objectives such as increasing efficiency, reducing costs, and maximizing profits. By contrast, sustainable innovation prioritizes reducing environmental impact and fostering societal well-being, often with less immediate emphasis on economic growth. These divergent logics can create tensions when integrating digital technologies into sustainability transitions. Therefore, for a coherent twin transition, digital innovation must be nested within the overarching aims of sustainability transitions. This argument draws parallels with the broader debate on innovation policy and transition policy (Foxon & Pearson, 2008), where Alkemade et al. (2011) call for a systemic approach to policymaking, in which "[sustainability] transition policy is considered an overarching policy goal that helps to align all other policies". Twin transition research frequently overlooks negative dynamics when it lacks such systemic sustainability considerations (Mouthaan et al., 2023).

Building on this foundation, the central focus of a twin transition policy framework should be to minimize misalignment by embedding digital technology goals within known sustainability objectives, while also addressing the broader systemic conditions of digital innovation systems. This means that policymakers must not only ensure that digital innovation contributes to, rather than diverges from, sustainability, but also attend to the socio-technical dynamics that shape alignment or misalignment in specific industry contexts.

## Characteristics of Digital Innovation Systems

Unpacking the nature and dynamics of Digital Innovation Systems (DIS) helps identify leverage points for policymakers to target their efforts. The analysis of Technological Innovation Systems (TIS) has been valuable in policymaking, helping identify intervention strategies within complex systems (Jacobsson & Bergek, 2011). The structural-function TIS approach helped trace barriers that hinder sustainable innovation activity (Negro et al., 2012). Additionally, the system-building of technological fields has also been studied using TIS, emphasizing that the drivers of TIS are not just emergent but often deliberately steered (Musiolik et al., 2012). Recently, the TIS has been used to study DIS (John et al. 2022, 2024; Yu et al., 2021).

The DIS shapes the development, diffusion, and use of all variants of digital innovation within "user" industry contexts, whether incremental or radical (see Fig. 16.1). Therefore, growing and steering the directionality of this innovation system is crucial for twin transition synergies. By definition, DIS includes twin innovations and non-sustainability-related digital innovations. The boundary-drawing discussion in innovation systems literature indicates that system-context boundaries are mostly drawn based on the study's analytical focus (Bergek et al., 2015; Markard et al., 2015). Applying this logic to twin transitions, boundaries may be drawn around the sustainability-focused DIS. However, it is impossible to differentiate between sustainable and non-sustainability-focused digital innovations due to the nature of pervasive innovations (John et al., 2022). Policies must address this dual nature, balancing risks and opportunities, and direct digital innovation towards sustainable goals.

DIS development within an industry often faces barriers that reflect the institutional misalignment between the logics of the DIS and those of the "user sector" (John et al., 2024; Wesseling et al., 2022). Due to its multi-system, pervasive, and general-purpose nature (Bekar et al., 2018; John et al., 2024; 2022a; Waßenhoven et al., 2023), DIS presents unique system-building characteristics not seen in the more typical (narrow) TIS cases that dominate empirical literature (John et al., 2022). Consequently, a well-functioning DIS must consider the extensive technological applications, the dynamic innovation activities, and the numerous participants in the innovation system.

## Towards a Twin Transition Policy Framework

Developing a coherent twin transition policy framework requires addressing both the technological and socio-institutional dimensions of digital and sustainable innovation. This section systematically develops four policy intervention *logics* that describe the dynamics by which policy interventions in the focal complex system could help meet twin transition policy goals of fostering digital innovation that contributes to deep decarbonization.

### Policy Intervention Logic 1: Reorienting Digital Technologies Towards Sustainable Pathways

Sustainability transitions typically involve the replacement of existing unsustainable technologies or systems with niches centred around sustainable innovation (Kemp et al., 1998). Transitions may involve several sustainable innovations that collectively or individually form technological trajectories or pathways (Geels & Schot, 2007).

As digital technologies have wide applicability through their general-purpose characteristics, they could provide second-order benefits in all kinds of application areas. Subsequently, in a twin transition context, one role of digital technologies could be to provide incremental improvement to sustainable innovation; however, digital technologies may also provide innovation possibilities for existing *un*sustainable technologies and systems (John et al., 2022; Mäkitie et al., 2023). Therefore, policy instruments need to be developed to deploy digital technologies towards sustainable pathways, while also steering digital technologies away from unsustainable application areas.

## Policy Intervention Logic 2: Deploying Digital Technologies to Overcome Barriers to Sustainable Innovation

Despite their transformative potential, sustainable innovations face significant, complex, and interrelated barriers that are technological, organizational, institutional, political, or socio-cultural in nature (Arthur, 1989; Foxon & Pearson, 2008; Geels & Schot, 2007). Overcoming such barriers to sustainable innovation is an effective way to intervene in transition dynamics (Negro et al., 2012).

Digital technologies could play a role in overcoming such barriers, as they may enable system innovation (Mäkitie et al., 2023) by providing novel linkages necessary for socio-technical change (Geels, 2005). However, it is necessary to explore the extent to which the changes in socio-technical systems brought about by digital technologies complement the changes required for advancing sustainable innovations (John et al., 2022). Subsequently, digital technologies need to be steered towards the complex barriers that hinder sustainable innovation.

It is important to clarify that barriers to sustainable innovation discussed in Intervention Logic 2 differ from the sustainable innovation pathways highlighted in Intervention Logic 1. For example, in the energy system transition, digital technologies that increase the efficiency of solar panels are different from grid innovations that change the structure of the industry towards flexible grid operations. The former involves digitally supported sustainable innovation, while the latter involves creating previously absent socio-technical linkages. What is common in the approach between Intervention Logic 1 and 2 is the technical nature by which digital technologies can be leveraged for sustainability transitions.

## Policy Intervention Logic 3: Overcoming Barriers Common to Both Digital and Sustainable Innovation

While a technological approach may lead digital technologies to form a direct part of the "solution" to sustainable innovation barriers (Intervention Logic 2), these barriers are inherently socio-institutional in nature and require a broader policy approach. Barriers to innovation are typically characterized as systemic problems within their overarching innovation systems (Klein Woolthuis et al., 2005; Wieczorek & Hekkert, 2012). Further, beyond the barriers unique to sustainable innovation, digital innovations also face barriers in their development and diffusion in user industries (John et al., 2024). These barriers often intersect with those hindering sustainable innovation. These overlapping barriers indicate overarching problems that hinder both digital and sustainable innovation rooted within the socio-institutional characteristics of the industry in question. For example, regulatory inertia or lack of skilled labour might equally hinder advancements in renewable energy systems and digital platforms. Policy interventions to overcome common barriers will have separate and synergetic effects on digital and sustainable innovation.

However, we argued previously that sustainable twin transition policies must prioritize overcoming barriers to sustainable innovation (see Section 2). Therefore, targeting digital innovation barriers that are linked to sustainable innovation barriers is essential, as this ensures that twin transition policy does not merely increase generic digital innovation activity but steers towards socio-institutional change that helps support sustainable innovation within the industry.

## Policy Intervention Logic 4: Creating Directionality Within DIS to Navigate Risks and Opportunities in Sustainable Innovation

In contrast to addressing systemic problems, Musiolik et al. (2020) have shown that strategic system building can be an additional intervention approach for policymakers. System-building policies intending to influence directionality can help steer away from risks towards opportunities within the innovation system.

Currently, DIS are typically still in their growth phase in most user industries undergoing sustainability transitions (John et al., 2022). Twin transition policy should leverage this potential to shape the socio-institutional system elements towards sustainable innovation. Notably, if left unchecked, DIS could (continue to) develop in a way that hampers sustainable innovation. We propose that the barriers to sustainable innovation can again be used as a measuring device to assess such opportunities and risks. Analysing system-building activities in DIS can help identify dynamics that might either contribute to overcoming or perpetuating sustainable innovation barriers. Subsequently, these dynamics can serve as leverage points for addressing barriers, avoiding risks, and embedding sustainability within DIS.

Overall, the resulting four policy intervention logics can be grouped into two overarching approaches. Policy Intervention Logics 1 and 2 constitute the *technological approach, which emphasizes leveraging digital technologies* to directly contribute to sustainability through sustainable applications (logic 1) or to indirectly contribute to sustainability transitions by overcoming barriers to sustainable innovation (logic 2). Policy Intervention Logics 3 and 4 comprise the *socio-institutional approach, which focuses on supporting the systemic conditions required to sustain twin transition synergies by overcoming shared barriers and fostering strategic directionality*. Together, these two complementary approaches provide a framework for developing industry-specific policy goals, intervention logics, and instruments, ensuring that the twin transition prioritizes sustainable outcomes while addressing the complexities of DIS. The next section outlines the methods used to apply this framework in the EPI case and illustrates the corresponding results.

## Constructing Twin Transition Policy Intervention Logics and Instruments for the Dutch Processing Industry

### Methods for Framework Application

This section details the methods used in the technological and socio-institutional approach for developing twin transition policies in the processing industry, illustrated in Fig. 16.2. The analysis builds on prerequisite questions that guide the identification of barriers, dynamics, and solutions, ensuring that the resulting policies are grounded in case-specific insights.

The technological approach and its subsequent policy intervention logics and instruments required an understanding of decarbonization pathways and the persistent barriers preventing the transition. To establish this foundation, we follow steps previously reported (John et al., 2022), drawing on existing deep decarbonization literature within the processing industry to list the various sustainable innovations involved with the transition. Next, we reviewed research on sustainability transitions that focused on the processing industry. This review helped us identify a list of five socio-technical barriers to sustainable innovation. These barriers also form the basis for other steps in the analysis, therefore we elaborate them in the next subsection.

To understand the technological impact of digital technologies on decarbonization pathways and the barriers to sustainable innovation, we build on previously reported research assessing the impact of artificial intelligence on the global steel industry (John et al., 2022). It is typical for the EPI literature to group processing industries together in case studies or policy discussions, given their common characteristics (Wesseling et al., 2017). Further, the research was based on interviews and an inventory analysis of AI tools. The inventory analysis is based on a systemic review of scientific publications, and involved mapping existing AI applications across the steelmaking value chain to assess their interaction with industry innovation activity. While AI in steel served

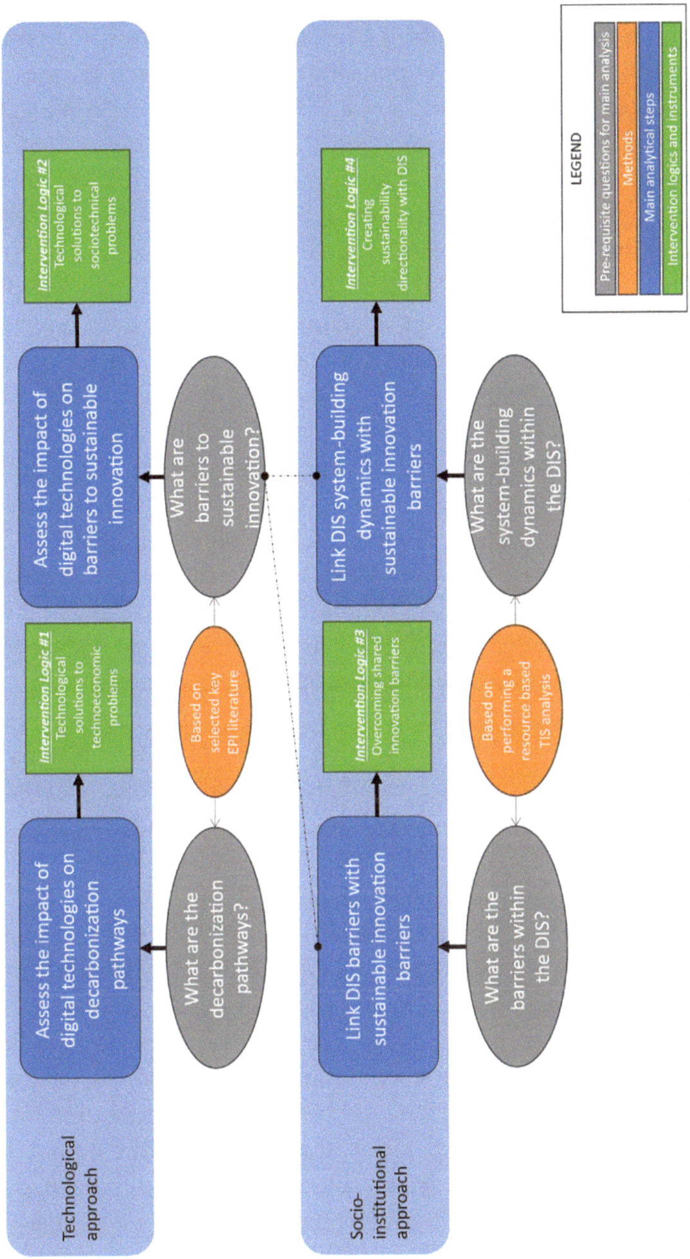


**Fig. 16.2** Framework application—Steps involved in the case study analysis

as a specific example, to allow for some generalizability, we incorporated observations made at various industry conferences at the regional level, where similar trends and applications of digital technologies were discussed across EPIs. Additionally, we used the interviews conducted as part of the DIS analysis to validate that findings from AI were applicable to digital innovation more broadly. Furthermore, the level of insights derived from the impact analysis, as will be demonstrated in the results, was sufficiently broad to extend to digital technologies as a whole.

Overall, this analysis enabled us to provide a generalized interpretation of the current impact of digital technologies on the decarbonization transition. Based on these insights, we constructed policies using Intervention Logics 1 and 2 on how to best deploy digital technologies for a twin transition. The socio-institutional approach first required assessing the DIS in the Dutch processing industry. We performed a structural-functional analysis and arrived at a comprehensive list of 27 barriers and 31 system-building dynamics within the DIS (John et al., 2024, 2022c).

To identify intervention points within DIS for the twin transition, we performed an analysis to establish the socio-institutional links between digital and sustainable innovation. This linking analysis was grounded in the resource-based characterization of sustainable innovation barriers. By examining which innovation resources (e.g., financial, knowledge-based, or institutional) were involved in each barrier and how TIS dynamics could influence these resources, we identified connections between the DIS and sustainable innovation challenges. The methods follow typical framework-building and indexing steps in qualitative policy research (Ritchie & Spencer, 2002). The five barriers to sustainable innovation—described in the technological approach—served as a priori themes for the thematic framework in the link analysis. Indexing was done for the digital barriers list, tagging each barrier according to the relevant sustainable innovation barrier link. Similar indexing was done for the list of system-building dynamics; however, the link was additionally characterized into either risk, opportunities, or both. The second author carried out a reliability test with deliberation sessions within the author team to mitigate interpretive differences. This resulted in a filtered list of TIS barriers and system-building dynamics that could be leveraged for policies using Intervention Logics 3 and 4.

## Background on Barriers to Sustainable Innovation in the Processing Industry

Sustainability transitions research has helped unpack the complex and interrelated barriers to sustainable innovation in the processing industry, addressing dimensions like the role of industry characteristics (Wesseling et al., 2017, 2022), infrastructure, lock-in (Janipour et al., 2020; Wesseling & Van Der Vooren, 2017), incumbents, markets, (Hansen & Coenen, 2017), networks (Bauer et al., 2018), materials, and policy (Bataille et al., 2018; Nilsson et al., 2021). We summarize these dimensions into five barriers shown in Table 16.1.

## The Technological Approach to Twin Transition Policy in the Processing Industry

### Effective Deployment of Digital Technologies for Decarbonization

Before developing interventions to deploy digital technologies towards sustainable pathways, the notion of sustainable and unsustainable pathways needs to be clarified within the context of the processing industry. Business-as-usual, carbon-intensive methods are currently prevalent in the processing industry. These technologies primarily rely on the use of fossil fuels such as natural gas, coal, and oil, leading to significant energy use and carbon emissions. The processing industry has several pathways to achieve decarbonization, which Bataille et al. (2018) categorized into three main options: dematerialization and recycling, retaining and modifying existing processes, and adopting radically new sustainable technologies.

We rely on two key assumptions to categorize sustainable innovation. First, considering the deep decarbonization goal of processing industries, a transition pathway around incremental improvements, like dematerialization and recycling, that provide efficiency improvements of fossil-based technologies, increases inertia and hinders the implementation of more radical sustainable innovation (Janipour et al., 2020; Wesseling

**Table 16.1** Barriers to sustainable innovation in processing industries

| Barrier to sustainable innovation in the process industry | Description |
|---|---|
| Lock-in and path dependency | There is a focus on incremental innovation due to decades of cumulative development. Compounding this barrier further is the unclear pathways due to misalignment between low-carbon innovations and the persistent problems related to phase-out. |
| System integration and flexibility challenges | It is a challenge to integrate new technologies because industries are highly routinized, integrated, and clustered, leading to high replacement costs and potential stranded assets. |
| Lack of demand and markets for sustainable products | Restrictive contracts and a lack of demand from downstream industries create market-related barriers for sustainable innovation. |
| Restrictive culture of risk averseness and safety | There is limited opportunity of testing and upscaling due to risk averse culture and high safety routines and standards in the industry. |
| Vested interests and lack of governance capacity | There is lack of institutional capacity and governance structures for the transition, with a risk of uneven actor power in directionality creation. |

et al., 2017). Second, from an innovation perspective (Wesseling et al., 2017), the option of "modifying existing processes" on many occasions will require adopting carbon capture or alternative novel energy sources, both of which are new technologies. Therefore, apart from dematerialization and recycling, decarbonization requires the development and diffusion of several new technologies and feedstock (ex)changes, including carbon capture, biomass, hydrogen, electrification, and industrial symbiosis. These innovative solutions fall under the category of sustainable innovation for the processing industry.

Turning to the twin transition, it is essential to first assess how digital technologies are currently being used within the processing industry. Our analysis of the applications of digital technologies shows that through prediction, traceability, automation, or optimization, these technologies add value to almost every major application area, ranging from process technologies to organizational functions. However, the percentage share of technologies found in each application area reveals disproportionately more innovation occurring for existing technologies rather than for advancing sustainable innovation. This indicates that the current digital technologies in the processing industry are significantly misaligned with decarbonization pathways. Fig. 16.3 represents a stylized summary of the problem.

Policy interventions to foster sustainable twin transitions must counter the identified technology trend to help break out from fossil fuel technology pathways and support applications that enable the decarbonization transition. We identify two examples of broad policy solutions for the technological twin transition policy approach involving sustainable pathways (Policy Intervention Logic 1). The first involves providing higher incentives for digital technologies that are applied to sustainable innovation areas. Industrial actors, including suppliers of digital technologies, end-users (e.g., steel and chemical manufacturers), and industry consortia projects can actively favour and procure digital innovation projects which are targeted towards hydrogen, bio, or carbon capture. The second helps complement the first through a fossil-fuel-disincentivizing strategy. This means that policies can aim to disincentivize digital technologies that are applied to fossil-based areas—or at least withdraw policy support for such applications—as these applications may delay the decarbonization transition. Given the high scale and potential business gains from efficiency projects of fossil-based technology, they may typically not even require any additional policy support to be implemented in the first place. Such an approach can allow for more resources to be directed towards projects with expected long-term sustainability benefits.

The intended result of these technology-specific twin transition interventions would be an altered technology trend, i.e., a shifting of the

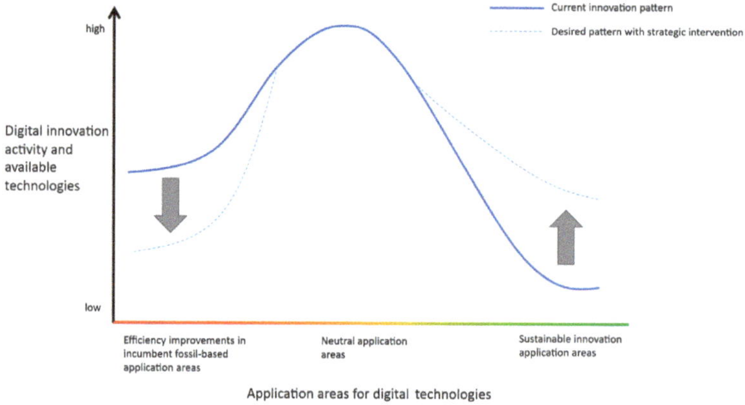


**Fig. 16.3** The pattern of digital innovation (stylized), based on John et al. (2022)

distribution as indicated by grey arrows and dotted new curve in Fig. 16.3.

It is important to nuance and clarify these interventions since it might not be directly evident whether digital projects and technologies fall into categories of "fossil-based" or "sustainable". To illustrate, if the digital application directly adds emissions, policymakers should actively disincentivize it. However, some industry actors may see efficiency gains for ongoing fossil-based technologies (such as improvements to the coal-based blast furnace in iron and steelmaking), an important "bridging" solution for the near term. Policymakers should critically consider (or commission expert assessments of) digital technologies' emission-reducing potential, while properly weighing in potential lock-in effects like sunk costs. Commissioning technology assessments and life cycle assessments, or requiring such studies as part of larger project funding applications, would help policymakers to reveal these balancing criteria. At the level of public policy, more interaction between the decarbonization and digitalization policymakers could provide the necessary capabilities for policy instruments towards such assessments.

Finally, the current innovation pattern notably shows that the largest share of digital technologies is typically in a grey zone, which is a "decarbonization transition neutral" area between fossil and low-carbon technologies. These include site operations and management activities, common technologies for all pathways like motor-driven systems, or safety and compliance areas. In such projects (as well as in unavoidable fossil-based projects), actors can focus on improving innovation system resources that increase the performance of the DIS within the industry. This is discussed further in other policy intervention logics of the policy framework which address the grey area more directly.

## Digital Technologies for Socio-Technical Barriers

Barriers to sustainable innovation are currently still present in the industry. In the inventory study, no specific examples of digital technologies emerged that currently solve those barriers. However, industrial actors have pointed to the potential of digital technologies for solving such barriers either in an emergent process over time through the proliferation of many digital technologies or through the development of a few specific digital technologies, like digital twins, that can directly solve some of these barriers. Further, the research shows that although digital technologies may yield energy efficiency benefits, they do not have a positive impact (overcoming effect) on *all* barriers to sustainable innovation; rather they may create an increased risk for perpetuating certain barriers to sustainable innovation in the processing industry:

1. Lock-in and path dependency: The focus on efficiency and scale, along with limited emphasis on deep decarbonization, means digital tools often support existing technologies as previously shown. However, tools like machine learning could enhance efficiency and reduce CO2-emissions cost-effectively therefore not adding sunk costs.
2. System integration and flexibility challenges: Digital technologies can provide advanced scenario testing and real-time data analysis, which might help industries simulate operations, identify disruptions, and balance flexibility with integration.
3. Lack of demand and markets for sustainable products: Digital technologies could facilitate data sharing and accessibility, potentially

improving articulation of sustainable market demand, enabling industries to adjust production and innovation strategies. For example, it may increase traceability of products' material sources and emissions, enabling associated premiums or standards for more sustainable products.

4. Restrictive culture of risk averseness and safety: Digital technologies for simulation and scenario analysis can help manage risks by allowing new processes to be tested virtually before implementation. They could also support new business models, like service-oriented strategies, that improve return on investment and might reduce hesitancy towards innovation.
5. Vested interests and lack of governance capacity: There is risk that large actors could retain power by investing in digital innovation. However, increased transparency from digital technologies might democratize data and clarify vested interests, aiding inclusive governance.

## The Socio-Institutional Approach to Twin Transition Policy in the Processing Industry

### Overcoming Barriers Common to both Digital and Sustainable Innovation

The link analysis between DIS and sustainable innovation barriers helps prioritize those barriers in the DIS that are most relevant to overcome for the twin transition. Of the 27 systemic problems in the DIS, 15 were excluded after the link analysis due to their specificity to the digital context. For instance, legitimacy creation and access to key decision-makers are challenges unique to digital innovation scaling and organizational mismatches within the industry. Resource mobilization for context-specific tools and the overvaluation of data are also limited to the DIS and do not impact sustainable innovation. The remaining 12 DIS barriers are distributed over the "sustainable innovation barriers links" as follows: Lock-in and path dependency (3), System integration and flexibility challenges (3), Lack of demand and markets for sustainable products (0), Restrictive culture of risk averseness and safety (2), Vested interests and lack of governance capacity (4). Examples of the barrier linking analysis and its outcomes are presented in Table 16.2. The full linking analysis, barriers, policy goals, and example solutions of Policy Intervention Logic 3 are provided in supplementary materials.

### Strategic DIS-Building for Sustainable Innovation

The link analysis between DIS system-building dynamics and sustainable innovation barriers helps to identify risks and opportunities towards sustainable twin transition synergies. Like the previous barrier link analysis, of the 31 system-building dynamics 8 were excluded which related to innovation resources highly specific to the DIS. The remaining 23 dynamics had various nuanced risks and opportunities associated with sustainable innovation. Due to the dual-directional nature of the link, there are many examples of more than one sustainable innovation barrier being affected by a single DIS dynamic. The distribution is as follows: Lock-in and path dependency (9), System integration and flexibility challenges (5), Lack of demand and markets for sustainable products (2), Restrictive culture of risk averseness and safety (8), Vested interests and lack of governance capacity (6). Examples of the system-building dynamics linking analysis and outcomes are presented in Table 16.3. The full results with subsequent policy goals and example solutions of Policy Intervention Logic 4 are provided in the supplementary materials.

## Concluding Remarks

### Reflections on the Technological and Socio-Institutional Approach

In answer to our question "How can specific policy goals, intervention logics and policy instruments be formulated towards sustainable twin transitions in the processing industry?", we found that twin transition policy goals require nuance. On the one hand, our technological

**Table 16.2** Illustration of barrier linking analysis

| DIS barrier | Sustainable innovation barrier link | Link description | DIS policy goal | Example policy instruments or solutions |
|---|---|---|---|---|
| Lack of supplier scalability and integration | Restrictive culture of risk averseness and safety; system integration and flexibility challenges | Scaling of innovations in the process industry is hard due to several factors including site complexity. | Improve scalability and integration support to overcome technical scalability challenges. | Support for scalable solutions, integration frameworks, and institutional assistance for small suppliers |
| Lack of applicable data or data quality | Lock-in and path dependency | Path dependence has meant that applicable data for low-carbon technologies is missing too. | Improve data standards and interoperability to enhance data consistency and integration. | Industry-wide data standardization initiatives and interoperability frameworks |
| Resources for individual tools not translating into combining of multiple tools | System integration and flexibility challenges | System integration and flexibility not yet present in DIS in the same way that multiple sustainable innovations are often challenging to make work together. | Promote interoperability to enhance flexibility. | Interoperability standards, integrated platforms, and cross-tool training programmes |

**Table 16.3** Illustration of DIS system-building linking for twin transitions

| DIS system-building dynamic | Sustainable innovation barrier link | Link description: risk | Link description: opportunity | DIS policy goal | Example policy instruments or solutions |
|---|---|---|---|---|---|
| New business models spinning off from pre-existing activities | Lock-in and path dependency; system integration and flexibility challenges | Potential for increased path dependence on non-sustainable-innovation application areas | New business models might increase flexibility and enable change | Encourage business model innovation that aligns with sustainable practices. | Business model innovation programmes, market expansion grants |
| Industrial firms engaging with consortia to access new knowl-edge | Vested interests and lack of governance capacity | No link | Could help increase accessibility to knowledge resources and increase awareness for SI through networks | Support consortia activities that enhance knowledge sharing for sustainable innovation. | Consortia funding, knowledge-sharing platforms |
| Firms providing resource access for internal mid to low-level actors | Restrictive culture of risk averseness and safety | No link | Opportunities for reducing risk, generating and testing new innovation knowledge | Ensure equitable resource access within firms to foster inclusive innovation practices. | Internal resource sharing programmes, intranet innovation portals |

approach illustrated that digital innovation can directly impact emission or energy reduction negatively (e.g., new oil field exploration), neutrally (improved HR systems), and positively (improved energy management analytics). Short-term twin transitions policy goals would demand positive direct impacts. However, our socio-institutional approach showed that digital tools with positive direct impacts may reinforce barriers to the sustainable innovations necessary for deep decarbonization on the long term. Vice versa, digital tools with negative short-term sustainable impacts may help overcome barriers to long-term deep decarbonization. Our findings enable distinguishing between short-term and long-term twin transition policy goals, see Table 16.4, with win-win solutions like digital twinning.

The above policy goals can be achieved via different, complementary intervention logics (i.e., dynamics by which policy interventions in complex systems may achieve twin transition goals) including:

1. Reorienting digital technologies towards sustainable pathways
2. Deploying digital technologies to overcome barriers to sustainable innovation
3. Overcoming barriers common to both digital and sustainable innovation
4. Creating directionality within DIS to navigate risks and opportunities in sustainable innovation

Intervention Logics 2, 3, and 4 require awareness of a) barriers to the innovation systems of sustainable innovation and b) technical potential of digital tools (#2) or DIS barriers (#3 and #4) and subsequently leveraging these with appropriate policy instruments. For instance, following Intervention Logic 4, we identified leverage points for creating sustainability spillovers for digital innovation application areas with no/neutral direct sustainability impacts. Examples include the emergence of new business models spinning off from existing digital activities, where internal tools are repurposed into marketable services with sustainability potential. Another example is leveraging consortia matchmaking intermediaries, who can strategically expand networks and support the inclusion of actors focused on sustainable innovation. These leverage

**Table 16.4** Long- and short-term dimensions of twin transition policy goals

| | | **Long term: impact on barriers to sustainable innovation** | | |
|---|---|---|---|---|
| | | Reinforces sustainable innovation barriers | Neutral | Overcomes sustainable innovation barriers |
| **Short term: sustainability impact** | Negative | - AI for increased fossil fuel use (e.g., enhanced oil recovery)<br>- Automation that increases production and energy use without sustainability constraints | | - AI models with high training-related emissions, used for climate scenario testing or other long-term transition planning |
| | Neutral | | | |
| | Positive | - Process control and energy efficiency tools that prolong fossil-based asset life<br>- Software for optimizing fossil-based logistics and operations | | - Digital twins for sustainable plant redesign and retrofitting<br>- AI for industrial symbiosis and resource sharing<br>- Blockchain for green traceability and market creation<br>- Platforms for coordinating cross-sectoral transitions (e.g., heat, hydrogen, power) |

points help direct policy attention towards shaping more favourable system conditions for sustainability alignment.

The complementary benefit of Intervention Logic 2 vis-a-vis the others is illustrated with the following example. We found little overlap between the barrier to sustainable innovation "Lack of demand and markets for sustainable products" and identified DIS dynamics (zero barrier links and two system-building links). This is not surprising as the barrier has little to do with socio-institutional aspects within the DIS; however, the technological approach shows that digital technologies can be technically leveraged to overcome this barrier. While such technology-driven solutions to social challenges are often part of the positivist narrative of digitalization, it is essential to contextualize which barriers are worth addressing using such solutions to avoid adverse effects.

## Limitations and Further Research

We identify some points of limitation and future research. First, our twin transition policy framework focused on assessing digital and sustainable innovation dynamics and leveraging theory to derive subsequent policy intervention logics. However, the intervention logics may not be exhaustive and additional logics should be explored. We mainly focused on implications of digital innovation for sustainability, arguing for the need to nest digital policy within transition policy. However, additional synergies can be gained by identifying "sustainable innovation system"-building dynamics that could synergize with DIS. Generalizability of our exploratory findings also require further testing.

Second, this chapter's contribution lies in identifying the intervention logics but the underlying examples of policy instruments require more attention. Co-design with policymakers and industry stakeholders can yield more feasible and effective instruments. Our policy framework can provide a basis for designing data collection workshops and communicating the complex interactions involved in twin transitions.

Finally, we focused on a hard-to-abate sector with substantial misalignment in the current twin transition. When applied to more aligned sectoral contexts, the technological approach may not reveal too many actionable leverage points. However, the socio-institutional approach is still valid as barrier prioritization and system-building dynamics will remain important leveraging points for policy.

**Supplementary Material** Refer to the files "SuppTable1_Barrier-link-analysis.xlsx" and "SuppTable2_System-building-link-analysis.xlsx".

The two supplementary tables provide the full linking analysis, describing the digital innovation system's barriers and system-building dynamics and the reasoning for including or excluding each of them on the basis of their link to the barriers to sustainable innovation. Subsequently, corresponding DIS policy goals and example policy solutions are highlighted.

**Acknowledgement** The research reported in this chapter was part of project ERGO—Energy reduction by Condition-Based Maintenance. This project is funded by TKI-E&I with the supplementary grant 'TKI-Toeslag', numbered I40-20-05, for Topconsortia for Knowledge and Innovation (TKI's) of the Ministry of Economic Affairs and Climate Policy, The Netherlands.

## References

Aazami, A. (2021). Topsector energie—Digitaliseringsagenda-Digitalisering en energie: méér dan de som der delen. T. energie.

Alkemade, F., Hekkert, M. P., & Negro, S. O. (2011). Transition policy and innovation policy: Friends or foes? *Environmental Innovation and Societal Transitions, 1*(1), 125–129. https://doi.org/10.1016/j.eist.2011.04.009

Andersen, A. D., Mäkitie, T., Steen, M., & Wanzenböck, I. (2024). Integrating industrial transformation and sustainability transitions research through a multi-sectoral perspective. Available at SSRN 4766925.

Arthur, W. B. (1989). Competing technologies, increasing returns, and lock-in by historical events. *The Economic Journal, 99*(394), 116–131. https://doi.org/10.2307/2234208

Bataille, C., Åhman, M., Neuhoff, K., Nilsson, L. J., Fischedick, M., Lechtenböhmer, S., Solano-Rodriquez, B., Denis-Ryan, A., Stiebert, S., Waisman,

H., Sartor, O., & Rahbar, S. (2018). A review of technology and policy deep decarbonization pathway options for making energy-intensive industry production consistent with the Paris agreement. *Journal of Cleaner Production, 187*, 960–973. https://doi.org/10.1016/j.jclepro.2018.03.107

Bauer, F., Hansen, T., & Hellsmark, H. (2018). Innovation in the bioeconomy–dynamics of biorefinery innovation networks. *Technology Analysis & Strategic Management, 30*(8), 935–947. https://doi.org/10.1080/09537325.2018.1425386

Bekar, C., Carlaw, K., & Lipsey, R. (2018). General purpose technologies in theory, application and controversy: A review. *Journal of Evolutionary Economics, 28*(5), 1005–1033. https://doi.org/10.1007/s00191-017-0546-0

Bennich, A. (2024). The digital imperative: Institutional pressures to digitalise. *Technology in Society, 76*, 102436. https://doi.org/10.1016/j.techsoc.2023.102436

Bergek, A., Hekkert, M., Jacobsson, S., Markard, J., Sandén, B., & Truffer, B. (2015). Technological innovation systems in contexts: Conceptualizing contextual structures and interaction dynamics. *Environmental Innovation and Societal Transitions, 16*, 51–64. https://doi.org/10.1016/j.eist.2015.07.003

Berkhout, F., & Hertin, J. (2004). De-materialising and re-materialising: Digital technologies and the environment. *Futures, 36*(8), 903–920. https://doi.org/10.1016/j.futures.2004.01.003

Bohnsack, R., Bidmon, C. M., & Pinkse, J. (2022). Sustainability in the digital age: Intended and unintended consequences of digital technologies for sustainable development. *Business Strategy and the Environment, 31*(2), 599–602. https://doi.org/10.1002/bse.2938

Erlinghagen, S., & Markard, J. (2012). Smart grids and the transformation of the electricity sector: ICT firms as potential catalysts for sectoral change. *Energy Policy, 51*, 895–906. https://doi.org/10.1016/j.enpol.2012.09.045

European Commission. (2019). Report from the Commission to the European Parliament,"Strategic report 2019 on the implementation of the European Structural and Investment Funds" (Document 52019DC0627). EUR-lex. https://eur-lex.europa.eu/legal-content/EN/TXT/?uri=CELEX%3A52019DC0627&qid=1720427744556

European Commission. (2020). Communication from the commission "new industrial strategy for Europe" (document 52020DC0102). *EUR-Lex*. https://eur-lex.europa.eu/legal-content/EN/TXT/?uri=CELEX%3A52020DC0102

Fouquet, R., & Hippe, R. (2022). Twin transitions of decarbonisation and digitalisation: A historical perspective on energy and information in European economies. *Energy Research & Social Science, 91*, 102736. https://doi.org/10.1016/j.erss.2022.102736

Foxon, T., & Pearson, P. (2008). Overcoming barriers to innovation and diffusion of cleaner technologies: Some features of a sustainable innovation policy regime. *Journal of Cleaner Production, 16*(1, Supplement 1), S148–S161. https://doi.org/10.1016/j.jclepro.2007.10.011

Geels, F. W. (2005). Processes and patterns in transitions and system innovations: Refining the co-evolutionary multi-level perspective. *Technological Forecasting and Social Change, 72*(6), 681–696. https://doi.org/10.1016/j.techfore.2004.08.014

Geels, F. W., & Schot, J. (2007). Typology of sociotechnical transition pathways. *Research Policy, 36*(3), 399–417. https://doi.org/10.1016/j.respol.2007.01.003

Hankel, A. (2014). Understanding higher order impacts of green ICT. In *ICT for sustainability 2014 (ICT4S-14)*. Atlantis Press. https://doi.org/10.2991/ict4s-14.2014.48

Hansen, T., & Coenen, L. (2017). Unpacking resource mobilisation by incumbents for biorefineries: The role of micro-level factors for technological innovation system weaknesses. *Technology Analysis & Strategic Management, 29*(5), 500–513. https://doi.org/10.1080/09537325.2016.1249838

Jacobsson, S., & Bergek, A. (2011). Innovation system analyses and sustainability transitions: Contributions and suggestions for research. *Environmental Innovation and Societal Transitions, 1*(1), 41–57. https://doi.org/10.1016/j.eist.2011.04.006

Janipour, Z., De Nooij, R., Scholten, P., Huijbregts, M. A. J., & De Coninck, H. (2020). What are sources of carbon lock-in in energy-intensive industry? A case study into Dutch chemicals production. *Energy Research and Social Science, 60*, 101320. https://doi.org/10.1016/j.erss.2019.101320

John, N., Frenken, K., & Wesseling, J. (2022). The digital innovation system of the Dutch processing industries. https://topsectorenergie.nl/documents/34/TKI_Energie__Industrie_-_Final_Report_Digital_Innovation_System_Dutch_Process__Igr5oNd.pdf

John, N., Wesseling, J., & Frenken, K. (2024). The accessibility and applicability of resources in innovation systems: Unpacking systemic problems within digital innovation systems. https://osf.io/28h3r_v1

John, N., Wesseling, J., & Frenken, K. (2022). Developing a resource-based analysis of technological innovation systems: the case of digital innovation in Dutch processing industries.

John, N., Wesseling, J. H., Worrell, E., & Hekkert, M. (2022). How key-enabling technologies' regimes influence sociotechnical transitions: The impact of artificial intelligence on decarbonization in the steel industry. *Journal of Cleaner Production, 370*, 133624. https://doi.org/10.1016/j.jclepro.2022.133624

Kemp, R. (1994). Technology and the transition to environmental sustainability: The problem of technological regime shifts. *Futures, 26*(10), 1023–1046. https://doi.org/10.1016/0016-3287(94)90071-X

Kemp, R., Schot, J., & Hoogma, R. (1998). Regime shifts to sustainability through processes of niche formation: The approach of strategic niche management. *Technology Analysis & Strategic Management, 10*(2), 175–198. https://doi.org/10.1080/09537329808524310

Klein Woolthuis, R., Lankhuizen, M., & Gilsing, V. (2005). A system failure framework for innovation policy design. *Technovation, 25*(6), 609–619. https://doi.org/10.1016/j.technovation.2003.11.002

Klerkx, L., Jakku, E., & Labarthe, P. (2019). A review of social science on digital agriculture, smart farming and agriculture 4.0: New contributions and a future research agenda. NJAS-Wageningen journal of. *Life Sciences, 90-91*, 100315. https://doi.org/10.1016/j.njas.2019.100315

Kovacic, Z., García Casañas, C., Argüelles, L., Yáñez Serrano, P., Ribera-Fumaz, R., Prause, L., & March, H. (2024). *The twin green and digital transition: High-level policy or science fiction?* Nature and Space. https://doi.org/10.1177/25148486241258046

Mäkitie, T., Hanson, J., Damman, S., & Wardeberg, M. (2023). Digital innovation's contribution to sustainability transitions. *Technology in Society, 73*, 102255. https://doi.org/10.1016/j.techsoc.2023.102255

Markard, J., Hekkert, M., & Jacobsson, S. (2015). The technological innovation systems framework: Response to six criticisms. *Environmental Innovation and Societal Transitions, 16*, 76–86. https://doi.org/10.1016/j.eist.2015.07.006

Meijer, A. (2024). Perspectives on the twin transition: Instrumental and institutional linkages between the digital and sustainability transitions. *Information Polity, 29*, 35–51. https://doi.org/10.3233/IP-230015

Mouthaan, M., Frenken, K., Piscicelli, L., & Vaskelainen, T. (2023). Systemic sustainability effects of contemporary digitalization: A scoping review and research agenda. *Futures, 149*, 103142. https://doi.org/10.1016/j.futures.2023.103142

Musiolik, J., Markard, J., & Hekkert, M. (2012). Networks and network resources in technological innovation systems: Towards a conceptual framework for system building. *Technological Forecasting and Social Change, 79*(6), 1032–1048. https://doi.org/10.1016/j.techfore.2012.01.003

Musiolik, J., Markard, J., Hekkert, M., & Furrer, B. (2020). Creating innovation systems: How resource constellations affect the strategies of system builders. *Technological Forecasting and Social Change, 153*, 119209. https://doi.org/10.1016/j.techfore.2018.02.002

Negro, S. O., Alkemade, F., & Hekkert, M. P. (2012). Why does renewable energy diffuse so slowly? A review of innovation system problems. *Renewable and Sustainable Energy Reviews, 16*(6), 3836–3846. https://doi.org/10.1016/j.rser.2012.03.043

Nilsson, L. J., Bauer, F., Åhman, M., Andersson, F. N. G., Bataille, C., de la Rue du Can, S., Ericsson, K., Hansen, T., Johansson, B., Lechtenböhmer, S., van Sluisveld, M., & Vogl, V. (2021). An industrial policy framework for transforming energy and emissions intensive industries towards zero emissions. *Climate Policy, 21*(8), 1053–1065. https://doi.org/10.1080/14693062.2021.1957665

Ometto, L., Challapalli, S., Polo, M., Cestari, G., Villagrossi, A., Sandri, M., & Pellegrini, E. (2019). Successful use case applications of artificial intelligence in the steel industry. Iron & steel technology (AIST). *Digital transformations, 2019*, 44–53.

Ritchie, J., & Spencer, L. (2002). Qualitative data analysis for applied policy research. In *Analyzing qualitative data* (pp. 173–194). Routledge.

Siemons, J., Katakwar, P., Croezen, H., Helmi, A., van der Pol, E., Blahusiak, M., Mahmoud, M., Haan, J. d., Flick, V., & Verkaik, N. (2020). Project 6–25 Technology Validation. https://www.6-25.nl/wp-content/uploads/2020/09/P6-25-Validation-Study_Final_RHDHV_PDC_01072020_2.pdf

Waßenhoven, A., Rennings, M., Laibach, N., & Bröring, S. (2023). What constitutes a "key enabling technology" for transition processes: Insights from the bioeconomy's technological landscape. *Technological Forecasting and Social Change, 197*, 122873. https://doi.org/10.1016/j.techfore.2023.122873

Wesseling, J., Kieft, A., Fuenfschilling, L., & Hekkert, M. (2022). How sociotechnical regimes affect low-carbon innovation: Global pressures inhibiting industrial heat pumps in The Netherlands. *Energy Research & Social Science, 89*, 102674. https://doi.org/10.1016/j.erss.2022.102674

Wesseling, J. H., Lechtenböhmer, S., Åhman, M., Nilsson, L. J., Worrell, E., & Coenen, L. (2017). The transition of energy intensive processing industries towards deep decarbonization: Characteristics and implications for future research. *Renewable and Sustainable Energy Reviews, 79*, 1303–1313. https://doi.org/10.1016/j.rser.2017.05.156

Wesseling, J. H., & Van Der Vooren, A. (2017). Lock-in of mature innovation systems: The transformation toward clean concrete in The Netherlands. *Journal of Cleaner Production, 155*, 114–124. https://doi.org/10.1016/j.jclepro.2016.08.115

Wieczorek, A. J., & Hekkert, M. P. (2012). Systemic instruments for systemic innovation problems: A framework for policy makers and innovation scholars. *Science and Public Policy, 39*(1), 74–87. https://doi.org/10.1093/scipol/scr008

Wittmayer, J. M., & Schäpke, N. (2014). Action, research and participation: Roles of researchers in sustainability transitions. *Sustainability Science, 9*, 483–496. https://doi.org/10.1007/s11625-014-0258-4

Wurm, D., & Wittmann, F. (2023). Mission-oriented innovation policies for the twin transition: A CO:DINA research report.

Yu, Z., Liang, Z., & Xue, L. (2021). A data-driven global innovation system approach and the rise of China's artificial intelligence industry. *Regional Studies, 1-11.* https://doi.org/10.1080/00343404.2021.1954610

Zolfagharian, M., Walrave, B., Raven, R., & Romme, A. G. L. (2019). Studying transitions: Past, present, and future. *Research Policy, 48*(9), 103788. https://doi.org/10.1016/j.respol.2019.04.012

# 17

# Multi-System Dynamics, Industrial Transformation, and Transition Policy: Future Research Priorities

Teis Hansen, Koen Frenken, Tuukka Mäkitie, Markus Steen, and Taran Thune

## Introduction

Arriving at the book's final chapter provides an opportunity for identifying future research priorities for multi-system dynamics and industrial transformation based on the insights from the preceding sixteen chapters. These chapters have not only shed light on the current state of knowledge, but have also revealed important limitations, emerging questions, and underexplored dimensions that warrant closer scrutiny. This concluding chapter is devoted to outlining such research priorities. It does not offer a definitive list, but our intention is to identify important topics that future work should pay attention to. The future

T. Hansen (✉) · T. Mäkitie · M. Steen
Department of Technology Management, SINTEF Digital, Trondheim, Norway
e-mail: teis.hansen@ifro.ku.dk

T. Hansen
Department of Food and Resource Economics, University of Copenhagen, Copenhagen, Denmark

K. Frenken
Copernicus Institute of Sustainable Development, Utrecht University, Utrecht, The Netherlands



M. Steen et al. (eds.), *Sustainability Transitions and Industrial Transformation*,
https://doi.org/10.1007/978-981-95-4893-4_17

research priorities outlined here are meant to serve as tentative suggestions that encourage engagement based on the status of current research and societal priorities.

In the following sections, based on contributions in this volume as well as other recent literature, we highlight avenues for future investigation, structured according to the three sections of the book (multi-system dynamics, firms and industries, and policy). We also include a section on digitalization, which is a cross-cutting topic addressed in multiple parts of the book. The final section addresses a broader outlook and recent concerns, and their impact on sustainability transitions.

## Multi-System Dynamics

As previously argued (see also Andersen & Markard, 2024), studies on multi-system dynamics of sustainability transitions have until now mainly focused on bi-lateral system interlinkages, for example, between the energy and transport sectors. Several chapters in this book ventured beyond such two-sector dynamics in their empirical enquiry. Mignon et al. (this volume), for example, explored interconnections between food, energy, and transport (and waste). Ateş et al. (this volume) analysed interlinkages between transport and energy with increasing electrification of the former and the role of digital technology, identifying digital technology and infrastructure as representing a key separate and enabling infrastructure system. Hanson and Mäkitie (this volume) explored how two different production methods for hydrogen production have very different sectoral configurations. Mäkitie et al. (this volume) focused on interlinkages between electricity and transport (in their case maritime transport), but in addition to 'horizontal' linkages of supplying and using electricity in shipping, the analysis also revealed 'vertical' interlinkages with various supplier sectors (such as technology and component

T. Mäkitie · T. Thune
TIK Centre for Technology, Innovation and Culture, University of Oslo, Oslo, Norway

producers) that had shaped the transition. This ‘broadening out’ surely comes with limitations. As more ‘systems’ (or sectors) are added, the number of linkages aspects easily increase exponentially. The resulting outputs may be heavy on description, but gloss over important nuances, and possibly lack explanatory purchase.

A key future research priority is therefore to continue the development of novel concepts and theoretical constructs to help reveal the interlinkages and inter-system dynamics that enable or constrain transition processes. As several chapters in this volume suggest, however, much can be gained from building further on previous work on, for example, large technical systems (Hughes, 1983) or sectoral couplings (Konrad et al., 2008). One under-conceptualized aspect is for instance how different sectors or systems involved in multi-system transitions may have very different temporal change dynamics and as such be in different stages of development (Ohlendorf et al., 2023). As exemplified in the chapter by Löhr (this volume), more theoretical attention is also needed on various types of agency that influence transition dynamics. This conceptual advancement will necessarily have to be accompanied by further empirical research (to test, verify, refine) in various multi-system sectoral and geographical contexts in both the Global North and Global South.

This brings us to our second suggested future research area. Transition research has been dominated by studies that are geographically confined to single countries or regions. This also applies to most of the chapters in this book, save for the chapter on institutional responses to the entry of Uber (Koutsimpogiorgos et al., this volume). We therefore call for more research that grapples with multi-system dynamics in different geographical (and sectoral, political, institutional, social) contexts. For example, new value chains are currently emerging around $CO_2$, hydrogen, or alternative fuels for ships, and these value chains in many instances transcend both national and sectoral boundaries, adding complexity, uncertainty, and risk to involved actors, and with often unclear implications for the places, peoples, and regions that are affected. This, we believe, also opens perspectives to important questions concerning both industrial transformation (i.e., affecting supplier or resource extraction and processing industries) and justice in multi-system transition dynamics.

Third, and connecting to the chapters on this volume on firms and industries in transitions, significant gaps persist in our understanding of how (singular) firms and other actors operate across and at the interfaces (Andersen et al., 2023) between different systems (Bakhuis et al., 2024; Löhr & Chlebna, 2023; Magnusson et al., 2025). This includes a range of firm types, including large utilities (in, e.g., energy, mobility, water, sanitation, waste), specialized firms (e.g., recyclers, digital platform providers), multi-industry supplier firms, or consultancies. These may all in different ways condition how transition processes unfold, for example by transferring distinct 'configurational templates' (Miorner et al., 2025) both across geographies as well as different socio-technical systems.

Our fourth suggestion for future research on multi-system dynamics in transitions concerns trade-offs and knock-on effects. To give but one example, while it is paramount to reduce the use of fossil-based plastics, there are arguably limits to how fast this can happen, and—given lack of feasible alternatives as well as the tremendous amounts of plastics being used in practically all sectors—it may therefore be argued that there is a need to bolster plastic recycling. However, this may create new lock-ins around the use of plastics, and potentially disincentivize stronger efforts to develop alternative solutions. We suggest that transition scholars (and other social scientists and organizational scholars interested in similar topics) continue to explore interdisciplinary collaboration with colleagues who may provide insight on for example the environmental outcomes of multi-system transitions.

Finally, we take note of methodological challenges. As demonstrated by the chapters in this book, transition studies are dominated by the use of qualitative methods, such as interviews and (manual) reading and analysis of documents, typically following a case study approach. Such methods can be powerful in uncovering the complex mechanisms of transitions, such as path dependencies and interaction effects shaping temporal transition processes (Geels, 2011). However, the broadening of analysis to multi-system dynamics requires in-depth analysis of interactions between evermore broader sets of actor networks, institutions, technologies, and events, challenging the analyst's capabilities to capture and analyse large sets of complex interactions. Moreover, qualitative case study designs may overemphasize context-sensitivity and thus oversee, for example, the transnational and international dynamics of transitions. Broadening of the methodological toolbox is thus necessary, including exploring Agent-Based Modelling and computational methods such as machine learning techniques.

## Firms and Industries in Transitions

The chapters on firms and industries in transitions have contributed to addressing salient research issues in the sustainability transitions field, extending theoretical perspectives and exploring new empirical contexts. The chapters have particularly responded to the call for further research that provide nuanced perspectives on incumbency (Kungl, 2024; Magnusson & Werner, 2023; Steen & Weaver, 2017; Turnheim & Sovacool, 2019), particular in sectors that are emission-intensive and hard to transform, and where incumbents likely play a key role both in progressing and stalling sustainability transitions. Here, we highlight five areas where the book chapters have made some contributions to address pertinent issues and discuss further research needs in light of these.

First, the interest in firms and industries in transitions is part of a turn to agency in transition studies (Farla et al., 2012), where there has been considerable interest in incumbent firms in sectors such as heavy transport, oil and gas industry, raw materials, and agri-food (Kungl, 2024). The chapters in the book have contributed to the understanding of incumbents in transitions in several ways, particularly by highlighting that incumbent organizations are part of complex inter-industry relations that span geographical and sectoral boundaries.

Large established firms are leading players that have considerable power over suppliers, collaborators, and customers, and exercise these through technological interdependencies, market positions, and institutional set-ups, such as standards. Sustainability transitions research has to date had a limited understanding of just how far the influence of large incumbents stretch (Bohnsack et al., 2020) and how powerful and dominant they are, and not only in their own sector or geographical locations (Rikap, 2021). The concentration of corporate power and its impact on innovation and transitions is an important topic as the global economy becomes increasingly dependent on and dominated by a handful of companies. The research aimed at "pluralising incumbency" (Turnheim & Sovacool, 2019) has to a limited degree considered the wide reach of large incumbents and how they connect multiple sectors and firms through their complex value chains (see for instance Ertelt (this volume) and Thune and Mäkitie (this volume) for examples of studies that address incumbents and value chains. We suggest that connecting insights from research on incumbents and multi-system dynamics in transitions could be a promising way forward to provide a better understanding of firms and industries in transitions.

Relatedly, the literature on incumbent and transitions has to limited extent addressed how firms and industries shape their institutional environments and the power and influence they have or develop in order to do so (Hoogstraaten et al., 2020; Kukk et al., 2016). For instance, how firms interpret regulatory changes and how they create industry standards is relevant for understanding firms in transitions, as well as how they shape behaviours of other firms and organizations through collective action or value chain domination (see Ertelt (this volume) and Thune and Mäkitie (this volume) for explorations into this topic). This also highlights the importance of connecting organizational and systems-oriented perspectives on firms and industries in transitions, as several of the chapters in this book do. Further knowledge about how firms and industries influence institutional frameworks (Ertelt, this volume), including standards, normative prescriptions, and regulations, is needed. Research that connects studies of policies and incumbent firms and industries in transitions would be valuable, as we need to deepen the understanding of how incumbents influence policy and vice versa.

Second, where sustainability transitions research traditionally focused on development and diffusion of new and environmentally friendly technologies and solutions, the management and organization literature rather focus on strategic adaptation and organizational behaviour related to sustainability (Kolk & Pinkse, 2005). The chapters with a firm level focus in this volume all draw on concepts from organization and management studies to study how firms engage with sustainability. As such, the chapters are part of a novel line of inquiry in sustainability transitions research that engages with topics like how firms understand, frame, exploit, and position themselves to address sustainability in proactive and reactive ways. Moreover, the chapters in this volume highlight how engaging with sustainability is a strategic adoption to a changed institutional environment, where incumbent firms become highly skilled at talking "sustainability talk", but less willing to "walk the walk" (Soppe & Schupfer, this volume). As several of the chapters in the book is based on empirical studies in the oil and gas industry, this is perhaps not so surprising. But how firms frame and adopt sustainability discourse and the loose coupling to actual transformation has been a key topic in research based on multiple firms and diverse industries (Hahn et al., 2018; Schupfer, 2024), the topic still warrants systematic investigation. Ullern and Damman (this volume) suggest paradox theory as an interesting lens to address such issues, and generally the rich conceptual apparatus in organization theory can be used to deepen the insights about firms and industries in transitions.

Third, the literature on firms in transitions has mainly focused on single firms. The chapters in this book rather highlight collective aspects of industry transformations. The different chapters underline how networks, formal and informal partnerships, corporate venturing, and value chain relationships are important channels for how firms interpret and approach sustainability (see Ertelt, Thune & Mäkitie; Ullern & Damman and Soppe & Schupfer; all in this volume). However, it is still unclear how collective interpretations and actions at the industry level contributes to sustainability transitions in firms, and further knowledge about the mechanisms or channels of influence that connect firms and industries is warranted. The considerable information about sustainability activities reported by firms, often developed by industry associations in collaboration with regulators, may be used for comparative studies or quantitative research, and can be exploited to extend insights across multiple firms and inside and across industries.

Fourth, it is a basic insight in innovation studies that firms are better able to implement new technologies and business practices that extend rather than transform their core business. More radical transformations, for instance strategic reorientation of firms or entering new markets, are a risk prone, costly and complex process, and as such it is a rare phenomenon. In line with the observation that firms often innovate to prolong status quo, some of the chapters in this book have looked at how digitalization and digital tools are implemented in firms to meet carbon emission reduction targets or reduce the use of energy and materials (examples include Ullern and Damman (this volume) and Hydle and Schupfer (this volume)). Such innovations are often implemented to extend existing (and potentially unsustainable economic activities), making it important to study the effects and consequences of such actions over time. Further research to address long-term effects of relatively incremental innovations in for instance hard-to-abate sectors, is needed, and this requires longitudinal studies.

Finally, in terms of methods and empirical approaches, studies of firms and industries in transitions have tended to focus on singular case studies, sometimes with a longitudinal approach to uncover developments over time. The focus on case studies is something that this research topic shares with sustainability transitions research in general (Zolfagharian et al., 2019). Incumbents in different sectors have by now been extensively studied via case studies (for exceptions, see, e.g., Steen & Weaver, 2017; Vormedal et al., 2023), and could benefit from approaches that look beyond singular cases. For example, studying multiple firms operating in value chains, innovation ecosystems, or sociotechnical systems could help identify more generic patterns of incumbent behaviour in sustainability transitions. Comparative case studies, for instance using qualitative comparative methods (an example of this approach is Hess and Mai (2014), mixed methods and quantitative research, would further enrich the field. This is in line with recent calls for more methodological diversity in sustainability transitions research (Hansmeier et al., 2021; Truffer et al., 2022).

## Policy

The chapters focused on policy share a joint interest in the interface between transformative characteristics and the more traditional policy emphasis on stimulating industrial development and competitiveness. While the chapters enrich our understanding of policies for multi-system transformations, we highlight three core areas for future research to address.

First, the importance of policy coordination is emphasized across all contributions. Policy coordination is a central challenge of transformative policymaking (Weber & Rohracher, 2012). While policy coordination is arguably also relevant to consider for more traditional policy approaches, the complexity of transitioning entire socio-technical systems requires greater coordination across policy fields, and thereby intensified efforts towards breaking down policy silos (Grillitsch et al., 2021). For multi-system transformation, the need for policy coordination is further accentuated.

Despite this unanimous insistence on the importance of policy coordination, we know surprisingly little on the functioning, opportunities, and limitations of different types of coordination mechanisms. Fruitful trading zones may be established in the future with research in political science and public administration on policy integration, i.e., different forms of cooperation by policy actors from different policy domains. Typical topics covered in these fields include motivation for policy integration, measures applied, influence on the policymaking process, and results of interventions (Tosun & Lang, 2017). As this literature sees policy integration as a process that depends on actors, institutional conditions, and relations between policy domains (Candel & Biesbroek, 2016), there are obvious conceptual touchpoints with relevant frameworks from innovation studies. Work on transformative policy coordination could start from the rich insights on the attainments and limits of approaches such as joined-up government and collaborative governance (e.g., Carey et al., 2015; Jochim & May, 2010; Jordan & Lenschow, 2010; Lang, 2019; Pelkonen et al., 2008; Ross & Dovers, 2008).

Second, the policy chapters operate at different levels of analysis. To exemplify, the chapter of John et al. (this volume) is at a more aggregate level, focusing on policy intervention logics, while the chapter by Andersson and Hellsmark (this volume) is primarily focused on a specific policy instrument for decarbonising the Swedish energy-intensive industry. Hansen et al. (this volume) discussed the return of industrial policy and its transformative (in)capacity, with empirical attention to Europe and the US. However, irrespectively of these differences in analytical level, there is little focus across the chapters on policy design (Howlett, 2023). As such, the chapters are illustrative of most work on transformative policies that analyse instruments in place, potentially offering suggestions for processual improvements, rather than analysing the actual policy formulation processes. However, *"a 'policy turn' in sustainability transitions"* (Rosenbloom, 2025, 1) would arguably necessitate greater attention to power relations and capacities involved in policy formulation processes within and between central policymaking organizations (Normann et al., 2024).

Third, geopolitical policy priorities, in particular security and defence-related priorities, have risen to the top of the policymaking agenda in many European countries since the commissioning of this book, which obviously has significant implications for the opportunities for implementing policies with transformative (from an environmental perspective) ambitions. Some of the policy chapters touch on this issue through references to security of supply in the chapter by Elzinga et al. (this volume) or attention to the geopolitical tensions that partly motivate the recent wave on green industrial policy (see the Hansen et al. (this volume) chapter). However, issues around technology sovereignty (Edler et al., 2023) and the security implications of transition processes (Kivimaa, 2022) were scarcely considered in the chapters. These issues are central to engage with in future research. From a multi-system perspective, national security priorities will likely influence developments in all core socio-technical systems over the coming years. This highlights the need for adequately considering this in our conceptual frameworks, beyond simply relegating developments in this area to 'contextual developments'. In addition, from a policy practice perspective, it becomes central to identify opportunities for policy to strengthen synergies between sustainability transitions and security priorities to avoid the risk of sidelining the transition agenda.

## Digitalization

The salient trend of digitalization in the economy cuts across all industries we have analysed. Indeed, digitalization has been highlighted as an important, emergent, and cross-cutting topic in sustainability transition research (Andersen et al., 2021; Bennich, 2025; Mäkitie et al., 2023). This agenda is urgently needed for at least two reasons. First, digital technology itself is consuming an exponentially growing among of energy and resources which renders the technology ecologically unsustainable, at least, in its current form and use (Andersen et al., 2021). Second, digitalization may support transition processes both as a landscape force providing opportunities for innovations and as a core component of new niche that may disrupt existing regimes. However, at the same time, digitalization also underpins many of the incremental innovation along established, unsustainable, technologies trajectories extending lock-ins (John et al., 2022).

A particular interesting feature of digitalization is platformization (Hydle & Schupfer, this volume), which can be understood as the main organizational aspect of digitalization that is accompanying technological changes that are subsumed under digitalization. Platforms, as an organizational model, can be applied in many ways with varying governance structure and (data) sharing arrangements. What is more, platforms can be applied in a business-to-business, consumer-to-business, and consumer-to-consumer contexts.

Another observation to be made holds that, from a transition perspective, digitalization can be driven by newcomers as well as by incumbents and can come from within a system or from outside the system. Understanding the role of incumbents in adopting digital innovation and developing these to fit into the industrial context at hand is particularly important (Soppe & Schupfer, this volume). Especially in hard-to-abate industries where incumbents are less likely to be disrupted, the digital innovations may come primarily from such large and experienced players. This also means that, despite the "affordance" that digital technologies and online platforms bring for radically new production methods and business models congruent with sustainability, the very same technologies and platforms are also being used to foster incremental innovation and sustaining the unsustainable status quo.

Taken together, the notion of a twin transition where digitalization and sustainability transitions would go hand and hand and be mutually reinforcing, should be treated with caution. It is equally possible, and in our view often more likely, that firms follow digitalization trajectories that enhance unsustainable lock-ins by further optimizing the technologies and production processes at hand. While such incremental digital innovation may well reduce energy use and enable lower waste production, they would not support a fundamental transition towards alternative and sustainable systems of production and consumption. This means that the notion of twin transition, now widespread due to EU's policy programmes building on this notion, should be further scrutinized. There is an urgent need to critically evaluate twin transition policies to understand what types of digital innovations would support sustainability transitions (John et al., this volume; Mäkitie et al., 2023).

## Broader Outlook and Emerging Concerns

This book has conveyed that we need to widen our conceptual lenses in order to develop better understandings of sustainability transitions and the strategies and policies required to accelerate them. The traditional focus on single systems, and relative neglect of for example geographical variation and value chains, needs to be broadened to grasp the uneven speeds and directions of transitions unfolding across and within production-consumption systems in different economies.

Taking a reflexive perspective, sustainability transition research itself has become an established field. While this has been a great achievement over the past decades, it also brings the risk of becoming less open to other academic fields and intellectual traditions. After all, as the sustainability challenges are mounting with planetary boundaries being crossed, many more scholars have taken up questions of sustainability and sustainability transitions, from quite different angles. In our view, the transition field would greatly benefit from more intensive and systematic engagement with other fields in its endeavour to develop research-based insights and actionable knowledge on multi-system dynamics, firms and industries, transformative policies, and the role of digitalization, as well as other pertinent transition research themes. Next to ongoing interaction with for instance economic geography (under the heading of 'Geography of Sustainability Transitions'), the contributions in this book call for strengthening and further development of interfaces with other research fields, such as industrial economics, organization studies, information systems research, and political science.

We further recognize that research is never static as it evolves in response to new discoveries, shifting societal needs, and technological advancements. This is also evident in this book, in that it does not capture concerns that have risen on the agenda since our initial call for contributions. For example, the book pays relatively limited attention to questions around technological sovereignty, security, and defence, which due to geopolitical circumstance has become a key policy priority in Europe and elsewhere. In relation to the issue of geopolitics, many countries and regions experience increasing contestation and pushback against the sustainability policy agenda. The reasons behind this development are complex and intertwined, and include the rise of right-wing populist political movements in several European countries, armed conflicts, and state budget increases to security and military spending, and a general sense of global geopolitical instability.

In addition, this development may also be fuelled by the relatively marginal results achieved through decades of climate policies and international agreements, which has implications for the legitimacy of policy efforts to stimulate transitions. This is perhaps especially felt in European countries, where climate policy and transformative innovation policy have been relatively progressive so far.

From a European point of view, as is implicit in the bulk of chapters in this book, there is certainly also a need for more research into how sustainability transition and industrial transformation processes are interlinked and have repercussions across other countries and continents. For example, China's rise to the fore as the global manufacturing powerhouse of several key 'cleantechs' (e.g., batteries, solar PV, EVs) is tilting the global economic landscape, with unclear implications for distribution of jobs and value-creating activities. Another important example is how sustainability transitions are driving extractive activities in many Global South contexts, with both positive and negative socio-economic and environmental outcomes (Hyldmo et al., 2025).

On a final note, we recognize that the viewpoints and arguments put forward in this volume all emanate from Global North perspectives, and primarily also from a limited number of countries in the North-West of Europe. This reflects that all chapter authors are based in universities or research institutes in this corner of the world. While there are obviously generic lessons to be drawn from the different chapters, reflecting in part that many first-of-their-kind transition processes do indeed unfold in these countries (e.g., decarbonisation of maritime shipping), we most certainly acknowledge the place-specificity and sectoral, political, social, and institutional context-dependency of sustainability transition dynamics (Ghosh & Mguni, 2025; Hansen et al., 2018; Truffer et al., 2015). It is our hope that this book volume will nonetheless provide food for thought and that it will spur subsequent research both in our and in many other parts of the world.

## References

Andersen, A. D., & Markard, J. (2024). *Multi-system dynamics in sustainability transitions: Introduction and outlook*. Cambridge Open Engage. https://doi.org/10.33774/coe-2024-x6x8n

Andersen, A. D., Frenken, K., Galaz, V., Kern, F., Klerkx, L., Mouthaan, M., Piscicelli, L., Schor, J. B., & Vaskelainen, T. (2021). On digitalization and sustainability transitions. *Environmental Innovation and Societal Transitions, 41*, 96–98. https://doi.org/10.1016/j.eist.2021.09.013

Andersen, A. D., Geels, F. W., Steen, M., & Bugge, M. M. (2023). Building multi-system nexuses in low-carbon transitions: Conflicts and asymmetric adjustments in Norwegian ferry electrification. *Proceedings of the National Academy of Sciences of the United States of America, 120*(47), Article e2207746120. https://doi.org/10.1073/pnas.2207746120

Andersson, J., & Hellsmark, H. (this volume). The scope and outcomes of transformative policy missions: A conceptual framework and empirical exploration of Swedish industry decarbonization. In M. Steen, K. Frenken,

T. Hansen, T. Mäkitie, & T. M. Thune (Eds.), *Sustainability transitions and industrial transformation: Multi-sector dynamics and policy perspectives.* Palgrave.

Ateş, A., Lovell, K., & Rogge, K. S. (this volume). Understanding multi-system interlinkages between energy and mobility systems and their governance in the UK. In M. Steen, K. Frenken, T. Hansen, T. Mäkitie, & T. M. Thune (Eds.), *Sustainability transitions and industrial transformation: Multi-sector dynamics and policy perspectives.* Palgrave.

Bakhuis, J., Kamp, L. M., Barbour, N., & Chappin, É. J. L. (2024). Frameworks for multi-system innovation analysis from a sociotechnical perspective: A systematic literature review. *Technological Forecasting and Social Change, 201*, Article 123266. https://doi.org/10.1016/j.techfore.2024.123266

Bennich, A. (2025). Untangling digitalisation: A topic of growing relevance for transition scholars. *Environmental Innovation and Societal Transitions, 57*, Article 101021. https://doi.org/10.1016/j.eist.2025.101021

Bohnsack, R., Kolk, A., Pinkse, J., & Bidmon, C. M. (2020). Driving the electric bandwagon: The dynamics of incumbents' sustainable innovation. *Business Strategy and the Environment, 29*(2), 727–743. https://doi.org/10.1002/bse.2430

Candel, J. J. L., & Biesbroek, R. (2016). Toward a processual understanding of policy integration. *Policy Sciences, 49*(3), 211–231. https://doi.org/10.1007/s11077-016-9248-y

Carey, G., McLoughlin, P., & Crammond, B. (2015). Implementing joined-up government: Lessons from the Australian social inclusion agenda. *Australian Journal of Public Administration, 74*(2), 176–186. https://doi.org/10.1111/1467-8500.12096

Edler, J., Blind, K., Kroll, H., & Schubert, T. (2023). Technology sovereignty as an emerging frame for innovation policy. Defining rationales, ends and means. *Research Policy, 52*(6), 104765. https://doi.org/10.1016/j.respol.2023.104765

Elzinga, R., Negro, S. O., & Janssen, M. (this volume). The more the merrier? A study on multiple mission-specific innovation systems in the Dutch energy sector. In M. Steen, K. Frenken, T. Hansen, T. Mäkitie, & T. M. Thune (Eds.), *Sustainability transitions and industrial transformation: Multi-sector dynamics and policy perspectives.* Palgrave.

Ertelt, S.-M. (this volume). (Re)interpreting circularity? Understanding the contested directionalities of the Swedish heavy-duty vehicle sector towards the Circular Economy. In M. Steen, K. Frenken, T. Hansen, T. Mäkitie, &

T. M. Thune (Eds.), *Sustainability transitions and industrial transformation: Multi-sector dynamics and policy perspectives*. Palgrave.
Farla, J., Markard, J., Raven, R., & Coenen, L. (2012). Sustainability transitions in the making: A closer look at actors, strategies and resources. *Technological Forecasting and Social Change, 79*(6), 991–998. https://doi.org/10.1016/j.techfore.2012.02.001
Geels, F. W. (2011). The multi-level perspective on sustainability transitions: Responses to seven criticisms. *Environmental Innovation and Societal Transitions, 1*(1), 24–40. https://doi.org/10.1016/j.eist.2011.02.002
Ghosh, B., & Mguni, P. (2025). *Sustainability transitions in the global south*. Cambridge Open Engage. https://doi.org/10.33774/coe-2025-c9v1q
Grillitsch, M., Hansen, T., & Madsen, S. (2021). Transformative innovation policy: A novel approach? In B. Godin, D. Vinck, & G. Gaglio (Eds.), *Handbook of alternative theories of innovation* (pp. 276–291). Edward Elgar. https://doi.org/10.4337/9781789902303.00028
Hahn, T., Figge, F., Pinkse, J., & Preuss, L. (2018). A paradox perspective on corporate sustainability: Descriptive, instrumental, and normative aspects. *Journal of Business Ethics, 148*(2), 235–248. https://doi.org/10.1007/s10551-017-3587-2
Hansen, T., Hansen, U. E., & Steen, M. (this volume). The transformative capacity of green industrial policy. In M. Steen, K. Frenken, T. Hansen, T. Mäkitie, & T. M. Thune (Eds.), *Sustainability transitions and industrial transformation: Multi-sector dynamics and policy perspectives*. Palgrave.
Hansen, U. E., Nygaard, I., Romijn, H., Wieczorek, A., Kamp, L. M., & Klerkx, L. (2018). Sustainability transitions in developing countries: Stocktaking, new contributions and a research agenda. *Environmental Science & Policy, 84*, 198–203. https://doi.org/10.1016/j.envsci.2017.11.009
Hansmeier, H., Schiller, K., & Rogge, K. S. (2021). Towards methodological diversity in sustainability transitions research? Comparing recent developments (2016–2019) with the past (before 2016). *Environmental Innovation and Societal Transitions, 38*, 169–174. https://doi.org/10.1016/j.eist.2021.01.001
Hanson, J., & Mäkitie, T. (this volume). The sectoral configurations of the hydrogen innovation system in Norway. In M. Steen, K. Frenken, T. Hansen, T. Mäkitie, & T. M. Thune (Eds.), *Sustainability transitions and industrial transformation: Multi-sector dynamics and policy perspectives*. Palgrave.
Hess, D. J., & Mai, Q. D. (2014). Renewable electricity policy in Asia: A qualitative comparative analysis of factors affecting sustainability transitions.

*Environmental Innovation and Societal Transitions, 12*, 31–46. https://doi.org/10.1016/j.eist.2014.04.001

Hoogstraaten, M. J., Frenken, K., & Boon, W. P. C. (2020). The study of institutional entrepreneurship and its implications for transition studies. *Environmental Innovation and Societal Transitions, 36*, 114–136. https://doi.org/10.1016/j.eist.2020.05.004

Howlett, M. (2023). *Designing public policies: Principles and instruments.* Routledge.

Hughes, T. P. (1983). *Networks of power: Electrification in western society, 1880–1930.* The Johns Hopkins University Press.

Hydle, K. M., & Schupfer, H. (this volume). Digitalization and platformization for sustainability: The case of the oil and gas industry. In M. Steen, K. Frenken, T. Hansen, T. Mäkitie, & T. M. Thune (Eds.), *Sustainability transitions and industrial transformation: Multi-sector dynamics and policy perspectives.* Palgrave.

Hyldmo, H. D. S., Wardhani, I. S., Kurniawan, N. I., Cahayati, D. D., Rye, S. A., & Vela-Almeida, D. (2025). Urgent transition, urgent extraction? Global decarbonization, national governance, and local impacts in the Indonesian nickel industry. *Environmental Research Letters, 20*(5), 054003. https://doi.org/10.1088/1748-9326/adc31c

Jochim, A. E., & May, P. J. (2010). Beyond subsystems: Policy regimes and governance. *Policy Studies Journal, 38*(2), 303–327. https://doi.org/10.1111/j.1541-0072.2010.00363.x

John, N., Wesseling, J. H., Worrell, E., & Hekkert, M. (2022). How key-enabling technologies' regimes influence sociotechnical transitions: The impact of artificial intelligence on decarbonization in the steel industry. *Journal of Cleaner Production, 370*, Article 133624. https://doi.org/10.1016/j.jclepro.2022.133624

John, N., Wesseling, J., & Frenken, K. (this volume). Navigating twin transitions: A policy framework to align digital innovation with sustainability, applied to the Dutch processing industry. In M. Steen, K. Frenken, T. Hansen, T. Mäkitie, & T. M. Thune (Eds.), *Sustainability transitions and industrial transformation: Multi-sector dynamics and policy perspectives.* Palgrave.

Jordan, A., & Lenschow, A. (2010). Environmental policy integration: A state of the art review. *Environmental Policy and Governance, 20*(3), 147–158. https://doi.org/10.1002/eet.539

Kivimaa, P. (2022). Transforming innovation policy in the context of global security. *Environmental Innovation and Societal Transitions, 43*, 55–61. https://doi.org/10.1016/j.eist.2022.03.005

Kolk, A., & Pinkse, J. (2005). Business responses to climate change: Identifying emergent strategies. *California Management Review, 47*(3), 6–20. https://doi.org/10.2307/41166304

Konrad, K., Truffer, B., & Voss, J. P. (2008). Multi-regime dynamics in the analysis of sectoral transformation potentials: Evidence from German utility sectors. *Journal of Cleaner Production, 16*(11), 1190–1202. https://doi.org/10.1016/j.jclepro.2007.08.014

Koutsimpogiorgos, N., Frenken, K., Herrmann, A. M., & Punt, M. B. (this volume). Pulling the brake: Worldwide institutional reactions to Uber's entry. In M. Steen, K. Frenken, T. Hansen, T. Mäkitie, & T. M. Thune (Eds.), *Sustainability transitions and industrial transformation: Multi-sector dynamics and policy perspectives*. Palgrave.

Kukk, P., Moors, E. H. M., & Hekkert, M. P. (2016). Institutional power play in innovation systems: The case of Herceptin®. *Research Policy, 45*(8), 1558–1569. https://doi.org/10.1016/j.respol.2016.01.016

Kungl, G. (2024). Challenges of the current discourse on incumbent firms in sustainability transitions. *Energy Research & Social Science, 108*, Article 103367. https://doi.org/10.1016/j.erss.2023.103367

Lang, A. (2019). Collaborative governance in health and technology policy: The use and effects of procedural policy instruments. *Administration & Society, 51*(2), 272–298. https://doi.org/10.1177/0095399716664163

Löhr, M. (this volume). Conceptualising multi-system configurations: The case of hydrogen in Germany. In M. Steen, K. Frenken, T. Hansen, T. Mäkitie, & T. M. Thune (Eds.), *Sustainability transitions and industrial transformation: Multi-sector dynamics and policy perspectives*. Palgrave.

Löhr, M., & Chlebna, C. (2023). Multi-system interactions in hydrogen-based sector coupling projects: System entanglers as key actors. *Energy Research & Social Science, 105*, Article 103282. https://doi.org/10.1016/j.erss.2023.103282

Magnusson, T., & Werner, V. (2023). Conceptualisations of incumbent firms in sustainability transitions: Insights from organisation theory and a systematic literature review. *Business Strategy and the Environment, 32*(2), 903–919. https://doi.org/10.1002/bse.3081

Magnusson, T., Onufrey, K., Werner, V., & Gillström, H. (2025). Inter-system linkage formation in multi-system transitions: Incumbents, asymmetries and learning cycles. *Research Policy, 54*(8), Article 105293. https://doi.org/10.1016/j.respol.2025.105293

Mäkitie, T., Andersen, A. D., & Steen, M. (this volume). The multi-sectoral interactions of electrification in Norwegian coastal shipping. In M. Steen, K. Frenken, T. Hansen, T. Mäkitie, & T. M. Thune (Eds.), *Sustainability transitions and industrial transformation: Multi-sector dynamics and policy perspectives*. Palgrave.

Mäkitie, T., Hanson, J., Damman, S., & Wardeberg, M. (2023). Digital innovation's contribution to sustainability transitions. *Technology in Society, 73*, Article 102255. https://doi.org/10.1016/j.techsoc.2023.102255

Mignon, I., Onufrey, K., & Thomson, L. (this volume). Mapping multi-system interactions in urban contexts—A study of the energy, transport and food systems in the municipality of Gothenburg. In M. Steen, K. Frenken, T. Hansen, T. Mäkitie, & T. M. Thune (Eds.), *Sustainability transitions and industrial transformation: Multi-sector dynamics and policy perspectives*. Palgrave.

Miorner, J., Schelbert, V., & Binz, C. (2025). How Transformative innovations travel: Tracing the diffusion of circular district-scale sanitation systems in Europe. *Economic Geography, 101*(2–3), 93–121. https://doi.org/10.1080/00130095.2025.2489965

Normann, H. E., Svartefoss, S. M., & Thune, T. (2024). Behind the scenes: Politics and pragmatism in formulating mission-oriented innovation policies in a national context. *Environmental Innovation and Societal Transitions, 52*, Article 100891. https://doi.org/10.1016/j.eist.2024.100891

Ohlendorf, N., Löhr, M., & Markard, J. (2023). Actors in multi-sector transitions—Discourse analysis on hydrogen in Germany. *Environmental Innovation and Societal Transitions, 47*, Article 100692. https://doi.org/10.1016/j.eist.2023.100692

Pelkonen, A., Teräväinen, T., & Waltari, S.-T. (2008). Assessing policy coordination capacity: Higher education, science, and technology policies in Finland. *Science and Public Policy, 35*(4), 241–252. https://doi.org/10.3152/030234208x308854

Rikap, C. (2021). *Capitalism, power and innovation: Intellectual monopoly capitalism uncovered*. Routledge.

Rosenbloom, D. (2025). Building momentum for a 'policy turn' in sustainability transitions: Lessons from Canada to consolidate strengths and bridge science-policy divides. *Environmental Innovation and Societal Transitions, 55*, Article 100956. https://doi.org/10.1016/j.eist.2024.100956

Ross, A., & Dovers, S. (2008). Making the harder yards: Environmental policy integration in Australia. *Australian Journal of Public Administration, 67*(3), 245–260. https://doi.org/10.1111/j.1467-8500.2008.00585.x

Schupfer, H. (2024). *Legitimacy challenges during sustainability transitions: How fossil fuel incumbents respond to rising sustainability pressures*. University of Oslo.

Soppe, B., & Schupfer, H. (this volume). Making oil and gas smarter, not greener: Fossil fuel incumbents' corporate venture capital investments in digital entrepreneurs during sustainability transitions. In M. Steen, K. Frenken, T. Hansen, T. Mäkitie, & T. M. Thune (Eds.), *Sustainability transitions and industrial transformation: Multi-sector dynamics and policy perspectives*. Palgrave.

Steen, M., & Weaver, T. (2017). Incumbents' diversification and cross-sectorial energy industry dynamics. *Research Policy, 46*(6), 1071–1086. https://doi.org/10.1016/j.respol.2017.04.001

Steen, M., Frenken, K., Hansen, T., Mäkitie, T., & Thune, T. M. (this volume). Introduction: Why this book and its contributions. In M. Steen, K. Frenken, T. Hansen, T. Mäkitie, & T. M. Thune (Eds.), *Sustainability transitions and industrial transformation: Multi-sector dynamics and policy perspectives*. Palgrave.

Thune, T. M., & Mäkitie, T. (this volume). Captains of industry? Value chain relations and sustainability reorientation of firms. In M. Steen, K. Frenken, T. Hansen, T. Mäkitie, & T. M. Thune (Eds.), *Sustainability transitions and industrial transformation: Multi-sector dynamics and policy perspectives*. Palgrave.

Tosun, J., & Lang, A. (2017). Policy integration: Mapping the different concepts. *Policy Studies, 38*(6), 553–570. https://doi.org/10.1080/01442872.2017.1339239

Truffer, B., Murphy, J. T., & Raven, R. (2015). The geography of sustainability transitions: Contours of an emerging theme. *Environmental Innovation and Societal Transitions, 17*, 63–72. https://doi.org/10.1016/j.eist.2015.07.004

Truffer, B., Rohracher, H., Kivimaa, P., Raven, R., Alkemade, F., Carvalho, L., & Feola, G. (2022). A perspective on the future of sustainability transitions research. *Environmental Innovation and Societal Transitions, 42*, 331–339. https://doi.org/10.1016/j.eist.2022.01.006

Turnheim, B., & Sovacool, B. K. (2019). Forever stuck in old ways? Pluralising incumbencies in sustainability transitions. *Environmental Innovation and Societal Transitions*. https://doi.org/10.1016/j.eist.2019.10.012

Ullern, E. F., & Damman, S. (this volume). Navigating paradoxes in sustainability transitions: How incumbent firms (can) foster circular economy. In M. Steen, K. Frenken, T. Hansen, T. Mäkitie, & T. M. Thune (Eds.), *Sustainability transitions and industrial transformation: Multi-sector dynamics and policy perspectives*. Palgrave.

Vormedal, I., Bjander, J., Larsen, M. L., & Lindberg, M. B. (2023). Technological change and the politics of decarbonization: A re-making of vested interests? *Environmental Innovation and Societal Transitions, 47*, Article 100725. https://doi.org/10.1016/j.eist.2023.100725

Weber, K. M., & Rohracher, H. (2012). Legitimizing research, technology and innovation policies for transformative change: Combining insights from innovation systems and multi-level perspective in a comprehensive 'failures' framework. *Research Policy, 41*(6), 1037–1047. https://doi.org/10.1016/j.respol.2011.10.015

Zolfagharian, M., Walrave, B., Raven, R., & Romme, A. G. L. (2019). Studying transitions: Past, present, and future. *Research Policy, 48*(9), Article 103788. https://doi.org/10.1016/j.respol.2019.04.012

# Index




M. Steen et al. (eds.), *Sustainability Transitions and Industrial Transformation*,
https://doi.org/10.1007/978-981-95-4893-4